Detection Challenges in Clinical Diagnostics

RSC Detection Science Series

Editor-in-Chief:
Professor Mike Thompson, *University of Toronto, Canada*

Series Editors:
Dr Subrayal M Reddy, *University of Surrey, Guildford, UK*
Dr Damien Arrigan, *Tyndall National Institute, Cork, Ireland*

Titles in the Series:
1: Sensor Technology in Neuroscience
2: Detection Challenges in Clinical Diagnostics

How to obtain future titles on publication:
A standing order plan is available for this series. A standing order will bring delivery of each new volume immediately on publication.

For further information please contact:
Book Sales Department, Royal Society of Chemistry, Thomas Graham House, Science Park, Milton Road, Cambridge, CB4 0WF, UK
Telephone: +44 (0)1223 420066, Fax: +44 (0)1223 420247
Email: booksales@rsc.org
Visit our website at www.rsc.org/books

Detection Challenges in Clinical Diagnostics

Edited by

Pankaj Vadgama
Queen Mary University of London, UK
Email: p.vadgama@qmul.ac.uk

and

Serban Peteu
National Institute for R&D in Chemistry and Petrochemistry,
Bucharest, Romania
Email: sfpeteu@umich.edu

RSC Publishing

RSC Detection Science Series No. 2

ISBN: 978-1-84973-612-1
ISSN: 2052-3068

A catalogue record for this book is available from the British Library

Published by The Royal Society of Chemistry,
Thomas Graham House, Science Park, Milton Road,
Cambridge CB4 0WF, UK

Registered Charity Number 207890

For further information see our web site at www.rsc.org

With deep appreciation to
Dixa, Reena, Rooshin and Preeya
Mihaela and Mihai
for all their support

Preface

There have been quite a number of topical texts and reviews published recently dealing with clinical diagnostics. So, one question to ask is why another book? This Preface hopefully provides a rationale for this. In a parallel vein, there was a belief in the idea of a "Fountain of Youth"; an aged-adult body enters the "Fountain" Service and exits with new body components, with as good-as-new regeneration. To bring this up to date, even if this was achievable, say by tissue engineering the question arises – would the insurance companies foot the bill? So not only are there residual uncertainties, even in utopia, but context is ever changing.

To return to the present topic, our aim was to bring the reader up to date within the context of rapidly evolving technology and to communicate this through the eyes of research leaders. A broad range of approaches is scoped and the diagnostics needs and bottlenecks surveyed. Both academic and industrial experts are included, all addressing robust tools for dealing with the world of *real* biological measurement – especially from the perspective of a commonly neglected expert: the end user.

Chapter 1 (by Thompson *et al.*) and **Chapter 2** (by Vadgama *et al.*) deal with Clinical Diagnostics, both in the laboratory and at the bedside, from the broader picture down to some details. There have been powerful advances in extralaboratory testing enabled by new solid-state technology encompassing reagent immobilisation and miniaturization. This is a difficult area to monitor and set standards for, because of the distributed nature of such testing across a variety of clinical sites and even the home. Laboratory analysis has taken on major advances with high throughput and small sample volumes, so polarisation between technologies that are aimed at laboratory testing *vs.* those for extra laboratory testing are inevitable.

RSC Detection Science Series No. 2
Detection Challenges in Clinical Diagnostics
Edited by Pankaj Vadgama and Serban Peteu
© The Royal Society of Chemistry 2013
Published by the Royal Society of Chemistry, www.rsc.org

Analytical specificity remains a vital issue, still not fully resolved, especially for low-concentration analytes, regardless of technique, and where sample separation cannot be part of the assay system, as in say *in vivo* sensors, quite significant effort is required to redesign the basic construct. This latter owes as much to materials science and engineering as to chemistry.

Selectivity in general is discussed in Chapter 1 (by Thompson *et al.*) with specific regard to Clinical Chemistry. In the *in vivo* context, sensors need to function selectively given their exposure to unmodified samples. There is the added complication of high local protein concentration and cellular ingress at the implant site. Returning to *the selectivity challenge*, there are helpful, published techniques for electrochemical bio/sensors reviewed in Chapters 1 and 6 (Thompson *et al.*; Peteu and Szunerits), including membranes to address solute transport and interfacing issues. Ultimately, one needs to be mindful of the *trade-off* between *analytical complexity* and *slower processing* as against the goal of *high specificity*.

This issue of damaging nonspecific adsorption (NSA), represents a true Achilles heel for direct contact sensors, and is examined in detail. With biological matrices constituting highly complex solute mixtures, it becomes clear they could well prevent the detection/quantification *of target analytes present at considerably lower concentration,* outlined in Chapter 1 (Thompson *et al.*).

Early advances in this regard, though not always seen as such, are the dry reagent systems developed for glucose as illustrated in **Chapter 3** (Wang and Hu). Here, unless the integrated laminates are not tailored to whole-device function and can be produced in mass numbers, the overall transduction value cannot be realized. This chapter examines the progress and challenges of the blood-glucose biosensors. Managing one's diabetes also decreases the occurrence of its serious complications such as nephropathy, neuropathy and retinopathy. The pathogenesis of diabetes and its complications seem to be correlated with the presence of nitro-oxidative species – including peroxynitrite – potentially implicated in beta-cells destruction, as highlighted in Chapter 6 (Peteu and Szunerits).

Chapter 4 (Gaspar *et al.*) critically assess recent progress and many challenges in electrochemical detection of disease-related diagnostic biomarkers. Mostly, we have relied on biomolecules, but *aptamer technology* shows how such synthetic structures can be harnessed to give stable "readers" for biochemical targets. These are early days still for the technology, and designer aptamers bred *via* SELEX (selective evolution of ligands by exponential enrichment), should extend their repertoire. It may also be that here and elsewhere, with use of arrays, absolute selectivity will not be a necessity.

As with sampling integrity, so with continuous use sensors for monitoring *in vivo*, Chapter 5 (by Meyerhoff *et al.*), shows there is great need to *reduce bioincompatibility and resultant surface fouling;* biofluids are not tolerant of foreign surfaces. *In vivo* **sensors** are, however, uniquely positioned to provide site-related information, in particular at specific extravascular, tissue locations, but face a huge safety and biocompatibility challenge. The fact that this has been resolved in some cases raises the possibility of broader forms of monitoring systems. Even using early proof-of-concept systems, it may yet be possible to pick up biological signatures that arise from wider disease sets.

Chapter 5 (Meyerhoff *et al.*) especially scrutinizes the challenges for **sensors long-term biocompatibility**. In spite of the great sensors advances *in vitro*, the commercial development of *implantable* chemical sensors has reached a bottleneck. So much so, that, even *with* the FDA-required recalibration the output of devices, is still *not considered reliable enough*. Special coating materials able to say *attenuate the activation of platelets* or materials able to *inhibit the inflammatory response* will become important.

In the case of more exotic short-lived radical species such as *peroxynitrite* (often ≤1 s lifetime) featured in **Chapter 6** (by Peteu and Szunerits) measurement in real environments will be difficult. For species such as peroxynitrite, quantification poses a whole new level of measurement uncertainties, yet they are important: cell signalling, reactivity and tissue damage are mediated by such short-lived radicals. Mitochondrial oxidation is itself a free-radical generator. Interestingly, *nitric oxide, superoxide,* the precursors of *peroxynitrite, and peroxynitrite* itself have been dubbed "the *good,* the *bad* and the *ugly*" because of their tissue-level effects. This chapter illustrates the chemical diversity of such reactive species and the way in which electrochemical interfaces and sensor chemistry could track some of these, as a glimpse into future clinical use.

The biomachinery resulting in ineffective haematopoiesis and augmented leukaemia risk in *myelodysplastic syndromes* is largely known. However, one major challenge illustrated in **Chapter 7** (by McNamara *et al.*) is how to correctly classify and "risk stratify" the patients. Molecular biology assessment could help as diagnostic tools. So often, the right diagnosis – offered early in the game – will affect morbidity and survival.

Chapter 8 (by Barr *et al.*) describes *Raman for noninvasive early cancer diagnosis*. Early histological appearances may be difficult to categorise, and there is less interobserver agreement. Some changes may not even be evident by traditional histology, moreover; diagnosis is expensive, time consuming and requires required tissue biopsy. For the specific case of oesophageal neoplasia, there is evidence that a Raman signature can identify molecular change prior to morphological aberrations.

Raman spectroscopy can deliver high sensitivity for degenerating, premalignant, oesophageal epithelium. The gain would be objectivity, speed and a real-time assessment for early removal of dysplastic tissue and follow up. Raman-based fibre-optic interrogation has potential as an *in situ* surgical adjunct.

Signal handling with arrays is well exemplified in **Chapter 9** (by Kendall *et al.*). Any opportunity to create multiple arrays should be taken, as this adds depth to measurement. Here, for volatiles analysis the power of pattern recognition is well demonstrated. Undoubtedly the principles can be extrapolated to other modes of measurement, including where selectivity and drift are challenges, be it *in vivo* or *in vitro*.

Another overarching challenge in clinical diagnostics: the critical need to provide *information not just data*. With ease of data generation, it could be said that there is too much data for the clinician to deal with, especially if it is real time as with *in vivo* monitoring devices. Herein lies the problem of what the data is really for; if it is a medical defence strategy it becomes a waste of finances, but if behind the data there is a genuine quest for what the bio/ pathological implications might be, then the data becomes of considerably greater value.

As with any scientific quest, one might consider this to be analogous to a person searching for their home keys under a streetlamp, even though these might have been dropped somewhere else, because "that's where the light is."... This book has tried to shed some light on some of the important detection challenges impeding future progress in clinical diagnostics. It may be the light is currently in the wrong place, but eventually the home keys will be found. Finally, we would like to thank the team at Royal Society of Chemistry who guided us so patiently through the publication maze and without whom there would be no book: Merlin Fox, Rosalind Searle and Alice Toby-Brant (Commissioning); Lois Bradnam and Sarah Salter (Production).

Pankaj Vadgama
London

Serban Peteu
Bucharest

Contents

RSC Detection Science Series No. 2
Detection Challenges in Clinical Diagnostics
Edited by Pankaj Vadgama and Serban Peteu
© The Royal Society of Chemistry 2013
Published by the Royal Society of Chemistry, www.rsc.org

**Chapter 3   Blood-Glucose Biosensors, Development
               and Challenges   65**
               Yuan Wang and Madeleine Hu

List of Contributors

L. Max Almond Surgical Fellow, Biophotonics Research Unit, Leadon House, Gloucestershire Royal Hospital, Great Western Road, Gloucester GL13NN, United Kingdom

Dr. Salzitsa Anastasova Post Doctoral Research Assistant, School of Engineering and Material Sciences, Queen Mary University of London, Mile End Road, E14N, United Kingdom

Prof. Hugh Barr Professor of Surgery, Consultant General & Gastrointestinal Surgeon, Biophotonics Research Unit, Gloucestershire Hospitals NHS Foundation Trust, Great Western Road Gloucester GL1 3NN & Cranfield Health, Cranfield University, Cranfield MK43 0AL, United Kingdom

Dr. Christophe Blaszykowski Research Associate, Department of Chemistry, University of Toronto, 80 St. George Street, Toronto, Ontario M5S 3H6, Canada

Dr. Megan C. Frost Associate Professor, Department of Biomedical Engineering, Michigan Technological University, 1400 Townsend Dr., Houghton, MI 49931-1295, United States of America

Dr. Szilveszter Gáspár Senior Researcher, International Centre of Biodynamics, 1B Intrarea Portocalelor, 060101 Bucharest, Romania

Madeleine Hu Premed Student, the College of New Jersey, 2000 Pennington Road, Ewing Township, NJ 08628, USA, madeleinehu@yahoo.com

Dr. Joanne Hutchings National Institute for Health Research Post-Doctoral Researcher, Biophotonics Research Unit, Leadon House, Gloucestershire Royal Hospital, Great Western Road, Gloucester GL13NN, United Kingdom

Dr. Catherine Kendall Senior Lecturer, Biophotonics Research Unit, Gloucestershire Hospitals NHS Foundation Trust, Great Western Road Gloucester GL1 3NN & Honorary Senior Clinical Lecturer Cranfield Health, Cranfield University, Cranfield MK43 0AL, United Kingdom

RSC Detection Science Series No. 2
Detection Challenges in Clinical Diagnostics
Edited by Pankaj Vadgama and Serban Peteu
© The Royal Society of Chemistry 2013
Published by the Royal Society of Chemistry, www.rsc.org

Prof. Naresh Magan Professor, Personal Chair, Applied Mycology, Research Director and Lead Academic for Doctoral Training Centre for Health and Biosciences, Cranfield Health, Vincent Building, Cranfield University, Cranfield MK43 0AL, United Kingdom

Dr. Christopher McNamara Consultant Haematologist, The Department of Haematology, Royal Free Hospital, London NW3 2QG, United Kingdom

Prof. Mark E. Meyerhoff Philip P. Elving Professor of Chemistry, Department of Chemistry, University of Michigan, 940 N. University, Ann Arbor, MI 49109-1055, United States of America

Dr. Serban F. Peteu Senior Scientist, National Institute for Research and Development in Chemistry and Petrochemistry, 202 Splaiul Independentei, 060021 Bucharest, Romania

Dr. Gavin Rhys-Lloyd Post-Doctoral Chemometrics, Biophotonics Research Unit, Leadon House, Gloucestershire Royal Hospital, Great Western Road, Gloucester GL13NN, United Kingdom

Dr. Alexander Romaschin Director, Clinical Biochemistry and Keenan Research Centre, St. Michael's Hospital, 30 Bond Street, Toronto, Ontario M5B 1W8, Canada

Prof. Wolfgang Schuhmann Professor, Analytische Chemie-Elektroanalytik & Sensorik, Ruhr-Universität Bochum, Universitätsstr. 150, Bochum D-44780, Germany

Sonia Sheikh Graduate Student, Department of Chemistry, University of Toronto, 80 St. George Street, Toronto, Ontario M5S 3H6, Canada

Prof. Neil Shepherd Professor of Histopathology, Biophotonics Research Unit, Leadon House, Gloucestershire Royal Hospital, Great Western Road, Gloucester GL13NN, Department of Histopathology, Gloucestershire Hospitals NHS Foundation Trust, Faculty of Bioscience and Medicine, Cranfield Health, Cranfield University, Bedfordshire, United Kingdom

Geeta Shetty Biophotonics Research Unit, Leadon House, Gloucestershire Royal Hospital, Great Western Road, Gloucester GL13NN, United Kingdom

Dr. Anna Spehar-Deleze Post Doctoral Research Assistant, School of Engineering and Material Sciences, Queen Mary University of London, Mile End Road, E14N, United Kingdom

Prof. Nicholas Stone Professor of Biomedical Imaging and Biosensing, Biophotonics Research Unit, Leadon House, Gloucestershire Royal Hospital, Great Western Road, Gloucester GL13NN, United Kingdom, Biomedical Imaging and Bio Sensing, School of Physics, University of Exeter, EX4 4QL, United Kingdom

Prof. Sabine Szunerits Professor, Institut de Recherche Interdisciplinaire, Université Lille 1, Parc de la Haute Borne, 50 Avenue de Halley, BP 70478, 59658 Villeneuve d'Ascq, France

Dr. Alison S. Thomas Clinical Research Fellow, The Department of Haematology, The Royal Free Hospital, London NW3 2QG, United Kingdom

Prof. Michael Thompson Professor of Bioanalytical Chemistry, Department of Chemistry and Institute for Biomaterials and Biomedical Engineering University of Toronto, Canada

Prof. Pankaj Vadgama Director of the IRC in Biomedical Materials; Professor of Clinical Biochemistry, Queen Mary University of London, Mile End Road, London, E14N, United Kingdom; Consultant Chemical Pathologist in the Department of Clinical Biochemistry at Barts Health NHS Trust

Dr. Alina Vasilescu Senior Researcher, International Centre of Biodynamics, 1B Intrarea Portocalelor, 060101 Bucharest, Romania

Yuan Wang Senior Scientist, Siemens Healthcare Diagnostics, 511 Benedict Avenue, Tarrytown, New York, NY 10591, United States of America

Alexander K. Wolf Graduate Student, Department of Chemistry, University of Michigan, 940 N. University, Ann Arbor, MI 49109-1055, United States of America

CHAPTER 1

Biosensor Technology and the Clinical Biochemistry Laboratory – Issue of Signal Interference from the Biological Matrix

MICHAEL THOMPSON,*[a] SONIA SHEIKH,[a] CHRISTOPHE BLASZYKOWSKI[a] AND ALEXANDER ROMASCHIN[b]

[a] Department of Chemistry and Institute for Biomaterials and Biomedical Engineering, University of Toronto, 80 St. George Street, Toronto, Ontario M5S 3H6, Canada; [b] Keenan Research Centre and Clinical Biochemistry, St. Michael's Hospital, 30 Bond Street, Toronto, Ontario M5B 1W8, Canada
*Email: mikethom@chem.utoronto.ca

1.1 Laboratory Clinical Biochemical Assays

Assays for biochemicals of clinical interest in biological fluids, tissues or feces can be arbitrarily divided into tests performed in either the clinical biochemistry laboratory or *via* a point-of-care device.[1] The latter technology often takes the form of a one-use structure that yields a result in a localized environment such as the home, hospital bedside or general practitioner's facility. The blood glucose and pregnancy assay devices, which involve discardable test strips, will be familiar to many. This type of structure has attracted a high level of interest

RSC Detection Science Series No. 2
Detection Challenges in Clinical Diagnostics
Edited by Pankaj Vadgama and Serban Peteu
© The Royal Society of Chemistry 2013
Published by the Royal Society of Chemistry, www.rsc.org

in recent years, not least from the commercial standpoint. Often quoted advantages are said to lie in convenience, speed of analysis and cost savings. There are of course attractive features in certain cases for procuring a measurement in a particular location. Obviously, biosensor devices might be expected to play a major role in this arena. However, the present discussion concentrates on the potential for use of biosensors in the dedicated central biochemistry laboratory, which is typically located in a major hospital for obvious reasons. Assays performed in this type of facility are very wide ranging in scope with samples originating from the areas of diagnosis of medical conditions, toxicology or monitoring of drug therapy. Sera or plasma dominate the types of samples analyzed but many assays also involve various tissues, feces and other bodily fluids such as urine.

Test technology associated with all the various aspects of medicine is a large and expensive operation. In terms of the cost of performing assays, the Province of Ontario, for example, spends well over 1 billion Canadian (CDN) dollars annually, the tests being conducted in both private laboratories and hospital facilities. A large hospital such as St. Michael's in Toronto performs a vast array of tests such as those involving immunology, hematology, bacteriology, mycology, cytology, genetics and various aspects of pathology in addition to biochemical measurements. With respect to the latter, some 3 million tests are performed annually with reagent costs alone being in the region of 2 million CDN dollars. Although not intended to be exhaustive, Table 1.1 provides an overview of the type of biochemical and immuno-assays conducted by the clinical biochemistry on an annual basis together with total numbers comprised of both in- and out-patient tests.[2] The "winner" is the thyroid stimulating hormone (TSH) test, which alone costs several million dollars annually. The equipment employed in the portfolio of clinical measurements ranges from conventional chromatography to sophisticated mass spectrometry, and combinations thereof. As would be expected given the sheer volume of samples, the laboratory is highly automated and robotized. Typically, a blood sample will be bar-coded then centrifuged to extract serum before entering an automated analytical train. Samples may then be directed automatically to their various analytical stations. A good example of a subsequent analysis would be the prevalent magnetic-bead enzyme-linked immunosorbent assay (ELISA) method for targets where immuno-assay is feasible. Despite the high level of automation incorporated into the analytical train, it should be emphasized that measurements are still performed in a batch-based fashion. Having said that, it is certainly the situation that a number of protocols offer multi-analyte capability, the aforementioned ELISA assay being a case in point.

Biosensor technology – in principle – offers rapid, label-free measurements, which can be conducted in highly automatable configurations. Furthermore, the possibility for multiplying tests using generic sensing physics is certainly an attractive strategy. However, the clear reality is that biosensor devices do not figure prominently in the clinical biochemistry laboratory. Use of ion-selective electrode technology for certain cations might represent a contradiction to this statement, but it could justifiably be argued that these devices do not constitute

Table 1.1 Overview of the type of biochemical and immuno-assays conducted by a clinical biochemistry laboratory on an annual basis (source: St. Michael's Hospital, Toronto, Canada).

Type of Test	Number performed (per year)
Acetone, quantitative*	448
Albumin, quantitative*	125 360
Alcohol, ethyl-quantitative*	5010
Alcohol, fractionation and quantification*	5889
Amylase*	21477
Barbiturates, quantitative*	13 898
Barbiturates, fractionation and quantification (serum) – includes other drugs requiring similar methodology (*e.g.* tricyclic antidepressants)*	704
Bilirubin, total*	72 000
Bilirubin, conjugated*	22 545
pH*	20
pCO_2, pO_2 and pH in combination*	62 757
Carbamazepine, quantitative (Tegretol)*	504
Chlordiazepoxide, quantitative (Librium)*	10
Calcium*	91 723
Calcium ionized*	9676
Catecholamines, fractionated*	2485
Chloride*	227 057
Cholesterol, total*	34 663
Acetaminophen*	5036
Carboxyhemoglobin*	4
CO_2 content, CO_2 combining power, bicarbonate (measured not calculated)*	219 783
Creatine phosphokinase*	63 914
Creatinine*	257 139
Creatinine clearance*	4884
Target drug testing, urine, qualitative or quantitative*	37 691
Diazepam, quantitative (Valium, Vivol)*	37
Drugs of abuse screen, urine*	48
Broad spectrum toxicology screen, urine – includes confirmatory testing*	7399
Electrophoresis, serum – including total protein	7556
Electrophoresis, other than serum – including total protein*	599
Glycosylated hemoglobin – Hgb A1*	19 394
Flurzepam, quantitative (Dalmane)*	0
Glucose tolerance test in pregnancy*	2694
Glucose tolerance test*	973
Gamma glutamyl transpeptidase*	7705
Glucose, quantitative (not by dipstick)*	188 863
Glucose, semiquantitative (dipstick if read with reflectance meter)*	46 538
High density lipoprotein cholesterol*	32 179
5H1AA quantification – U*	1554
Hemoglobin A2 by chromatography*	2030
Iron, total – with iron binding capacity and percent saturation*	21 054
Lactic acid (lactate)*	13 733
Lactic dehydrogenase (L.D.H), total*	19 245
Lipase	1475

Table 1.1   (*Continued*)

Type of Test	Number performed (per year)
Lipoprotein, electrophoresis*	3
Lipoprotein, ultracentrifugation*	499
Lithium*	866
Lidocaine*	86
Methotrexate (amethopterin)*	115
N-acetylprocainamide*	74
Magnesium*	84 408
Metanephrines, total – U*	2712
Methemoglobin*	77
Myoglobin, quantitative – U*	176
Occult blood*	4293
Osmolality (osmolarity)*	10 044
Oxalic acid (Oxalate) – U*	3104
Phenothiazines, quantitive – U*	69
Phosphatase, alkaline*	82 037
Phosphatase, alkaline fractionation*	62
Phosphorus (inorganic phosphate)*	86 956
Plasma hemoglobin*	44
Potassium*	235 045
Protein, total*	110 937
Primidone, quantitative (Mysoline)*	420
Procainamide*	71
Quinidine*	18
Salicylate, quantitative*	4875
SGOT (AST)*	88 095
SGPT (ALT)*	80 304
Sodium*	237 209
Thiocyanates*	15
Triglycerides*	33426
Urea Nitrogen (B.U.N.)*	201 706
Uric Acid (urate)*	23 038
Urinalysis, routine chemical (any of S.G., pH, protein, sugar, hemoglobin, ketones, urobilinogen, bilirubin, leukocyte esterase, nitrite)*	37 534
Urinalysis microscopic examination of centrifuged specimen*	11 497
Valproic acid (valproate)*	1154
VMA, Vanillylmandelic acid (Vanillylmandelate)*	1212
Biochemical assays not included above*	7844
Prostate specific antigen (PSA), total	1063
Alpha Glycoprotein Subunit	179
Amylase isoenzymes	535
Apolipoproteins A and B	12 478
FK506 (Tacrolimus)	8577
Lipoprotein (a)	792
Prostate specific antigen (PSA), total	4258
Troponin	44 477
Homocysteine	1585
Chylomicrons	4
Oligoclonal banding	1437
Citrate	2967
Sirolimus (Rapamycin)	526

Table 1.1 (*Continued*)

Type of Test	Number performed (per year)
Natriuretic peptide – Brain (BNP)	3201
Foetal fibronectin	16
Bioavailable testosterone	12 886
Aldosterone*	3059
Cortisol*	3470
Aminoglycosides (*e.g.* Gentamicin, Tobramycin)*	1108
Androsternedione*	1215
Digoxin*	1103
ACTH (andrenocorticotrophic hormone)*	1520
Folate, in red cells, to include hematocrite and if requested, serum folate*	7710
Estradiol*	1914
FSH (pituitary gonadotropins)*	2807
Growth hormone*	2343
HCG (human chorionic gonadotropins)*	9848
Hepatitis associated antigen or antibody immuno-assay - per assay (*e.g.* hepatitis B surface antigen or antibody, hepatitis B core antibody, hepatitis A antibody)*	11 469
Aminophylline (Theophylline)*	167
Anti-DNA*	1822
Diphenylhydantoin (Phenytoin), quantitative (Dilantin)*	2424
Insulin*	802
LH (luteinizing hormone)	3091
Ferritin*	17 240
Parathyroid hormone*	7256
Progesterone*	908
Prolactin*	3104
17-OH Progesterone*	308
IgE* - not to be billed for RAST test	1131
T4, free - absolute (includes T-4 total)*	11 888
Testosterone	12 767
TSH (thyroid stimulating hormone)*	30 406
Phenobarbitone*	363
Vitamin B12*	10 897
C-peptide immunoreactivity*	1084
Dehydroepiandrosterone sulfate (DHEAS)*	664
25-hydroxy vitamin D*	7703
T-3, free*	6350
Thyroglobulin*	2466
Alphafetoprotein	942
Hormone receptors for carcinoma (to include estrogen and/or progesterone assays)*	793
Total	3 274 901

biosensors anyway! The dearth of biosensor devices in the clinical biochemistry laboratory also appears to be matched by the lack of employment of so-called "lab-on-a-chip"[3,4] devices, which are generally characterized by microfluidic sample handling. Time will tell but at present neither of these technologies appears to have fulfilled the promise offered. This chapter attempts to evaluate possible reasons for this including a detailed look at one major problem – that of interference by adsorbed species from the biological matrix. In order to provide a backcloth to discuss the issues at stake, we take a prior concise look at biosensor technology. The chapter has something of an emphasis on label-free methodology.

1.2   Biosensor Technology

1.2.1   Biosensor Architecture

A biosensor is composed of chemical recognition sites (probes) attached to a substrate surface that, in turn, is in close proximity and union with a trans-ducer. Ideally, this configuration responds sensitively and selectively to the presence of (bio)chemicals, we term the target or analyte, usually present in the liquid phase. The technology is based on the recognition of species through selective binding of the target to the probe at the substrate surface of the device with such surface presence being converted into an electrical signal. The overall architecture is depicted in Figure 1.1. Note that probes are invariably composed of biological entities such as cells or biochemicals such as antibodies. The sample itself may or may not be of biological origin. An important aspect of biosensor technology is the nature of the response of the device with respect to time. The field is often considered to include the aforementioned "one-test" disposable systems, where there is no attempt to conduct a measurement over a period of time. This type of device is more often than not at the heart of the point-of-care assay. From the clinical biochemistry point of view with regard to the central laboratory, both time-responsive and one-shot structures might be applicable. This issue will be considered in more detail later in the chapter.

The signal obtained from a selective biosensor response will take the following form (Eq. (1.1)):[5]

$$X_n = S_{n1}C_1 + S_{n2}C_2 + \cdots\cdots + S_{nn}C_n \tag{1.1}$$

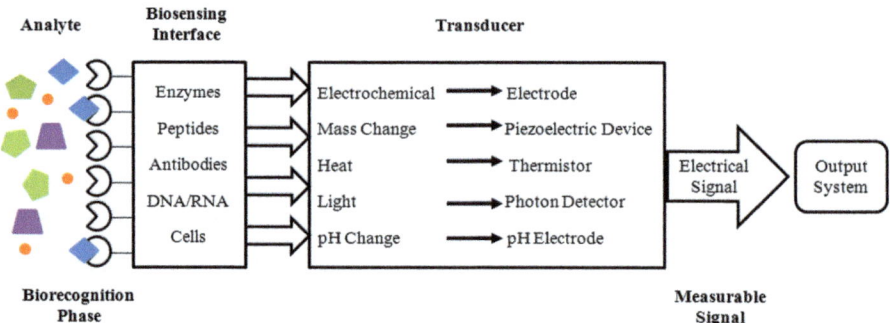

Figure 1.1   Schematic representation of a biosensor architecture featuring the biosensing interface, transducer and output system.

where C and S values represent the concentration of analyte *and* interferents in proximity to the device, and the response sensitivities, respectively. An enormous number of components would be expected to be involved in this equation in terms of samples such as serum or urine. A sensitive response implies maximization of one S value and, for a selective signal, a minimization of all other S values. Clearly, the sensor signal will be a *composite* of the chemistry of the attachment process and the physical perturbation caused by the probe–analyte complex. In this respect, it is crucial that, when designing a sensor for particular applications, the physics of the complete transduction process be thoroughly characterized and understood. In fact, some devices simply respond to the presence of the analyte, whereas others detect structural shifts caused by target–probe binding. The distinction between these mechanisms is not always evident.

1.2.2　Probe Attachment to Device Surfaces

A mandatory aspect of biosensor fabrication is the attachment of the biochemical probe to the surface of the transducer of choice. Over the years, a host of methods have been developed with this goal in mind.[6,7] A number of criteria are relevant to this component of sensor production:

- It will be vital to retain the binding activity of the probe in terms of its ability to bind the target. It is anticipated that certain proteins, for example, may be partially denatured and lose affinity upon attachment to the sensor surface, whereas others such as antibodies may be more robust.
- The proper spatial orientation of the probe with respect to the plane of the device surface will be an important element to ensure exposure of binding sites for target capture.
- In terms of sensitivity, it is crucial to maximize the density of probe molecules per surface area unit. Simply stated, the more probe available on the device surface, the greater will be the sensor signal.
- The spatial characteristics of the probe with regard to the surface plane are a factor that is important but not studied widely. If probes are in too close proximity on the surface, this may result in steric hindrance to target binding. In practice, there exists a maximal (optimal) effective probe loading.
- There are other more mundane factors such as the cost of reagents and the potential longevity of the modified sensor surface. These will clearly be critical from a commercial standpoint.

Some examples of methods for probe attachment follow. This short compendium is not intended to be an exhaustive treatment.

1.2.2.1　Noncovalent Attachment

The probe of interest can be chemically or physically adsorbed from solution directly, in most cases, on the substrate surface of the device. Various protocols are employed to achieve such an effect such as dip casting, painting, spraying and spin coating, where these terms will be self-explanatory to the reader. The interaction of the probe with the surface will be characterized primarily by hydrogen bonding and/or hydrophobic interactions. Adsorbed molecules can also act as a linker for eventual biomolecule attachment. A system that is widely employed is avidin/streptavidin/neutravidin chemistry.[8] Avidin is a tetrameric protein that contains four binding sites for the ligand, biotin. Interaction of the protein with this molecule results in a particularly strong binding ($K_d = 10^{-15}$ M). Accordingly, various biomolecules can be modified with relative facility by biotin incorporation in order to link them to surface-attached avidin or a sister molecule such as the aglycosylated version, neutravidin. An analogous strategy uses protein A.[9] The latter is a polypeptide (MW ~ 42 kDa) isolated from *Staphylococcus Aureus* that is capable of binding specifically to the F_c region of various antibody molecules and has thus been employed in immunosensor technology.

Another noncovalent approach is the trapping of the recognition molecule within the three-dimensional structure of a specific chemical matrix. The matrix may simply act as a "holding" moiety or be modified in some way to take part in the transduction process. For the former, polymers such as polyacrylamide have been employed in a number of experimental protocols. However, there are a number of potential disadvantages to this type of approach for placing the probe on the device surface including the possibility of biomolecule leakage and/or denaturation. An additional consideration is the necessity for the target to diffuse into the polymer matrix in order for the biochemical interaction to take place. An analogous procedure uses encapsulation by sol-gel technology,[10] wherein a solution of a monomer such as an alkoxide (the sol) is induced to polymerize into a biphasic configuration (the gel) that incorporates both liquid and solid. A typical monomer among many would be tetraethylorthosilicate (TEOS – EtO_4Si), which is readily hydrolyzed by water to produce a siloxane-based polymeric structure with a gel consistency. A cavity can be formed around the probe of interest in somewhat the same manner as for the polymers mentioned above.

1.2.2.2   Covalent Binding

Attachment *via* covalent bonds has been by far the most used approach. A very wide variety of chemistries have been employed with modest success in terms of the criteria outlined above.[11,12] Many functional groups, whether directly present on the device substrate or obtained by modification, have been utilized to form a probe partial monolayer. Groups are available on biomolecules to instigate the surface link and examples of these for proteins are presented in Table 1.2.

A common protocol to bind enzymes, antibodies or molecular receptors to the substrate is to initially functionalize the surface, which is then followed by a second reaction of activation. In essence, this process simply allows a convenient, highly reactive "linker" moiety to be introduced to the system.

Table 1.2 Target functional groups for biosensor immobilization of proteins.

Functional Group	Amino Acid
Thiol	Cysteine
Amine	Lysine (side chain)
Phenol	Tyrosine
Imidazole	Histidine
Indole	Tryptophan
Guanidine	Arginine
Carboxylic acid	Glutamic acid, Aspartic acid

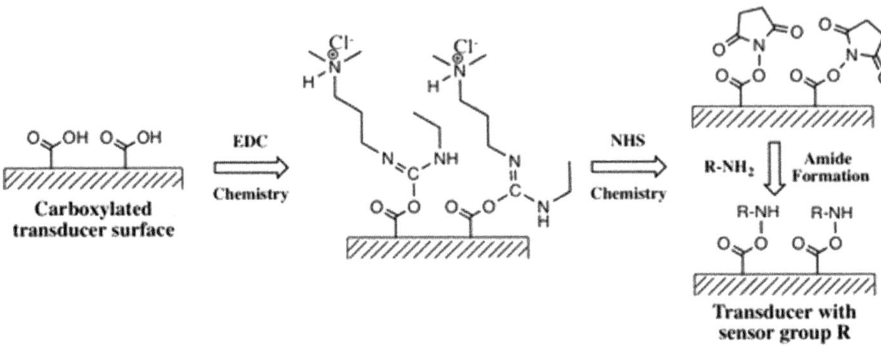

Figure 1.2   Schematic representation of a common strategy (EDC/NHS chemistry) to
covalently biofunctionalize surfaces *via* a preactivation process.
(Reprinted with kind permission of the Royal Society of Chemistry.)

The resulting interface is then normally allowed to react in turn with one of the
protein nucleophilic groups mentioned above. The literature is replete with
many examples of this sort of approach and, in Figure 1.2, we provide a
schematic of one strategy. If groups already evident on the device surface are
not used directly, functionalization of surfaces is often achieved with species
such as aminopropyltriethoxysilane (APTES), which reacts with interfacial
hydroxyl groups on whatever substrate they are present.

1.2.2.3   Self-Assembled Monolayer Chemistry

The introduction of close-packed monolayers has been used widely in biosensor
development. Two very different strategies are employed the first being the
Langmuir–Blodgett film technique.[13] In this experiment, close-packed monolayers are imposed on a surface by transferring lipid or lipid-like (amphiphilic)
films from the Langmuir trough, under the correct surface pressure conditions,
by a dipping process. In principle, several layers can be deposited in sequence
using this dipping approach. The very important advantage offered by this
technique is the possibility to combine artificial lipid membrane configurations
directly and *in situ* with integral membrane proteins (IMPs). Such membrane
systems are, of course, widely prevalent as signaling devices in biology.

An important chemistry, which has been significantly developed in recent
years, is the self-assembled monolayer (SAM). This approach relies on the use
of linking molecules that are engineered to *spontaneously* form ordered molecular assemblies on solid substrates. In this case, the most common strategy
has been the assembly of relatively long chain thiols on the surfaces of clean
gold substrates,[14,15] as shown in Figure 1.3. The distal end of bifunctional thiols
can then be employed for conventional covalent binding of proteins and oligonucleotides as described above (Figure 1.2).[16] An alternative technique is the
use of trichlorosilanes rather than thiols, which can be bound to surfaces that
have been functionalized with or naturally possess –OH groups.[17] Examples of

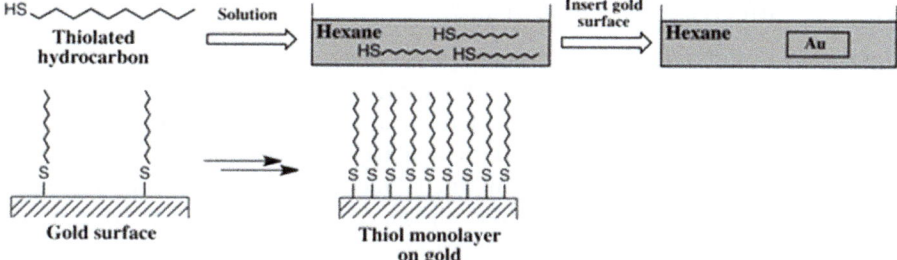

Figure 1.3 Long-chain alkyl thiol deposition onto a gold substrate to generate a self-assembled monolayer.
(Reprinted with kind permission of the Royal Society of Chemistry.)

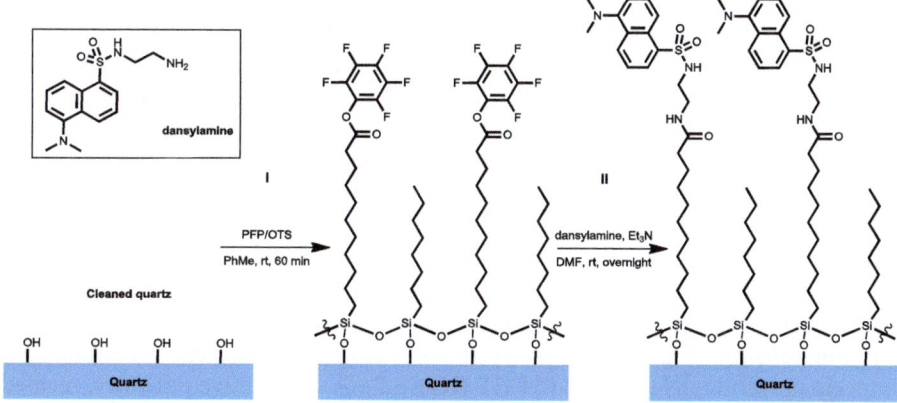

Figure 1.4 Preparation of an organosilane adlayer on quartz and subsequent covalent functionalization.
(Reprinted with kind permission of the American Chemical Society.)

substrates in this case would be silicon dioxide (*e.g.* quartz – Figure 1.4)[18] and indium-tin oxide (ITO).[19]

1.2.3 Devices and Transduction

A very large amount of work has been performed on a variety of biosensor structures. A bibliography detailing this effort up to the end of 2012 is provided herein.[20] Three distinct areas of physics have been employed with respect to the transduction process, these being electrochemistry, acoustic wave technology and electromagnetic radiation. There is considerable variety in terms of the nature of measurement, for example, possible label-free operation and secondly dipstick or response with time methodology. Moreover, a number of adjunct techniques have been employed in order to enhance the level of information obtained from the sensor determination.

1.2.3.1 *Electrochemical Devices*

A number of different approaches are possible *via* electrochemistry,[21,22] which involves the transfer of electrons or charge at solid/liquid or liquid/liquid interfaces. Those involving solid structures unsurprisingly dominate the field, which is summarized in Table 1.3. A particularly useful feature of many of these techniques is that they can be combined "naturally" with various forms of integrated electronic circuitry. Indeed, the necessary chemistry can be imposed directly on the surface of such an electronic device (see below).

Historically, one of the earliest devices generated from the world of electrochemistry was the sensor based on *potentiometry*. In this technique, an indicating electrode for the analyte of interest is incorporated in an electrochemical cell (with reference electrode) for which minimal or no current is passed. In this category, the ion-selective (indicating) electrode (ISE) forms the basis of systems, which are capable of detecting ions and other species of biochemical interest.[23,24] By far the most common in use is the glass electrode, which is sensitive to changes in hydronium-ion concentration. Other systems available for selective ion detection include electrodes that employ crystalline matrices, *e.g.* LaF_3 for F^-, or incorporate liquid ion exchangers in polymer matrices such as systems for sensing Ca^{2+}. The ISE has also been the basic structure used to develop enzyme-based molecular sensors where analytes are allowed to interact with immobilized enzymes and transform into products detected by the electrode.

An alternative transducer to the conventional electrode outlined above is the field-effect transistor (FET), which was introduced by Bergveld[25] in the 1980s for ion sensing. Figure 1.5 shows the structure of a typical n-channel enhancement mode insulated-gate FET (IGFET). A conductive pathway, termed a channel, can be instigated between n-type regions. One of the n-type contacts is called the source and the other the drain; a potential is applied that drives current in the channel. The contact on the surface of the center part of the insulating layer is called the gate. It is in this location that the device is employed to detect charge changes connected to the chemistry conducted at the gate/solution interface (Figure 1.6). Note that certain regions of the device must be encapsulated and that the system as mentioned above needs an adjunct reference electrode. The device has been employed in various formats including

Table 1.3 Electrochemical techniques available for application in biosensor technology.

Technique	Methodology
Amperometry	Measures current *versus* concentration; potential is controlled
Conductometry	Measures conductance (= 1/resistance) at controlled concentration
Coulometry	Measures change over time with controlled potential
Potentiometry	Measures electrochemical potential at current = 0
Voltammetry	Measures current at applied potential; controlled concentration of electroactive species

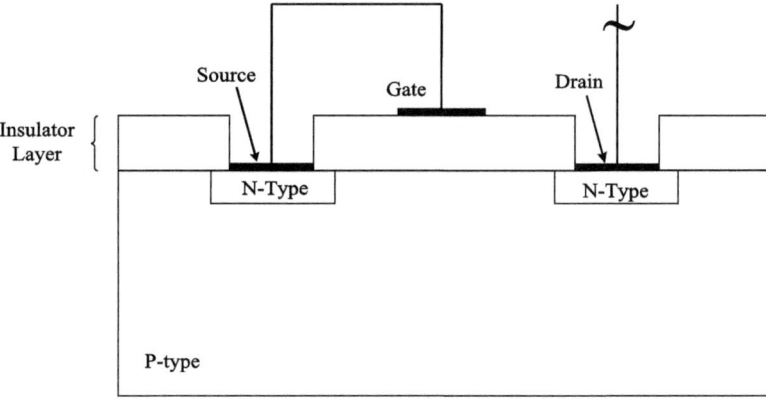

Figure 1.5   The n-channel enhancement mode insulated-gate field effect transistor –
IGFET.
(Reprinted with kind permission of the Royal Society of Chemistry.)

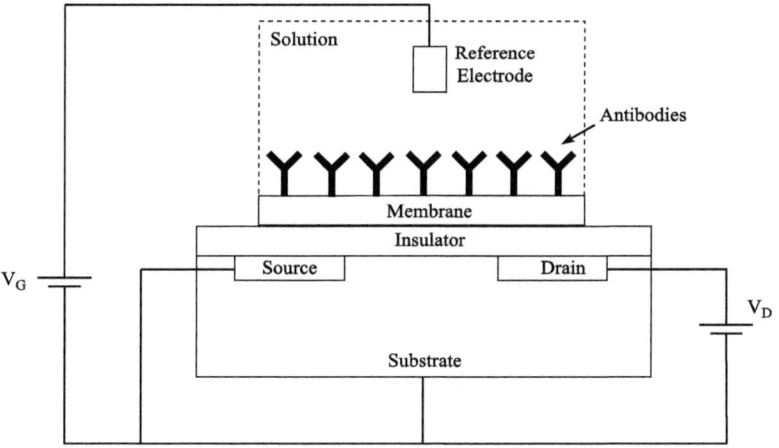

Figure 1.6   IGFET typical immunosensor set-up. Note the insulating protection of
source and drain contacts.
(Reprinted with kind permission of the Royal Society of Chemistry.)

the detection of ion concentrations (ISFET)[26] or immunochemical interactions
(IMMUNOFET).[27]

Techniques that are based on measurement of the resulting current *versus*
applied potential are called *voltammetric* methods. Anodic stripping and cyclic
voltammetry would be two examples of such techniques. *Amperometry* has
many definitions in the literature but is more often than not simply regarded to
involve the measurement of current associated with an oxidation–reduction
reaction at an electrode. It is worth noting that the genesis of biosensor
technology was the oxygen-mediated electrode for the amperometric assay of
glucose. Over the years, the electrochemical detection of this analyte, given its
importance, has been the subject of literally thousands of studies, the history of

which has been reviewed by Wang.[28,29] One of the most important advances in the technology was the introduction of organometallic mediators such as ferrocene. This iron π-arene complex composed of a cyclopentadiene sandwich of Fe acts as a convenient electron acceptor for glucose oxidase, thus avoiding the dependence on the role of O_2.

The detection of nucleic acid interactions and duplex formation at an electrode surface by an amperometric protocol has been the subject of intensive research for some time.[30] One example among several is the use of the redox properties of, for example, ruthenium complexes to enhance the electroactivity of nucleic acid species (the intrinsic redox behavior of nucleic acid moieties is insufficiently sensitive for analytical purposes). The electrochemistry community often describes this protocol as a "label-free" detection strategy, which is obviously nonsense given the necessary use of the Ru adjunct agent.

Finally, we mention *electrochemical impedance spectroscopy (EIS)*. In this technique, a sinusoidally-varying voltage is applied to an electrochemical system and the resulting current is measured, usually over a range of angular frequencies. Recording of these parameters can be employed to compute the real and imaginary components of the electrical impedance (Z). Primarily, the system has been used to detect biomolecular recognition events taking place at an electrode much as the situation with the other electrochemical arrangements outlined above.[31]

1.2.3.2 Acoustic Wave-Based Biosensor Technology

Acoustic wave sensors are generally very much associated with piezoelectric physics. (It should be noted, however, that a number of available structures do not employ piezoelectric components in a direct sense.) Piezoelectricity is the electric polarization produced by mechanical strain in crystals belonging to certain classes, the polarization being proportional to the strain and changing sign with it.[32] Of course, we now recognize that the reverse is true, that is, a specific crystal can be mechanically strained when subjected to an electric polarization. The devices used in biosensing have in common the deformation caused on a piezoelectric crystal by the application of an electric field. The origin of this effect lies in the interaction between electric charge and elastic restoring forces in the crystal. All importantly, the phenomenon *cannot* take place in a crystal possessing *central symmetry*. In order to maximize the coupling of electrical and mechanical effects, it is conventional for practitioners to use particular slices of crystals or *cuts*. For example, with respect to the predominant piezoelectric material, quartz, these are AT-, ST- and BT-cuts. Movement of particles in a piezoelectric crystal caused by an oscillating electric field-induced stress, restored by elastic forces, leads to standing or travelling waves in the material and the phenomenon of piezoelectric resonance.

Several types of acoustic wave devices have been used as biosensors, but the simple thickness–shear mode (TSM) has been by the most employed for detection purposes. The TSM is composed of an electroded (often gold)

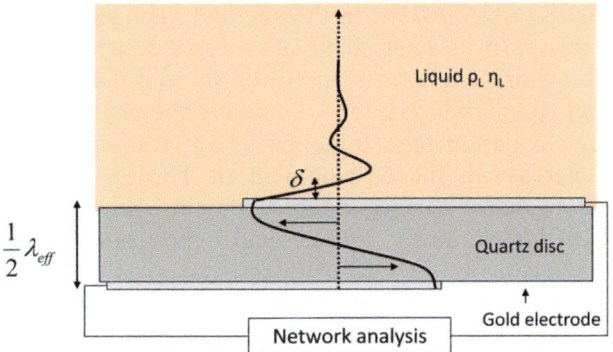

Figure 1.7 Working principle of the thickness-shear mode (TSM) acoustic wave device featuring the standing wave in the substrate with propagation of a damped acoustic shear wave into the liquid.
(Reprinted with kind permission of the Royal Society of Chemistry.)

piezoelectric wafer (usually AT-cut quartz), as depicted in Figure 1.7. An oscillating electrical potential is applied *via* the electrodes in order to drive mechanical motion in the device. A resonant acoustic shear wave is generated, which travels through the piezoelectric material with little energy dissipation. The wave is reflected at the device/surroundings interface in order to maintain a standing wave (Figure 1.7). In chemical and biosensor applications, material is generally added to the device surface, for example, species involved in bio-molecular interaction. On a simple theoretical basis, this scenario was the subject of very early work by Sauerbrey,[33] who showed that when material is deposited on the device surface the resonant frequency is changed according to his famous equation that is expressed for quartz as the piezoelectric material (Eq. (1.2)):

$$\Delta f = -\frac{2f_0^2}{A\sqrt{\rho_q \mu_q}}\Delta m \tag{1.2}$$

where Δf is the frequency change, f_0 is the primary resonant frequency of the sensor, A is the effective surface area associated with the piezoelectric process, ρ_q is the density of quartz, μ_q is the shear modulus of the particular cut of quartz employed and Δm is the change in mass. The idea of mass detection has spawned the ubiquitous term, *quartz crystal microbalance or QCM*. Unfortunately, this argument has been widely extended to operation of the device in the liquid phase and has become something of a dogma, especially in the community of electrochemists. In such a medium however, it is a reality that acoustic energy is transferred to the surrounding liquid resulting in a damped wave *via* the effects of bulk phase viscosity (Figure 1.7).

Experimentally, a number of approaches have been employed in order to measure the resonant frequency and other acoustic parameters, in some case in a static fashion, and in others, in flowing liquid through a flow-injection con-figuration. The most rigorous approach is that provided by what is often termed *acoustic network analysis*. In this method, the magnitude and phase of impedance of the sensor is determined at a set of frequencies under resonance conditions. A network analyzer is employed to record the Butterworth–Van Dyke equivalent circuit, which essentially relates the physical properties of the device to electrical parameters. In terms of applications, recent years have seen a rapid increase in the development of TSM technology, which has been employed to detect surface-induced protein conformational changes,[34] immunochemical interactions,[35] nucleic acid hybridization[36] and cell-surface attachment.[37]

A recent acoustic wave device development is the introduction of the electromagnetic piezoelectric acoustic sensor (EMPAS) structure.[38] In this technology, acoustic waves are instigated in an electrodeless quartz wafer by an electromagnetic field produced in close proximity to the wafer by a flat spiral coil (Figure 1.8). The secondary electric field associated with the coil drives the piezoelectric effect in the device. The configuration possesses a number of important advantages:

- It is not necessary to operate the device with contact metal electrodes in place and with electrical connections. Acoustic resonance is driven remotely. This renders advantages in terms of flow-through design and surface chemistry can be studied directly in an unhindered fashion.
- Crucially, it is possible to operate the sensor at ultra-high frequencies, *e.g.* 1 GHz, *via* bulk acoustic wave overtones. This leads to higher analytical sensitivity.
- It is possible to tune the device with ease to specific frequencies, which could potentially lead to important interfacial chemical information.
- The surface chemistry for biomolecule attachment involves SiO_2, which is a more developed area of chemistry than is the case for binding to metals such as gold.

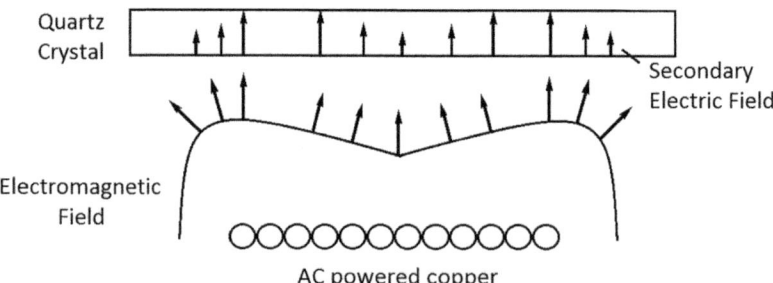

Figure 1.8   Working principle of the electromagnetic piezoelectric acoustic wave sensor (EMPAS): acoustic resonance is remotely driven in the electrode-free quartz substrate by a secondary electric field induced by the electromagnetic field associated with an AC-powered flat spiral coil.
(Reprinted with kind permission of the Royal Society of Chemistry.)

Surface-launched wave devices have also been employed in biosensing, where waves are generated in a piezoelectric substrate by transducers that are placed on the surface of the material. Particle movement, which is generally detected by separate transducer(s) imposed on the same surface, is often restricted to the "near" surface of the substrate, unlike the case for the TSM discussed previously. For an introduction to several of these devices, please refer to reference 39. The *surface acoustic* or *Rayleigh wave* (SAW) sensor has an interdigital transducer (IDT) fabricated on the piezoelectric substrate (*e.g.* ST-cut quartz) as shown in Figure 1.9. The chemistry related to the sensing processes is conducted on the area between the IDTs and it is generally the case that the electrodes have to be isolated from the liquid if the device is operated in that medium, in a similar fashion to that described above for FETs. An example of

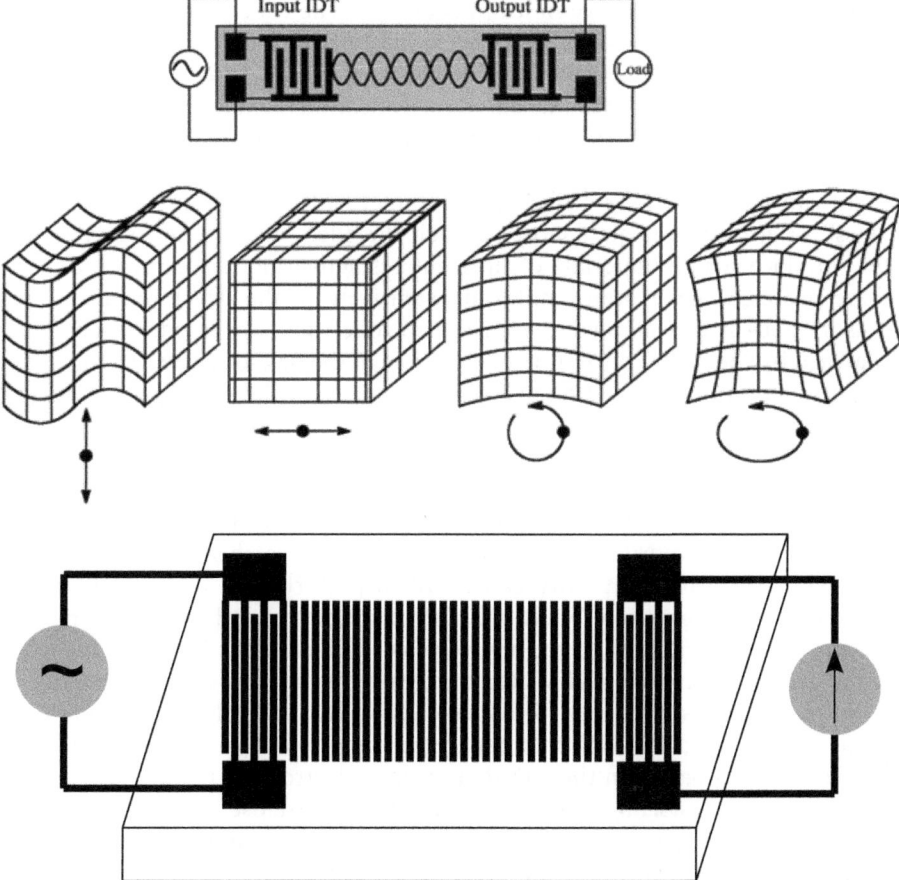

Figure 1.9 Upper: schematics of a surface acoustic wave device. Center: Rayleigh waves contain both shear and compressional components. Lower: schematics of a surface transverse sensor.
(Reprinted with kind permission of the Royal Society of Chemistry.)

another device employed *via* this technology is also portrayed in Figure 1.9. This is the *surface transverse wave* (STW).[40] The use of these biosensors, which all involve very similar biomolecule immobilization strategies to those outlined above, have been reviewed by Rapp and coworkers.[41]

1.2.3.3   Electromagnetic Radiation and Optical Biosensor Technology

Virtually, the full gamut of physics offered by optical science has been employed over the years for the detection of fundamental biophysical processes, biochemical binding events or species of bioanalytical interest such as biomarkers for disease. Techniques include those based on measurements of absorption, luminescence, interference, reflectance, scattering (including Raman spectroscopy) and refractive-index phenomena. The wide variety of techniques employed in optical sensing have been nicely reviewed.[42] One of the main features of this area has been the concentration on label-free detection.[43]

One of the earlier devices was based on optical waveguide technology and, in particular, the optical-fiber sensor. In essence, light transmission along a fiber is produced *via* a guided wave through an integral process of internal reflection. Light can transmit along a fiber with complete internal reflection or *via* some "loss" of electromagnetic energy through the effect of reflection/refraction at the interface where materials of different refractive index are involved. Both these processes have been employed extensively in the development of biosensors and in this context the term "optrode" has appeared, as obviously spawned by the older concept of an electrode in the field of electrochemistry. In the first methodology, the fiber is simply used as a delivery system for light interaction with an optical configuration placed at the distal end of the structure,[44] as shown in Figure 1.10a. In the second scenario called the *intrinsic* structure, some light energy finds its way under certain conditions into the medium outside the fiber in the form of a penetrative evanescent wave (Figure 1.10b). Importantly, the interaction of surface chemistry at this interface with the evanescent wave can result in perturbation of the light phase, intensity and polarization. There are many examples of such an arrangement in biosensor detection.[45] This type of device has been employed successfully in the assay of "real" samples such as for the detection of pathogens.[46]

Surface plasmon resonance (SPR) has become one of the leading technologies with respect to detection in bioanalytical and biophysical chemistry. The physics effect is based on the fact that valence electrons of metals such as gold and silver exhibit oscillations in their density. If these oscillations, in the form of surface waves, are present at the interface of the metal with a material of a different dielectric constant, they can be excited by the introduction of electromagnetic radiation. The classical approach to the study of this phenomenon was initially introduced by Kretchmann[47] and is shown schematically in Figure 1.11a. The incident radiation is reflected at the metal/dielectric boundary resulting in an evanescent wave in the metal, much as described above for optical fibers. At the correct (resonant) angle, this wave can couple in a *resonant* fashion with the frequency of the incoming radiation, resulting in the excitation of electron density oscillation mentioned above, leading to the term *surface plasmon resonance*. Experimentally, in sensor operation, biochemical binding events are conducted at the metal surface and detected through measurement of shifts in the SPR angle, observed wavelength of absorption or change in the position of reflectivity. A resonant transfer of electromagnetic energy into the surface plasmon wave is observed through a minimum in the plot of incident angle *versus* reflected intensity (Figure 1.11b).

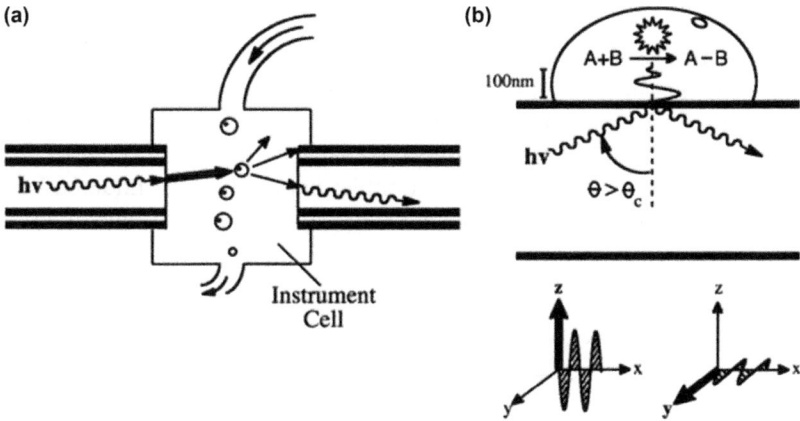

Figure 1.10   (a) Schematics of extrinsic fiber-optic delivery of radiation to a cell for absorption measurement. (b) Evanescent radiation penetrating to the exterior of a fiber in an intrinsic configuration.
(Reprinted with kind permission of the Royal Society of Chemistry.)

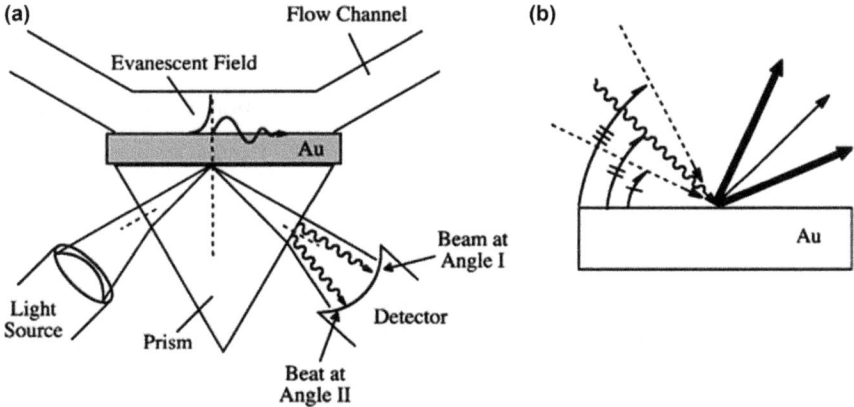

Figure 1.11   (a) Kretschmann configuration for surface plasmon resonance (SPR). (b) Incidence angle *versus* reflected intensity of radiation as a result of SPR.
(Reprinted with kind permission of the Royal Society of Chemistry.)

A simplified schematic of a typical SPR experiment is shown in Figure 1.12. This type of instrument designed for work in the biosensor arena is generally capable of analyzing several channels (gold surfaces) using microfluidic sample introduction in a real-time, label-free fashion through flow-injection technology. The system generates a response plot, which is often referred to in the field as a "sensorgram", presumably in the light of chromatograms and the like! With respect to applications, the SPR technique has been employed in a very wide variety of cases such as adsorption of proteins, cells and nucleic acids on gold and modified metal surfaces, detection of biomolecular interactions such as found in immunochemistry and nucleic acid duplex formation.[48–53] The method has proven to be particularly helpful for epitope determination in the former area.

Finally in this section, we mention the technique of *interferometry*. The latter is based on the superposition of, usually, electromagnetic waves that yields information concerning the original nature of the waves. In biosensing technology, several types of device designs have been employed in order to attempt the detection of biochemical species, examples being Mach–Zehnder's, Young's and Hartman's sensors (these are depicted in Figures 1.13a–c). The first of these

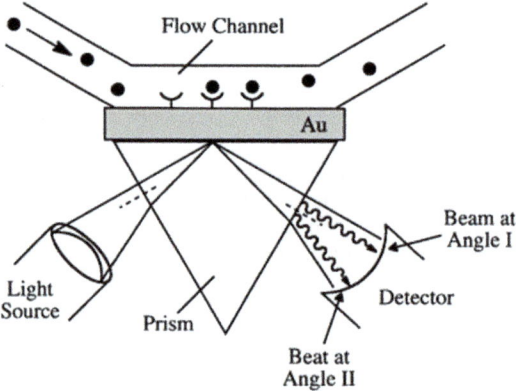

Figure 1.12 Typical SPR experiment involving flow injection of target analyte into the flow cell.
(Reprinted with kind permission of the Royal Society of Chemistry.)

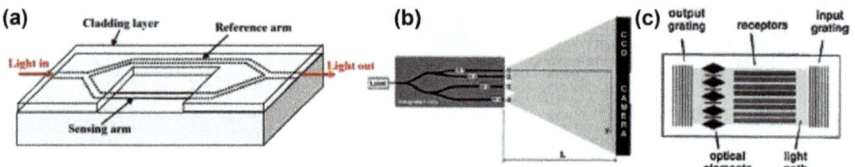

Figure 1.13 (a–c) Mach–Zehnder's, Young's (multichannel) and Hartman's interferometer sensors, respectively.
(Reprinted with kind permission of Elsevier.)

involves two different light pathways, as is typical in interferometry, with one being subjected to passage through the sample. The laser radiation is then combined with a reference beam resulting in the interference-based signal. The analytical arm incorporates an evanescent field, which interacts with, in this application, a biochemical moiety. This technology was employed in the earlier years for the detection of protein chemistry, where analytical detection limits of around 50 pM were found.[54]

An additional interference-based structure is Young's device.[55] The overall principle is similar to the sensor mechanism employed in Mach–Zehnder's device, the detection mode is different, however. In this case, the output from the sample and reference channels is combined to form interference fringes on a CCD screen. The method involves Fourier transform of the spatial intensity measure at the detector screen. In Hartman's interferometer,[56] electromagnetic radiation is coupled into a waveguide using gratings, which allows the interference phenomenon to occur between reference and sample strips. As with the other two devices outlined above, protein chemistry was detected as imposed on the waveguide strip.

1.3 Biosensors and Measurement of Clinical Targets

It is very evident from Table 1.1 that clinical biochemical detection and measurements involve a vast array of targets in biological matrices. Although it is not the aim of this chapter to detail the fundamental requirements of the laboratory – these are provided elsewhere[57] – we do summarize some of the main necessary criteria. These are then compared in the context of what is offered by biosensor technology. Essential properties and protocols of clinical determinations are:

- *High throughput.* As outlined above, the clinical biochemistry laboratory faces literally thousands of samples on a daily basis, which may have one particular analyte determination required or perhaps several on the same sample. The sheer numbers involved mandate a very high level of automation and robotic technology. This means that considerable technical effort is required to keep equipment in operation. For the most part, modern-day clinical analysis incorporates the capability to handle samples in a batch-wise process.
- *Speed of analysis.* Obviously, the requirements in this criterion will vary enormously. However, it would be desirable to possess the capability to perform rapid analysis, for example, as requested by a physician in an emergency situation.
- *Generic methodology.* Where possible, it is advantageous to possess the possibility to perform the determination of different targets employing the same detection physical chemistry. It goes without saying that this aspect, by the very nature of target analyses required, will be somewhat limited. However, there exist several technologies possessing this attribute, such as magnetic-bead ELISA.

- *Validation of dose–response and limit-of-detection (LOD) characteristics.* The former parameter refers to the meaningfulness of the signal obtained from the target in a *biological matrix* in terms of its concentration. This would be more familiar to the analytical chemist as the "calibration curve" for the particular method involved. The LOD would also be well known to the analyst being the minimum concentration value that a method can detect (in analytical chemistry, generally 3 times the standard deviation of the background noise).
- *Cost per analysis.* Given the great expense involved in the operation of a central clinical laboratory, it is hardly surprising that the calculated cost per assay will always be a factor under consideration. Factored into this are also the costs associated with salaries of operators, reagents, equipment and its maintenance.
- *Functionality in complex media.* It is clearly crucial for a particular method to function in samples such as blood, serum, plasma or urine. We have left this aspect to the end for reasons that will become apparent.

We now turn to the "ideal" biosensor. The relevance of these criteria to the clinical chemistry operation will be apparent to the reader.

- High sensitivity with resulting low value of limit-of-detection. This parameter will be governed by the physics of the detection technique employed – the S values outlined above.
- High accuracy in terms of concentration measurement for the analyte. Signaling conducted with high reproducibility – a precision issue.
- The dynamic range with respect to analyte response should be as wide as possible. For certain targets, indeed, clinical values may vary considerably.
- Signal approach – dipstick/batch or real-time sensing. It would be advantageous for a particular device to be capable of operation in the dipstick/batch manner or as a real-time sensor. The speed of response will obviously be important in both situations although for different reasons. Real-time operation offers the potentially attractive possibility to incorporate the biosensor into an automated flow-injection scenario. This is not the well-known segmented situation, but a technique based on analyte dispersions allowed to flow past the device surface. The flow-injection analysis (FIA) will require that the biosensor signal be reversible or the target can be washed from the surface within the flow system.
- Device response calibration in terms of concentration. This is mandatory and especially important for the real-time sensor. This feature has constituted an extremely difficult problem when it comes to, for example, the operation of a corporeal implantable structure. Early solutions to this issue may well lie in a strategy of device self-referencing.
- Selectivity or even specificity (see reference 5 for an excellent definition of these parameters) with respect to response to the analyte. In this respect with specific regard to clinical chemistry, it is essential that the device

function selectively in media that contain very high concentration of various biomolecules such as proteins and even cells. These species will tend to adsorb to the sensor surface potentially interfering with the required signal. This is one of the reasons why the legendary closed-loop implantable glucose device is not available despite many years of research. *The issue of nonspecific adsorption (NSA) represents a true Achilles Heel for application of biosensor technology in clinical biochemistry.*

Finally, it is worth noting that there have been several attempts to assess the capability of biosensors to detect molecules of clinical interest. Table 1.4 shows an example where the authors place performance criteria in the context of classical analytical chemistry.[58]

1.4 Signal Interference and the Non-specific Adsorption Problem

In biosensor technology, the undesired "non-specific adsorption" (NSA) of adversary species (sometimes also referred to as "non-specific binding") – as opposed to the "specific adsorption" of target analytes – is a serious and prevailing concern for many detection platforms intended to perform analysis in extremely challenging biological samples, more often than not blood serum or plasma. In fact, even cleared of the cellular components of blood, these biological matrices still consist of highly complex mixtures of potentially interfering biomolecules – in particular various types of proteins at high concentration $(60–80 \, g/L)$[59] – that have the propensity to adsorb non-specifically to the sensing surface of devices thereby preventing the detection, not to mention the quantification, of target analytes present at considerably lower concentration (down to ng/L or a difference of nine orders of magnitude).[60] Indeed, during the biorecognition phase (wherein analytes are expected to site-specifically bind to complementary receptors immobilized on the sensing platform), non-specifically adsorbing species also generate physicochemical stimuli that are indiscriminately detected by the biosensor and interfere with the specific response of the target analyte (Figure 1.14).

The most unfortunate immediate consequence of NSA – besides altering the performance of the biosensor, may it even be operational in biological samples in the first place – is the occurrence of "false positives", *i.e.* a response incorrectly interpreted as a genuine binding event, which understandably renders these biosensors irrelevant for real-world applications.[61] Arguably, NSA is the single most important reason why biosensors still have not found a prominent place as alternative diagnostic tools in clinical analysis at the beginning of the 21st century despite tremendous promise notably in terms of cost/ease of operation and miniaturization for "point-of-care" applications – the ultimate aim of biosensor technology.[62] More generally, NSA generates a high, often overwhelming background signal that may lower the sensitivity of the biosensor (poor signal-to-noise ratio) to clinically irrelevant levels.[60,63]

Table 1.4 Analytical figures of merit of different types of biosensor detecting various molecules of clinical interest. (Adapted from reference 58.)

Analyte detected (matrix)	Sensor type	Linear range	LOD	Sensitivity (slope)	Additional information
Glucose (buffer)	CNT enzyme-based biosensor	Up to 10 mM	0.25 μM	30.2 ± 0.5 μA/mM	$r^2 = 0.999$; $n = 10$
Glucose (blood serum)	Amperometric enzyme-based biosensor	1.0 μM–0.8 mM	0.5 μM[a]	2.3 mA/M	$r^2 = 0.999$; $n = 7$ CV(%) = 4.8
Glucose (human serum)	Amperometric enzyme-based biosensor	1.0 μM–1.6 mM	0.69 μM[a]	69.26 mA/M.cm²	$r^2 = 0.999$; $n = 25$ CV(%) = 1.1
Glucose (buffer)	Amperometric CNT-based biosensor	0.005–0.3 mM	3 μM[a]	80 ± 4 mA/M.cm²	$r^2 = 0.996$; $n = 5$ CV(%) <5%
Glucose (buffer)	CNT enzyme-based biosensor	Up to 2 mM	4 μM	0.33 μA/mM	$r^2 = 0.998$
Glucose (buffer)	CNT enzyme-based biosensor	15.0 μM–6.0 mM	7 μM[a]		$r^2 = 0.992$; $n = 8$ CV(%) = 1.6
Glucose (buffer)	Amperometric enzyme-based biosensor	0.02–5.7 mM	8.2 μM[a]	8.8 ± 0.2 mA/M.cm²	CV(%) = 2.1
Cholesterol (human serum)	Amperometric CNT-based biosensor	0.004–0.7 mM	1.0 pM[a]	1.55 μA/mM	$r^2 = 0.9954$; $n = 8$ CV(%) = 4.2%
Cholesterol (human serum)	CNT enzyme-based biosensor	0.004–0.10 mM	1.4 μM[b]		$y = 14.23x + 0.2602$ $r^2 = 0.999$; $n = 11$ CV(%) = 2.5
Cholesterol (buffer)	Amperometric enzyme-based biosensor	1.2 μM–1 mM	0.12 mM[c]		$r^2 = 0.9998$; $n = 11$ CV(%) <4.0
Cholesterol (buffer)	Amperometric enzyme-based biosensor	Up to 12 mM	0.18 mM	1.61 nA/mM	
Cholesterol (buffer)	CNT amperometric biosensor	0.2–6.0 mM	0.2 mM[a]	0.559 μA/mM	
Cholesterol (buffer)	Impedance enzyme-based biosensor	50–400 mg/dL	25 mg/dL		$r^2 = 0.9946$
Cholesterol (buffer)	Impedance enzyme-based biosensor	25–400 mg/dL	25 mg/dL	7.76×10^5 Abs/mg.dL	$y = 0.03546x + 7.76 \times 10^{-5}$
Cholesterol (human serum)	SPR enzyme-based biosensor	50–500 mg/dL	50 mg/dL	1.04 m°/mg.dL	
Cholesterol (human blood)	Voltammetric enzyme-based biosensor	0.01–0.06 mM		0.85 nA/μM	$r^2 = 0.99$ $y = 0.6054x + 0.2411$

Analyte (matrix)	Sensor	Linear range	Detection limit	Sensitivity	Regression/statistics
Glutamate (buffer)	CNT enzyme-based biosensor	0.2–250 μM	0.01 μM[a]	433 μA/mM.cm²	r² = 0.9984
Glutamate (buffer)	Amperometric enzyme-based biosensor	0.5 μM–500 mM	0.01 μM[a]	100 ± 10 mA/M.cm²	CV(%) = 4.8; r² = 0.991
Glutamate (buffer)	Amperometric enzyme-based biosensor	0.25 mM	2.0 μM[a]	80 ± 10 nA/μM.cm²	n = 5
Glutamate (buffer)	Amperometric enzyme-based biosensor	10 μM–1.5 mM	5 μM	88.8 nA/mM	y = 88.832x; r² = 0.9945; n = 3
Glutamate (buffer)	Amperometric enzyme-based biosensor	5 μM–0.5 mM	5 μM	0.62 mA/mM.cm²	r² = 0.9968
Glutamate (buffer)	Amperometric enzyme-based biosensor	20 μM–0.75 mM	20 μM		y = 5.0145x + 0.1638; r² = 0.9950; n = 3; CV(%) = 9.6
Glutamatc (buffer)	CNT enzyme-based biosensor	25 μM	25 μM	1.29 ± 0.18 nA/μM.cm²	y = 1.29x + 0.088
CEA (human serum)	Amperomctric immunosensor	0.5–5.0 ng/mL	0.2 ng/mL[a]		
CEA (human serum)	SPR immunosensor		0.5 ng/mL		S/N = 2.7
CEA (human serum)	Potentiometric immunosensor	4.4–85.7 ng/mL	1.2 ng/mL[c]		y = 24.36 log x − 10.37; r = 0.997; n = 5
CEA (human serum)	Electrochemical immunosensor	2–14 U/mL	1.73 U/mL[a]	3.2 ± 0.1 mL/U	CV(%) = 2.2; r² = 0.9975
Nitric oxide (buffer)	Amperometric CNT-based sensor	0.2–150 μM	0.08 pM		
Nitric oxide (buffer)	Hemoglobin-based electrochemical biosensor	1–50 μM	0.003 pM[a]	0.0424 μA/μM	y = 0.0424x + 2.441; r = 0.9978
Nitric oxide (buffer)	Hemoglobin-based amperometric biosensor	0.01–1.0 μM	5.0 pM[a]		y = 1.356x + 1.168×10⁻²; r² = 0.9989; n = 8; CV(%) = 4.2
Nitric oxide (buffer)	Hemoglobin-based amperometric biosensor	0.04 nM–5.0 μM	20 pM[a]		
C-Reactive Protein (human blood)	Magnetic immunosensor	0.025–2.5 μg/mL (28.5 pM–2.9 nM)	25.0 ng/mL[d]		y = 0.2108x + 72.31; r² = 0.989

Table 1.4 (*Continued*)

Analyte detected (matrix)	Sensor type	Linear range	LOD	Sensitivity (slope)	Additional information
C-Reactive Protein (human blood)	SPR Immunosensor	2–5 μg/mL (2.3–5.75 nM)	1 μg/mL[e]		$y = 10.675x - 16.995$ $r^2 = 0.9867$
C-Reactive Protein (human blood)	Capacitive immunosensor	0.025–1 μg/mL			$y = 2.78x + 1.29$ $r^2 = 0.9255$ CV(%) = 1.72
C-Reactive Protein (human blood)	SPR immunosensor	5–25 μg/mL			
Catecholamines: DA, EPI, NEPI (buffer)	Amperometric enzyme-based biosensor	DA: up to 40 μM EPI: up to 55 μM NEPI: up to 55 μM	0.2 μM[a] 0.4 μM[a] 0.3 μM[a]	75 nA/μM 45 nA/μM 60 nA/μM	CV(%) = 5.3 CV(%) = 6.0 CV(%) = 6.1
Catecholamines: DA, EPI, NEPI (human urine and plasma)	Optical fiber enzyme-based biosensor	5–125 pg/mL	DA: 2.1 pg/mL[c] EPI: 2.6 pg/mL[c] NEPI: 3.4 pg/mL[c]		$y = 0.344x + 0.4$; $r^2 = 0.9998$ $y = 0.140x + 0.04$; $r^2 = 0.9996$ $y = 0.252x + 0.38$; $r^2 = 0.9997$
EPI (buffer)	Fiber-optic enzyme-based biosensor	0.2–0.9 μM			$y = -0.016533x - 1.30882$ $r^2 = 0.988$
NEPI (buffer)	DNA-based biosensor	0.5–80 μM	5 nM		$y = 0.85x + 8.87$; $r^2 = 0.993$
DA (buffer)	Electrochemical enzyme-based CNT biosensor	1.0–30.0 μM	400 nM[a]		$y = 0.61x - 1.19$
DA (buffer)	Amperometric enzyme-based biosensor	5–120 μM		68.6 mA/mM.cm^2	$r^2 = 0.999$
Acetylcholine (buffer)	CNT-ISFET sensor	10–10000 μM	0.01 μM	378.2 μA/decade	
Acetylcholine (buffer)	Amperometric enzyme-based biosensor	1–1500 μM	0.6 μM[a]		
Acetylcholine (buffer)	Carbon fiber-based biosensor	1–100 μM	1 μM[f]		r = 0.998

Analyte (sample)	Biosensor	Linear range	LOD		Regression / Notes
Mycobacterium tuberculosis (simulation sample)	Bulk acoustic wave impedance biosensor	2×10^3–3×10^7 cells/mL	2×10^3 cells/mL		$\log y = 11.23457 - 0.00763x$; $r^2 = 0.9458$; $n = 5$
Mycobacterium tuberculosis (saliva)	QCM immunosensor	10^5–10^8 cells/mL	10^5 cells/mL		
Mycobacterium tuberculosis (buffer)	QCM sensor	10^2–10^7 cfu/mL	10^2 cfu/mL		$y = -2.90x - 64.29$; $r^2 = 0.997$
Human ferritin (serum)	QCM immunosensor	0.1–100 ng/mL			$y = 89.52 \log x + 117.68$; $r^2 = 0.9944$; CV(%): 4.97–20.4
Human ferritin (serum)	SPR immunosensor	0.2–200 ng/mL			$y = 152.13 \log x + 168.16$; $r^2 = 0.9997$
Choline (buffer)	Carbon fiber-based biosensor	1–100 μM	1 μM[f]		$r^2 = 0.9837$
Diphtheria antigen (buffer)	Potentiometric CNT-based immunosensor	24–1280 ng/mL	7.8 ng/mL		$y = 74 \log x - 98.6$; $r = 0.9978$; $n = 13$; CV(%) = 2.3
hCG (human serum)	Electrochemical CNT-based immunosensor	0.5–5.0 mIU/mL	0.3 mIU/mL[c]		$r = 0.9998$
hCG (human serum and urine)	QCM immunosensor	2.5–500 mIU/mL	2.5 mIU/mL[a]	5.38 ± 0.10 mIU/mL	
Insulin (serum)	SPR immunosensor	6–200 ng/mL	6 ng/mL		CV(%) < 5.0
Vibrio cholerae (buffer)	SPR immunosensor	3×10^5–3×10^9 cells/mL	10^5 cells/mL		$r = 0.9984$

CNT, Carbon nanotube; CEA, Carcinoembryonic antigen; DA, Dopamine; EPI, Epinephrine; LOD, Limit of detection; NEPI, Norepinephrine; QCM, Quartz crystal microbalance; SPR, Surface plasmon resonance; hCG, Human chorionic gonadotropin.

[a] LOD is three times the signal-to-noise ratio.
[b] LOD is $3s/k$ with s the standard deviation of 11 determinations and k the slope of calibration plot.
[c] LOD is three times the residual standard deviation.
[d] LOD is 10 times the background noise.
[e] LOD is higher than four times the background (0.5 AU, Arbitrary Units).
[f] LOD is approximately twice the noise level.

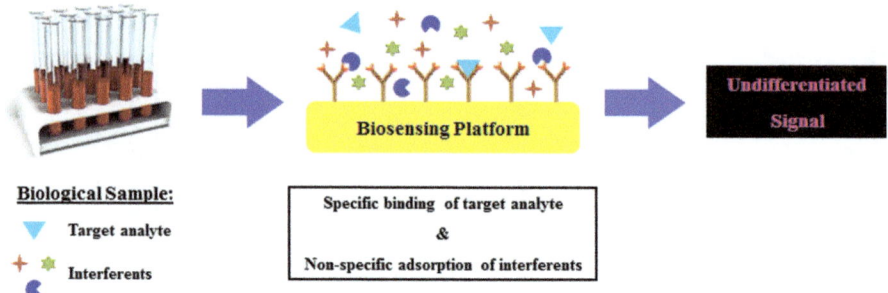

Biological Sample:

▼   Target analyte

Interferents

Specific binding of target analyte
&
Non-specific adsorption of interferents

Figure 1.14   The recurring problem in biosensor technology: the non-specific adsorption (NSA) of adversary species interfering with the specific target analyte response.

Admittedly, the conventional "enzyme-linked immunosorbent assay" (ELISA) – which displays remarkable sensitivity down to the low pM range[64] but remains a non-domestic test to be performed in a laboratory setting – still is promised a bright future.

To reliably alleviate the detrimental effect of NSA, it is necessary to better understand the complex mechanism(s) and dynamics of interaction involved once the biosensing surface comes into contact with the biological environment and the different types of potentially interfering species present within (*e.g.* proteins). In essence, the process of non-specific protein adsorption is quite complex and governed by several different factors, which includes: (i) the nature of the proteins contained in the biological fluid notably with respect to their structure (*e.g.* globular), size (molecular weight) or distribution of charge/polarity (isoelectric point); (ii) the physicochemical properties of the biosensing surface (*e.g.* charge, topography and morphology, surface energy); and (iii) the environmental conditions (*i.e.* pH, ionic strength and temperature).[65] To add to the complexity of the situation, proteins are flexible entities that can assume a variety of different conformational states, which results in the possibility for them to unfold and adopt the adequate geometry to fit the underlying surface, decrease the energy of interaction and irreversibly adsorb.[66] Furthermore, the energetics of adsorption itself is typically composed of diverse molecular forces (hydrophobic/electrostatic interactions, hydrogen bonding) that interplay to various degrees in a complex overall manner.[65] In multicomponent samples, proteins compete for surface binding sites, which triggers a series of collisions and results in a cascade of adsorption–desorption/exchange processes,[66] governed by the aforementioned conformational phenomena that occur at the more fundamental molecular level. Empirically, it has long been observed that protein adsorption from mixtures follows a temporal pattern, wherein higher mobility, more abundant proteins first adsorb transiently before being gradually replaced by less motile, scarcer ones with higher surface affinity.[66] This general, well-established phenomenon of sequential protein adsorption is known as the "Vroman effect".[66]

Undoubtedly, many factors affect the interactions of proteins with materials (of which the NSA of adversary species to biosensing surfaces is a particular case) and excellent reviews – pertaining to both the theoretical and experimental efforts made to model/probe the complex molecular events at play – can be found elsewhere in the literature.[66,67] Nevertheless, despite the mechanistic complexity of (non-specific) protein adsorption, some general rules relating the properties of proteins with their ability to adsorb to surfaces have been devised. For instance, the larger a protein, the more likely it is to possess multiple adhesion sites and readily adsorb to surfaces.[65] The same holds true when considering the flexibility of a protein (taken here as its ability to unfold), especially when (more) binding sites are exposed during conformational remodelling. This can occur in particular when a hydrophobic pocket (may it be a large inner protein core or simply a more limited surface domain) – previously hidden from the surrounding aqueous medium (wherein folded proteins reside fully hydrated) to minimize energetically unfavored polar/nonpolar interactions – is revealed upon adsorption.[66]

NSA is a particular case of a more general, ubiquitous phenomenon – surface fouling – that spontaneously occurs when exogenous biomaterials are exposed to biological environments. For many situations plagued by fouling – that extend well beyond the field of biosensors – tremendous efforts have been devoted over several decades to engineer antifouling surfaces, most traditionally through the imposition of organic films able to efficaciously resist protein adsorption.[68] As will be discussed in some more detail in the following section, numerous types of such coatings have been reported in countless literature publications and ultralow fouling ($<5\,ng/cm^2$) has now been achieved – for biotechnologically irrelevant, single-protein buffered solutions.[68] Regrettably, few coatings present/retain in actuality such a remarkable level of performance when exposed to highly complex biological media such as blood serum or plasma (even diluted), these being otherwise more demanding.[68] Fortunately, recent times have witnessed an increase in studies entirely focused on minimizing – or ideally eliminate entirely – fouling for real-world biological samples.[68] Assuredly, this trend should benefit the field of biosensor technology and help eradicate the NSA plague.

1.5 A Look at Surface Chemistries to Solve the NSA Issue

In Section 1.1, we have seen that the actual surface presented by a given biosensor structure to a biological fluid can take many forms. This will include inorganic material such as silica or indium-tin oxide in addition to polymers and organic linker molecules. It is a reality that the NSA or fouling discussed in the previous section can occur on all these materials to varying degrees. Accordingly, it is clear that there will be no panacea in terms of avoiding or even reducing the adsorption phenomenon. Although there has been a level of research aimed at the NSA issue with biosensors, it is certainly the case that most *relevant* work has been published on the study of material–biological fluid interactions. This topic has recently been reviewed in detail by the present authors.[68] The majority of the research described therein deals with adsorption of simple protein solutions and the like, with a much more modest effort devoted to biological fluids. A brief scan at fouling follows with some emphasis on our own efforts to avoid the phenomenon.

By far the majority of past related work describes the use of a great variety of organic coatings such as peptides, polyethylene glycols (PEG), zwitterionic sulfo- and carboxybetaines, methacrylates, acrylamide and biomimetic surface chemistries.[68] A vast literature on such coatings, often relatively thick in dimension, deals with the minimization, as mentioned above, of protein adsorption from simple buffer solutions. Whether in the latter type of solutions or more complex media, the PEG molecular family constitutes one of the most studied, but despite the high level of interest in this system the mechanism that lies behind the "PEG effect" remains obscure. One prevalent theory is that a kosmotropic water "barrier" to biological macromolecule adsorption is instigated by the polymer.[68]

In our research, we have focused on attempts to combine antifouling behavior with crosslinkers that are capable of immobilizing biochemical probes.[17] Covalently-bound adlayers based on silanization methodology, much as described in Section 1.2, have been employed in conjunction with the ultra-high frequency (GHz) EMPAS sensor. This is in order to study the surface effects caused when devices are subjected to a dispersion of goat serum.[69] The adlayer-forming molecules are custom-designed with several properties in mind. The molecules contain the ether linkage as present in PEG; they form ultrathin films ($\sim$ 5-Å thick) and are capable of merging with sister crosslinking molecules that form similar surface covalent bonds. The latter contain a distal functionalizable moiety that can be used to attach probes in a subsequent, preactivation-free step.[17] The behavior of MEG-OH was compared directly with related molecules with respect to fouling by the components of serum (Figure 1.15).[69] The remarkable result from this work is that a *single* ether oxygen atom in the monoethylene moiety of MEG-OH very significantly alters the surface behavior of the sensor in terms of interfacial adsorption from this matrix. On a speculative level, we believe that an intercalated water-based structure stabilized by the distal -OH groups is responsible for this effect. Further studies are underway both from a fundamental point of view and also with employment of the antifouling effect in biosensor experiments.

1.6   A Final Comment

A glance into a clinical biochemistry laboratory in a major hospital – such as St. Michael's located in Toronto, Canada – reveals groups of large instruments connected together with sizeable trains composed of robotic equipment. Much of this is orchestrated around batch-based assays and technology. The set-up is very far removed from biochip, biosensor and lab-on-a-chip devices, all of which have been touted for potential application in clinical assays. As outlined

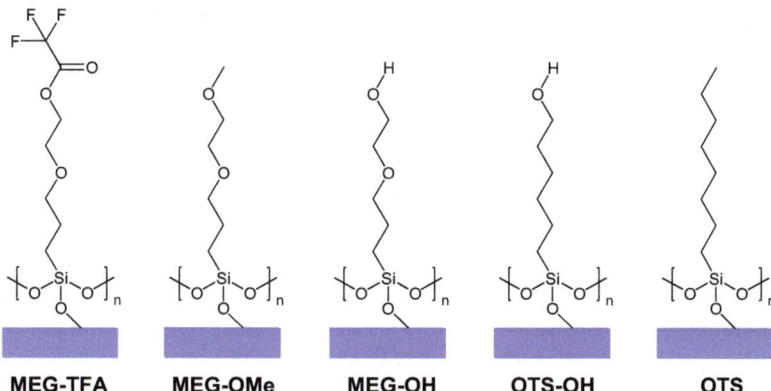

Figure 1.15 Modification of quartz with various structurally related organosilane surface modifiers.
(Reprinted with kind permission of the Royal Society of Chemistry.)

herein, the main reasons for this appear to lie in speed of analysis, convenience and saving with respect to reagent cost, *etc.* The latter appears to be a non-issue since such cost constitutes a minor component for the laboratory budget. In view of the requirements for the clinical laboratory, one may wonder how biosensor technology could realistically contribute to the operation of this sort of facility. In this respect, biosensor devices can function particularly well in a flow-injection scenario, especially those systems based on label-free detection involving acoustic wave and SPR physics. This approach can lead to the avoidance of batch measurements and lengthy trains of machines. Another advantage would be the possibility to perform generic detection, that is, the sensing of different targets with the same aforementioned physics/technologies. However, there exist two main problems that remain to be solved. The first would be to render a particular sensing platform reusable in the FIA configuration. A second would be to find a solution to the all-prevalent fouling problem encountered with biological fluids as discussed in some detail above.

Lastly, we would like to mention a final word about microfluidics and lab-on-a-chip devices, which appear to be ideal for combination with biosensor detection. These technologies seem to be more suited for point-of-care systems rather than the central clinical laboratory. However, despite the frequently emphasized promises offered by these devices, there is no doubt that NSA still remains a serious obstacle to be tackled. For example, NSA will not only occur at the detector employed in the device but also in channels of the microfluidic set-up.[70]

Acknowledgements

The authors are grateful to the Collaborative Health Research Program of BSERC and CIGR for support of their research. SS thanks the Province of Ontario for the award of an Ontario Graduate Scholarship.

References

1. C. P. Price, A. St. John and J. M. Hicks (ed.), *Point-of-Care Testing*, 2nd edn. American Association for Clinical Chemistry Press, Washington DC, 2004.
2. Source: St. Michael's Hospital, Toronto, Canada.
3. K. E. Herold and A. Rasooly (ed.), *Lab-on-a-Chip Technology – Vol. 1: Fabrication and Microfluidics*, Caister Academic Press, Norfolk, UK, 2009.
4. See *Lab-on-a-Chip*, Published by the Royal Society of Chemistry, Cambridge, UK.
5. H. Kaiser, *Pure. Appl. Chem.*, 1973, **34**, 35.
6. V. Dugas, A. Elaissari and Y. Chevalier, *Recognition Receptors in Biosensor* M. Zourob (ed.), Springer-Verlag, Berlin, Heidelberg Publisher, 2010, pp 47–134.
7. D. C. Kim and D. J. Kang, *Sensors*, 2008, **8**, 6605.
8. E. A. Bayer and M. Wilchek (ed.), *Avidin-Biotin Technology – Vol. 184*, Academic Press Inc., San Diego, CA, 1990.
9. G. P. Anderson, M. A. Jacoby, F. S. Ligler and K. D. King, *Biosens. Bioelectron.*, 1997, **12**, 329.
10. R. Gupta and N. K. Chaudhury, *Biosens. Bioelectron.*, 2007, **22**, 2387.
11. R. A. Williams and H. W. Blanch, *Biosens. Bioelectron.*, 1994, **9**, 159.
12. K. Jans, K. Bonroy, G. Reekmans, R. De Parma, S. Peeters, H. Jans, T. Stakenborg, F. Frederix and W. Laureyn, *Sensors for Environment, Health and Security* M.-I. Baraton (ed.), Springer-Verlag, Berlin, Heidelberg Publisher, 2009, pp 277–298.
13. A. Ulman, *An Introduction to Ultrathin Organic Films From Langmuir-Blodgett to Self-Assembly*, Academic Press Inc., San Diego, CA, 1991.
14. J. C. Love, L. A. Estroff, J. K. Kriebel, R. G. Nuzzo and G. M. Whitesides, *Chem. Rev.*, 2005, **105**, 1103.
15. C. Vericat, M. E. Vela, G. Benitez, P. Carro and R. C. Salvarezza, *Chem. Soc. Rev.*, 2010, **39**, 1805.
16. H. Häkkinen, *Nature Chem.*, 2012, **4**, 443.
17. S. Sheikh, J. C.-C. Sheng, C. Blaszykowski and M. Thompson, *Chem. Sci.*, 2010, **1**, 271.
18. C. Blaszykowski, S. Sheikh, P. Benvenuto and M. Thompson, *Langmuir*, 2012, **28**, 2318.
19. C. Blaszykowski, L.-E. Cheran and M. Thompson, *Can. J. Chem.*, 2011, **89**, 1512.
20. M. Thompson, L.-E. Cheran and S. Sadeghi. *Sensor Technology in Neuroscience*, Royal Society of Chemistry, Cambridge, UK, 2013.
21. N. J. Ronkainen, H. B. Halsall and W. B. Heineman, *Chem. Soc. Rev.*, 2010, **39**, 1747.
22. J. Wang, *Biosens. Bioelectron.*, 2006, **21**, 1887.
23. J. Korita and K. Stulik, *Ion Selective Electrodes*, 2nd edn. Cambridge University Press, 1983.

24. C. H. Fry and S. E. N. Langley, *Ion-Selective Electrodes for Biological Systems*, CRC Press, 2002.
25. P. Bergveld, *Biosensors*, 1986, **2**, 15.
26. P. Bergveld, *Sens. Actuators B-Chem.*, 2003, **88**, 1.
27. M. J. Schoning and A. Poghossian, *Analyst*, 2002, **127**, 1137.
28. J. Wang, *Electroanal.*, 2001, **13**, 983.
29. J. Wang, *Chem. Rev.*, 2008, **108**, 814.
30. J. Wang, *Perspectives in Bioanalysis – Vol. 1*, Elsevier, 2005, pp 175–194.
31. P. De Marco, A. Ng and D. Panduwinata, *Electroanalysis*, 2008, **20**, 313.
32. W. G. Cady, *Piezoelectricity – Vol. 1*, Dover Publications, New York, 1964, p. 4.
33. G. Sauerbrey, *Z. Phys.*, 1959, **155**, 206.
34. X. Wang, J. S. Ellis, P. Sundaram, E.-L. Moore and M. Thompson, *Molec. Biosys.*, 2006, **2**, 184.
35. D. H. Ather and V. Relpa, *J. Biophys. Chem.*, 2012, **3**, 211.
36. H. Su, K. M. R. Kallury, M. Thompson and A. Roach, *Anal. Chem.*, 1994, **66**, 769.
37. X. Wang, J. S. Ellis, C.-D. Kan, R.-K. Li and M. Thompson, *Analyst*, 2008, **133**, 85.
38. S. M. Ballantyne and M. Thompson, *Analyst*, 2004, **129**, 219.
39. M. Thompson, S. M. Ballantyne, L.-E. Cheran, A. C. Stevenson and C. R. Lowe, *Analyst*, 2003, **128**, 1048.
40. M. Thompson and D. C. Stone, *Surface-Launched Acoustic Wave Sensors: Chemical Sensing and Thin Film Characterization*, Wiley & Sons Inc., 1997.
41. K. Lange, B. R. Rapp and M. Rapp, *Anal. Bioanal. Chem.*, 2008, **391**, 1509.
42. X. Fan, I. M. White, S. J. Shopova, H. Zhu, J. D. Suter and Y. Sun, *Anal. Chim. Acta*, 2000, **620**, 8.
43. B. T. Cunningham, *Label-Free Biosensors – Techniques and Applications* M. A. Cooper (ed.), Cambridge University Press, 2008.
44. Y. Li, Y. Fan, L. Zhang, S. Ma and J. Zheng, *Proc. SPIE*, 2000, **4077**, 242.
45. P. A. E. Piunno, U. J. Krull, R. H. E. Hudson, M. J. Damha and H. Cohen, *Anal. Chem.*, 1995, **67**, 1635.
46. A. M. Valasex, C. A. Lana, S. Tu, M. T. Morgan and A. S. K. Bhunia, *Sensors*, 2009, **9**, 5819.
47. E. Kretschmann, *Z. Phys*, 1971, **241**, 313.
48. J. Homola, *Surface Plasmon Resonance Based Sensors*, Springer-Verlag, Berlin, Heidelberg Publisher, 2006.
49. J. Homola, S. S. Yee and G. Gauglitz, *Sens. Actuators B-Chem*, 1999, **4**, 3.
50. X. D. Hoa, A. G. Kirk and M. Tabrizian, *Biosens. Bioelectron.*, 2007, **3**, 151.
51. R. L. Rich and D. G. Miszka, *Curr. Opin. Biotech.*, 2000, **11**, 54.
52. X. Guo, *J. Biophoton.*, 2012, **5**, 483.
53. J. Homola, *Anal. Bioanal. Chem.*, 2003, **377**, 528.
54. R. G. Heideman, R. P. H. Kooyman and J. Greve, *Sens. Actuators B-Chem.*, 1993, **10**, 209.

55. A. Brandenburg, *Sens. Actuators B-Chem.*, 1997, **39**, 266.
56. B. H. Schneider, E. L. Dickinson, M. D. Vach, J. V. Hoijer and L. V. Howard, *Biosens. Bioelectron.*, 2000, **15**, 13.
57. M. L. Turgeon, J. J. Linnâe and K. M. Ringsrud, *Linne & Ringsrud's Clinical Laboratory Science: the Basics and Routine Techniques*, 5th edn. M. L. Turgeon (ed.), Mosby Elsevier, 2007.
58. C. I. L. Justino, T. A. Rocha-Santos and A. C. Duarte, *Trends Anal. Chem.*, 2010, **29**, 1172.
59. J. N. Adkins, S. M. Varnum, K. J. Auberry, R. J. Moore, N. H. Angell, R. D. Smith, D. L. Springer and J. G. Pounds, *Mol. Cell. Proteomics*, 2002, **1**, 947.
60. M. J. A. Shiddiky, P. H. Kithva, D. Kozak and M. Trau, *Biosens. Bioelectron.*, 2012, **38**, 132.
61. J. Tamayo, P. M. Kosaka, J. J. Ruz, Á. S. Paulo and M. Calleja, *Chem. Soc. Rev.*, 2013, **42**, 1287.
62. J. Breault-Turcot and J.-F. Masson, *Anal. Bioanal. Chem.*, 2012, **403**, 1477.
63. S. Choi and J. Chae, *J. Micromech. Microeng.*, 2010, **20**, 1.
64. D. A. Giljohann and C. A. Mirkin, *Nature*, 2009, **462**, 461.
65. K. C. Dee, D. A. Puleo and R. Bizios. *An Introduction to Tissue-Biomaterial Interactions*, Wiley & Sons Inc., Hoboken, NJ, 2002, pp 37–51.
66. R. A. Latour, Jr., *Encyclopedia of Biomaterials and Biomedical Engineering*, Taylor & Francis, 2005, pp 1–15 and references therein.
67. J. J. Gray, *Curr. Opin. Struc. Biol.*, 2004, **14**, 110.
68. C. Blaszykowski, S. Sheikh and M. Thompson, *Chem. Soc. Rev.*, 2012, **41**, 5599 and references therein.
69. S. Sheikh, D. Y. Yang, C. Blaszykowski and M. Thompson, *Chem. Commun.*, 2012, **48**, 1305.
70. R. Mukhopadhyay, *Anal. Chem.*, 2005, **77**, 429A.

CHAPTER 2

Integrated Chemistries for Analytical Simplification and Point of Care Testing

PANKAJ VADGAMA,* SALZITSA ANASTASOVA AND ANNA SPEHAR-DELEZE

Queen Mary University of London, Mile End Road, E14N, London, United Kingdom
*Email: p.vadgama@qmul.ac.uk

2.1 Introduction

In an era of individualised therapy and patient empowerment, it is a natural progression for assays that were once the sole preserve of the central laboratory to now fall within reach the of the nonlaboratory professional or the patient with assays typically undertaken out with the laboratory. The area of extra-laboratory testing has, accordingly, seen major expansion in recent years, but has also become the subject of considerable debate as regards its true value to the patient. It is certainly a more demanding and expensive way of undertaking tests, but the champions of such testing "immediacy" are clear that a wide range of benefits accrue. These include a more rapid return of data to the clinician to institute earlier therapy, reduced hospital stay, reduced medical complication rates, especially in chronic disease such as diabetes, and an overall improvement in quality of life at the population level.[1]

This form of decentralised testing, more commonly referred to as point-of-care testing (POCT), largely remains under the aegis of laboratory medicine.

RSC Detection Science Series No. 2
Detection Challenges in Clinical Diagnostics
Edited by Pankaj Vadgama and Serban Peteu
© The Royal Society of Chemistry 2013
Published by the Royal Society of Chemistry, www.rsc.org

Moreover, through technological advances, it is evolving rapidly both in analytical scope and the diversity of clinical application; monitoring of both ambulatory and nonambulatory patients is now readily feasible. POCT applications range from basic blood glucose measurement to complex enzyme assays. The organisational benefits include the elimination of specimen transport and sample preparation and certainty over when the assay data is available.[2] POCT may be undertaken using systems as contrasting as single-use hand-held devices and bench-top analysers, the latter being more allied to traditional laboratory analysers. Whilst the growth of POCT has had considerable impact on *how* monitoring is undertaken, it cannot replace clinical judgment. It is best used when a clinician is in a position to respond rapidly to test results.[3]

The wider patient pathway and organisational issues fall outside the scope of this review. However, the ability to roll out a measurement capability to the POCT setting in the first place, especially for complex assays, relies heavily on the harnessing of new technologies. These especially involve the integration of biochemical reagents as immobilised components of measuring platforms. Such platforms might be sensor based, creating a combined, monolithic, bio-/ artificial structure, or involve an independent biological phase and detector. Overall, quite disparate technologies need to be functionally integrated into bioengineered, ergonomically simple, constructs.[4] This review covers design issues for the various subcomponents and how these can satisfy the needs of selective, sensitive measurement, excluding *in vivo* monitoring. The broad topics are: routes to sample presentation using controlled flow, the way multiple fluid movements are achieved, *e.g.* in porous structures using lateral flow, how biochemical reactive and recognition systems are configured and how this biochemically mediated response triggers the measurement pathway. In the quest for operationally simple and miniaturised point of care systems, there has been considerable leveraging of bioreagent immobilisation chemistries and MEMS (microelectromechanical systems) technologies as well as a re-engineering of optical and electrochemical transducers. Along with this have been leading-edge developments in nanotechnology, to achieve extreme miniaturisation,[5] and better sample interfacing materials to control what actually reaches the measuring element of the analytical system. The latter is essentially a form of packaging to protect the vulnerable functional elements.[6] Typically, whole blood is used for analysis, and it may be necessary to use barrier layers to screen out the formed elements of blood for say an optical or colorimetric assay.

2.2 Fluidics for POCT

More complex assays for POCT, or where bench-top analysers are used, necessitate some form of sample and/or reagent flow in a fluidic channel. The adaptation of traditional fluidics has been particularly evident in the case of multiparameter blood gas and electrolyte systems. This contrasts with single-parameter reagent strips such as for urine albumin and blood glucose that require only the application of sample to the measuring surface. Bench-top analysers, as smaller versions of laboratory analysers, simplify vulnerable operator-dependent steps, mainly through automation.

MEMS may be used to achieve flow miniaturisation and sample conservation, and though microfluidics dimensions are not reached, other than in experimental systems, many of the logistics of fluid handling are now established. In the Alere Triage® System, which comprises a meter and various test panels, *e.g.* for cardiac marker measurement (CK-MB, myoglobin, Troponin-I),[7,8] an immunoassay reaction mixture, after controlled incubation, flows over different capture surfaces for selective assay. The i-STAT meter for blood gases, ions and metabolites has individualised fluidics in self-contained cartridges that include of the i-STAT calibrants; in one interesting veterinary use, synovial fluid was assayed for pH, glucose and lactate.[9] In a system where sample self-containment is paramount, the Pima CD4^{+} analyser for HIV testing, the sample is taken up by an integrated capillary with all subsequent immuno-binding and fluidic movement occurring within a cartridge.[10] High-number parallel assays would be difficult to integrate with a fluidic system. In the Evidence microchip immunoassay array, more than 20 anthelmintic drugs could be screened simultaneously, but each component constituted a single well into which preset reagents were added as the route to a simplified assay.[11]

Advances in printing, lamination and micromoulding technology have enabled greater miniaturisation along with the creation of more complex and versatile flow conduits that can carry sample and carrier phases to a detector phase in a time-controlled manner. In one experimental configuration, a microfluidic circuit was formulated with an open flow component serving as sampler for precision sample uptake and delivery.[12] However, the majority of flow devices utilise fully closed flow configurations. As well as sample conservation, an additional benefit of microfluidics is the generation of laminar flow that can allow formation of stable liquid/liquid interfaces within a single channel with only diffusion and convection determining solute exchanges. This is an ideal way of transferring an analyte from a sample stream to a mobile recipient phase. The principle of this is shown in Figure 2.1[13] where stable parallel flows are seen. Where the liquids flow together, solute exchange occurs at a rate determined by molecular weight, aiding solute selection; indeed this can be extrapolated to the separation of colloidal and cellular elements from the sample.

An immunoassay has been demonstrated where drug binding to a selective antibody retards its transport across the liquid/liquid interface thereby providing a measure of drug concentration. The assay principle here is that of competitive binding between sample and labelled drug with optical tracking.[14] The principle of size-based separation has been further developed in a protein separation assay for salivary MMP (matrix metalloproteinase) (Figure 2.2).[15] Here, the target protein binds to a fluorescently labelled antibody, and separation by size-based selectivity achieved using a porous gel and a size exclusion membrane; sample was moved electrophoretically rather than by pressure driven flow. We have developed *in situ* formed protein based membranes at

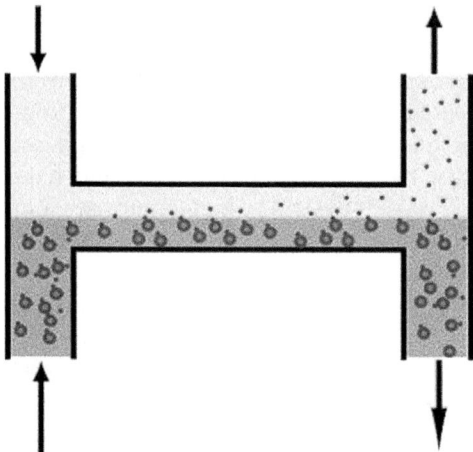

Figure 2.1 Laminar flow in an H-Filter® system (Micronics). Sample, *e.g.* whole blood, is introduced from the bottom left and flows in parallel with a receiver solution, *e.g.* buffer, introduced through top left. Smaller sample molecules diffuse from the sample to the receiver stream to be harvested from the top right outlet, while bigger molecules and particles are harvested from bottom right.
(Reproduced from ref. 13 with permission.)

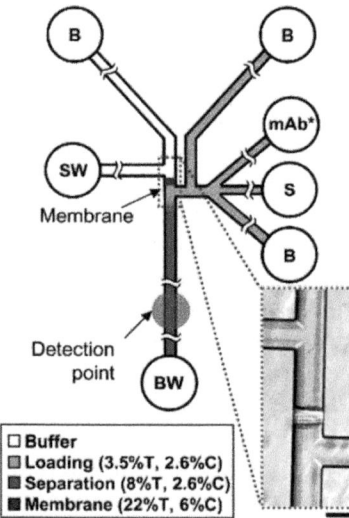

Figure 2.2 Microchip electrophoretic immunoassay (μCEI) device for the detection of MMP-8 from saliva samples. S-sample, B-buffer, SW-sample waste, BW-buffer waste, mAb*-fluorescently labelled monoclonal antibody to MMP-8. Polyacrylamide gel composition variation is indicated by grayscale shading (%T and %C are percentage of total acrylamide and bis-acrylamide crosslinker, respectively). Inset shows ×40 size-exclusion membrane bright-field image, scale bar 100 μm.
(Reproduced from ref. 15 with permission.)

parallel fluid flow interfaces using a rapid crosslinking with terephthaloyl chloride that then established a stable barrier between parallel streams;[16] such an arrangement holds the prospects for pore size-based control of lateral solute transport as well as serving as a support for bioreagent immobilisation.

A practical consideration with fluid flows is how to generate the initial flow. Precision pumps have been developed with appropriate miniaturisation, however, even with traditional pumps, complex patterns of microflow can be generated. Versatility of flow design was demonstrated in one example where agglutinated red cells were detected during blood crossmatching; here, agglutinated cells reduced light transmission across a fluidic channel as compared with non-agglutinated cells.[17] Micropumps can have drawbacks of reliability as well as the added need for powering and machining; passive pumps may offer a simpler alternative. In one example, an array of hexagonal microfabricated pillars with defined spacing achieved controlled fluid filling giving pumpless flow.[18] The system had minimal internal dead space and flow resistance and allowed stable flow during a one-step immunoassay. This particular flow configuration also incorporated various flow constraints and delay elements coupled with sample collection. With integration of heterogeneous components in this way, it becomes feasible to build lab-on-a-chip capabilities to achieve sequential sampling, separation, dilution, reagent mixing/incubation and readout steps that go to make up a typical clinical assay procedure for a laboratory analyser.

Particularly at low cross section channels, as exist in microfluidic systems, the lack of surface compatibility of cell- and protein-loaded fluids with standard biomaterials, polymers and inorganics such as silicon, remains a problem. One way to mitigate this is to coat or surface modify the materials with known colloid tolerant interfaces such as those based on poly(ethylene oxide) or biomimetic phosphoryl choline. An alternative has been the use of microdroplet flow where the relevant biocomponent is retained within a containment phase surrounded by an oil or air continuum.[19] The discretisation of both reagent and sample in this way can also reduce volume requirements and allow manipulative steps with reduced need for complex channels and valves. Electric field generation at insulator covered electrode arrays can allow microdroplet transport to be readily controlled (electrowetting). By this means, also, necessary microdroplet division and fusion can be instituted to allow for interactive reactive processes. An ambitious target reported for this approach was multiparameter newborn screening.[20] A dedicated laboratory for this requires to undertake immunoassays, enzyme assays and even DNA identification, but a droplet-manipulating platform capable of dispensing and washing droplets after binding/reactive steps appears to show promise with multiple model targets. Sample volume requirements here dropped to ~1% and high throughput assays for screening, *e.g.* for hypothyroidism and G-6-PD deficiency looked feasible It may be possible through innovations in materials for microfluidics to be able to customise and simplify production; using hot embossed cyclic polyolefin Bhattacharrya and Klapperich[21] were able to simplify microchannel fabrication for the immunoassay of C-reactive protein, an inflammation marker.

2.3   Lateral Flow Techniques – From Simple Colorimetric Strips to 3D Flow

2.3.1   Colorimetric Strips

Lateral flow provides the least technologically complex and demanding platform for the qualitative or semiquantitative analysis of an analyte. It has its origins in paper-based regent strips where an adsorbed reagent enables single-use colorimetric measurement, typically by just visual inspection and comparison with a colour chart In the production process, the reagent is deposited on adsorbent paper later split into square pads, in the case of urine testing providing test panels on a single dipstick. The core assays here were for urine albumin, ketones, white cells, specific gravity and nitrite. Though simple to use, they did depend in some instances on reaction timing, and of course the visual discrimination of the observer to assess the sometimes marginal colour changes. A switch to instrumentation with use of light-emitting diodes (LEDs), charge-coupled device (CCD) arrays and CCD cameras has allowed for greater objectivity and the avoidance of observer error. Later refinements included the use of enzymes, glucose oxidase being the classic example, and the mitigation of interference through added reagent, *e.g.* to scavenge or identify an interferent in the sample.[22] The transition from urine to blood chemistries utilised traditional regents, but in addition to the basic functional elements of a fluid sample uptake layer, a reagent layer and a light-reflecting layer,[23] there now appeared masking, trapping and separation layers. This was especially necessary given the extension to enzyme-based assay (*e.g.* for glucose, urea) and to the assay of enzymes (*e.g.* creatine kinase, amylase).

Quite sophisticated layer formatting has been possible as evidenced by an early phenytoin immunoassay.[24] In this instance, antibody to phenytoin constituted a dry reagent affinity component. Labelled and sample phenytoin competed for binding to the antibody; the label used was FAD, and only as a part of unbound phenytoin was it able to attach as a prosthetic group to activate glucose oxidase apoprotein. Resulting holoenzyme activity was then measured in the strip by a conventional colorimetric indicator reaction. An innovative use of a nonpolar phase coexisting with a polar (sample) phase involved use of the ionophore valinomycin as a K^+ binder within the nonpolar layer.[25] The valinomycin was associated within the apolar phase with a pH indicator dye; chelation of K^+ by valinomycin led to its becoming cationic, with ejection of H^+ into the polar phase from the indicator dye and a resulting dye colour change.

Though not a chromogen, the lanthanide ion terbium(III) forms a usable luminescent complex with p-aminobenzoic acid (PABA). This was exploited in a paper-strip format to assay the local anaesthetics benzocaine and procaine; their hydrolysis in alkali generated the necessary PABA end product leading to a luminescent chelate. The reagent strip is more stable than one using bio/organic reagents, but does require special luminescence detection apparatus. Albumin dye binding and the resulting colour change has been the mainstay of urine albumin measurement by a dry reagent pad. However, there is a serious sensitivity limit that precludes detection in the microalbuminuria range relevant to detecting and monitoring diabetic nephropathy. A tetrabromosulfonphthalein dye (DIDNTB) was found to have a specially high affinity for albumin[26] and allowed its detection in urine down to 10 mg/L; the constituted reagent strip maintained an optimum pH for the dye using impregnated citrate buffer. A counterpart to the high albumin affinity was the reduced binding competition from IgG/IgA and immunoglobulin light chains found in urine. A non-chemistry strip, but one that is highly successful is that used to monitor anticoagulant treatment. In a recent commercial embodiment, the CoaguCheckXS,[27] the test strip takes up venous or capillary blood, hydrating a thromboplastin layer that generates a blood-coagulation sequence to generate thrombin from the sample. The thrombin as indicator of the pro-thrombin time then cleaves a peptide substrate; here an electrochemical rather than an optical end point was monitored.

In contrast to the vertical transmission of sample through a porous phase, lateral flow devices involve horizontal flow and have the facility for topographic resolution of bound *vs* free analyte, and provided a visible label is used, allow for instrument free detection.[28] Their primary use has been in the detection of proteins using antibody binding. The first and foremost use has been detection of urine human chorionic gonadotrophin (hCG) for pregnancy testing. Essentially, two epitopes are recognised on a protein by preloaded antibodies. After the liquid sample is applied to a loading pad, it moves laterally within the porous support and is exposed to these antibodies in turn. An absorbent pad at the opposite end of the analytical strip draws up the liquid sample to drive the controlled lateral flow. First, the target protein binds to labelled antibody, and the two move onto an immobilised capture antibody zone that in turn immobilises the antigen labelled antibody complex to generate a visible, concentrated zone. Visibility is generally achieved by particulate labelling of the first antibody, *e.g.* using gold nanoparticles or latex beads.

2.3.2   Nanoparticle Labels

Scaling of sizes, the availability of new materials, variation of surface profiles and novel bulk properties open up new assay avenues with nanoscale structures. In one example, magnetic nanoparticles were used as immobilisation surfaces for antibody;[29] In this model, it was shown that orientation of antibody was highly relevant to sensitivity; particle binding through the carbohydrate component of the antibody, away from the binding site, allowed for freer access to hCG target antigen. Magnetic particle binding has also been used for sample washing and separation of free from bound antigen.[30] In this example, a possible basis for a commercial device, cardiac troponin I was assayed by first mixing antibody-loaded particles with sample, then activated electromagnets were used to drive the particles to a second antibody-binding surface, and finally diametrically positioned electromagnets separated unbound particles from the antibody-bound layer (Figure 2.3). Frustrated total optical internal reflection finally enabled imaging of the antibody-bound particulate layer.

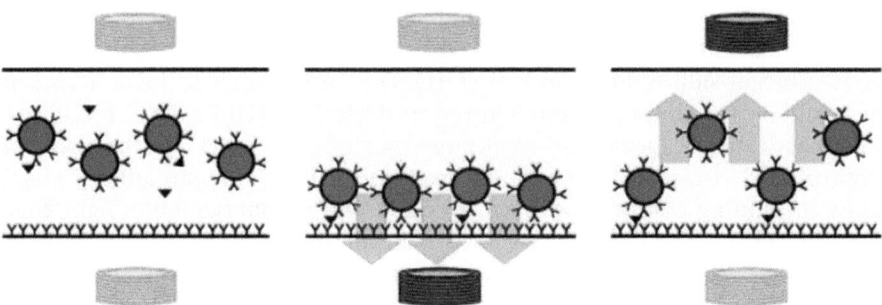

Figure 2.3   A schematic of magnetic-bead-based immunoassay: analyte binding by
                antibody-functionalised magnetic nanoparticles with top and bottom
                magnets off (left), nanoparticles bind to the sensor surface, lower magnet
                on (middle) and magnetic removal of free and weakly bound nanoparti-
                cles, upper magnet on (right).
                (Reproduced from ref. 30 with permission.)

Further labels include selenium, carbon, liposomes, phosphors and quantum
dots to add to the repertoire of traditional fluorescent and bioluminescent dyes.
With limited antibody, it is possible to devise competitive immunoassays where
the labelled and unlabelled antigens compete for the antibody binding.[31] In one
assay of serum albumin, a fluorescent label was used, where competition was for
antibody immobilised on a strip. Increased analyte concentration reduced label
binding and therefore fluorescent dye accumulation at the strip surface;[32] here,
however, a fluorescence reader was necessary for the assay, adding to instrumentation complexity. In order to detect multiple analytes simultaneously, it is
necessary to have individually distinguishable labels. Thus, to detect antibodies
to Hepatitis A, B and RSV, respectively, metallic nanoparticles capable of
surface-enhanced Raman scattering were differentially functionalised to give
resolvable Raman signatures and activated by specific antibodies to different
targets;[33] lateral flow to a single capture strip holding a second antibody for each
target then enabled selective, simultaneous quantification.

Carbon nanoparticles have proven to be versatile surfaces for label as well as
bioreagent binding, *e.g.* nucleic acids, antibodies and other proteins, potentially
conferring detection sensitivity greater than gold or latex particles.[34] Rayev and
Shmagel[35] preadsorbed affinity protein and subsequently crosslinked this to
give a stabilised binding surface enabling the attachment of observable carbon
nanoparticles to topologically localised target biomolecules. Thus, by coating
with protein G, which has a general affinity for immunoglobulin, it was possible
to identify antibodies from seropositive samples, where antibody had bound to
HIV antigen retained at a porous support. A lateral flow arrangement was also
demonstrated for hCG detection by pre-loading with antibody to the β-subunit
of hCG. A nucleic acid lateral flow assay has also been developed exploiting
this readily visible label. In one example, a *Plasmodium falciparum* DNA
template in PCR was labelled with primers bearing two tags, digoxigenin and
biotin;[36] carbon nanoparticles loaded with avidin bound to the biotin end and

the digoxigenin end was captured by immobilised antibodies. Amplified nucleic acid sequences have been detected using dye-loaded nanoscale liposomes; here lateral flow of a target sequence led to its capture by immobilised complementary strand DNA; this was subsequently made visible by binding to a reporter sequence on the surface of a liposome[37] and practical illustration has been given, with RNA amplification, of the detection of different serotypes of the Dengue fever virus.[38] For these serotypes, the sensitivity achieved ranged from the detection of 50 000 down to just 50 RNA molecules. Also, with an optical readout, quantitation was possible. A further affinity system was demonstrated by Kulagina *et al.*[39] who immobilised antimicrobial peptides (AMPs) as capture agents to confirm the targeting of model pathogens. Venezuelan equine encephalitis virus, vaccinia virus and *Coxiella burnetti* (an intracellular bacterium) were successfully identified with an effectiveness that proved even greater than with selective antibody. Further extension of sensitivity for this fluorescence-based assay could be possible using nanoparticles.

2.3.3 New Opportunities

In addition to single planar flow, 3D multiphase flows have been explored by Whitesides' Group[40] using paper/fibrous supports. These microfluidic paper analytical devices (μPADs), channel solutions horizontally as well as vertically and could deliver to multiarray reagent pads. The principle of the system was demonstrated for glucose and protein measurement.

With newer porous materials, which allow for reduced nonspecific binding and greater control over the lateral flow rate, improved selectivity and concentration resolution could eventually be achieved. In this way, the quantitative capabilities of lateral flow will become manifest. The key question will be how far it will be necessary to integrate optical or other quantitative readers into the newer systems. While a definitively quantitative measurement may be obtainable, this will also add costs to any commercial device, reduce portability and detract from its being a consumer item for home use. Unlike home meters for say glucose, where quantitative measurement is imperative, in many lateral flow applications, the usually macromolecular target need not be determined with exact quantitation. What is much more important is the ability to differentiate so-called normal from abnormal, or the qualitative identification of antibodies to pathogens. Nevertheless, if very cheap or even disposable readers can be developed, it may become possible to combine disposability with reliable quantitation, a further, important outcome would be the avoidance of observer error.

2.4 Biological Recognition

2.4.1 Immobilisation Techniques

Biological modification of a device or the transformation of a soluble bioreagent into a solid phase has innumerable analytical benefits. Such approaches have been the formal basis of biosensor construction as well as of standalone bioreactor and bioaffinity phases. The "biomodification" approach has allowed exploitation of the special binding selectivity and affinity of enzymes, antibodies and aptamers, respectively. However, the immobilisation of such recognition systems needs careful tailoring; without this the bioaffinity molecule may either not survive the retention process or be so poorly orientated at a surface that it is unable to achieve recognition capability. In the case of readily diffusible targets such as enzyme substrates, orientation may not be an issue as the target molecule can reach the active site, but for macromolecular targets, it may not be possible to have effective binding unless the binding subcomponent, *e.g.* the Fab region of an antibody, has sufficient exposure at the sample interface, and is not buried in the subjacent solid phase. The standard repertoire of physical adsorption, chemical crosslinking, covalent attachment and polymer entrapment remains the core means of immobilisation to this day, but of relevance to this chapter, these have been adapted to diagnostic platforms. However, the special needs of analysis have meant that there has been significant refinement of the basic immobilisation repertoire.

Porous supports provide high surface area matrices for biomolecule adsorption and retention. As such, they are attractive for direct adsorptive deposition of biomolecules. Whilst leaching is a potential issue, for single-use devices, an analysis can be completed well before such loss of the bioactive moiety. The usual physisorption binding forces are in operation, notably hydrophilic and hydrophobic interactions, van der Waals forces and ionic and hydrogen bonding. The local influence of these aggregated forces can be to distort biomolecule shape and so render the biomolecule inactive, more of a problem with enzymes than with antibodies. Kong and Hu[41] have reviewed the general strategies for biomolecule immobilisation on paper surfaces. They highlight the fact that cellulose is slightly anionic and so binding tends to involve cationic domains on biomolecules. Inevitably, the solution environment (pH, pI) modifies the effectiveness of binding. Special interactions also play a part, such as binding that is mediated through tyrosine residues to the edge plane of cellulose crystals.[42] Horseradish peroxidase was used as a general adsorptive model by Di Risio and Yan[43] who showed that with this globular protein, net positive charge at near-neutral solution conditions promoted adsorption, but there was also a further dominant effect, an enhancement of adsorption at hydrophobic surfaces. These hydrophobic interaction forces, however, also led to protein unravelling with loss of activity.

The attractive forces posed by a surface are, of course, not selective for the bioreagent, but furnish a locus for any extraneous protein deposition. This leads to masking of the measurement surface, sensitivity drift and loss of signal. Surface-blocking agents such as milk and bovine serum albumin (BSA) are well established in standard multiwell immunoassays for blocking nonspecific binding and could be used in lateral flow assays, but there is also the possibility of incorporating agents such as surfactants and BSA into the solution phase. An immunoassay where a liquid crystal was the reporter system showed reduced nonspecific binding through addition of Tween 20.[44]

Provided the biomolecule is not denatured on initial contact with a surface, it is likely to be stabilised. Thus, for alkaline phosphatase and horseradish peroxidase with adsorption on paper, storage at elevated temperatures, even up to 90 °C, demonstrated retained activity for many hours compared with activity loss in solution within minutes.[45] Degradation dynamics indicated two sequential, first-order, reactions, suggesting two distinct pathways, the second, slower one dominated half-life. In practical terms, alkaline phosphatase showed retention of 63% activity at 14 days storage at 23 °C. The study also showed the resistance of enzyme to shear forces generated during printing.

Inkjet printing of enzymes has been a route to high precision, rapid deposition, facilitating array formation. Computer-controlled printing, delivering nanolitre volumes can avoid crossover and the process also allows the biomolecule to be in solution; these are attractive features for device miniaturisation. Commercial printers have been used, for example, to deposit to a feature size of 100 micrometres with deposition of up to four different solutions.[46] Inkjet printing uses thermal or mechanical forces to drive solution droplets from a nozzle, while e-jet printing uses electric-field-gradient generation between the delivery nozzle and the substrate, driving out minute drops, depending on nozzle size. Thus, it has been possible to deliver droplet sizes in the 0.1 micrometre range;[47] with streptavidin as a model protein, activity was retained, based on the measured affinity behaviour of biotin.

Antibodies are generally more robust than enzymes. A study with anti-rabbit IgG adsorbed on cellulose paper[47] showed that at 40 °C there was an activity half-life of 10 days with no added degradative effect of humidity. Interestingly, precoating the paper with a crosslinked polyamide-epichlorohydrin film, conferring a cationic property to the cellulose, did not significantly enhance the loading of this anionic protein, nor did it improve thermal stability, possibly because thick layers of protein, not in direct contact with the surface, were involved. The implication of this study was that strip manufacture with hot-roll processing is feasible. Loading of antibodies has been increased through use of high porosity, high surface area structures. Porous silicon has the interesting dual effect of providing a low fouling, highly hydrophilic surface at the molecular level, while creating a hydrophobic macrostructure due to its nanoporosity, thereby helping to retain polar solution during assay conditions.[48] As an alternative to a cavity, a high-curvature microsurface using particles can enable antibody adsorption by simple electrostatic attraction; as an illustration of this, sulfonate functionalised polystyrene microspheres have been used and co-attachment of quantum dots as labels enabled design of a fluoroimmunoassay.[49]

Protein A binds to the Fc region of an antibody, and so preadsorbed protein A has been frequently used to orientate antibodies, allowing for an outfacing Fab region for optimum binding exposure. In one report, Protein A preadsorbed to nanogold enabled both a greater surface area for antibody binding and the desired antibody orientation for a higher sensitivity immunoassay, targeted to haptoglobin.[50] The impact of antibody orientation will

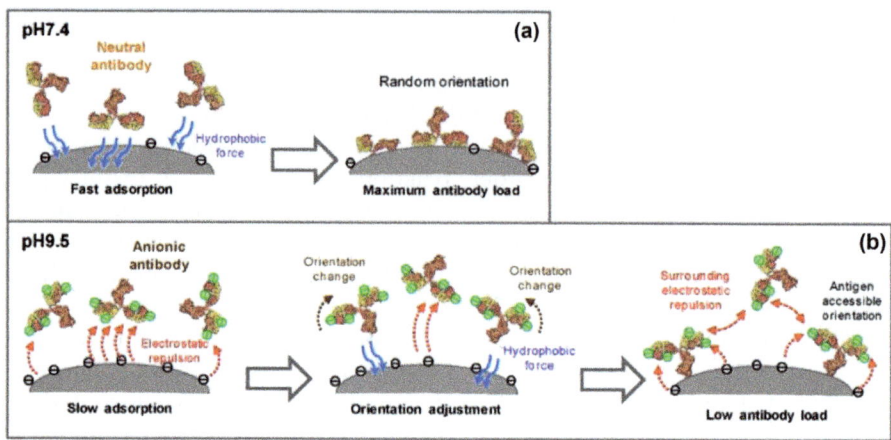

Figure 2.4 Proposed mechanism for electrostatic, preadsorptive antibody orientation at polystyrene due to ionisation under alkaline conditions: (a) random orientation at pH 7.4 in the absence of charge repulsion, (b) antibody reorientation due to charge repulsion between the Fab' region and polystyrene surface at pH 9.5.
(Reproduced from ref. 54 with permission.)

depend both on the nature of the solid support and that of the antibody. Thus, in a study of antibody and antibody fragments (Fab') using selective region bridging *via* biotin to surface-immobilised streptavidin, orientation led to increased binding,[51] with up to 6–10-fold greater binding capacity following the orientation, depending on the surface used. PEG was the anchor point at the support surfaces, and the length of this tether affected the binding sensitivity to orientation, with longer PEG chains generating greater orientation dependence. Orientation effects partly worked through delivering a higher surface packing density, further reinforced by high packing of streptavidin at the surface. The additional step of creating Fab' fragments may not justify the effort unless a further chemistry is designed to give surface orientation. Thus, in a study of Fab' and whole antibody binding to an IgG target, intact antibody loading was already greater than for the smaller Fab', unless the latter had been orientated.[52]

Regardless of whether covalent immobilisation or adsorption are the end outcome of the antibody retention process, noncovalent (*e.g.* hydrophobic, electrostatic) interactions at the interface are important in driving some of the initial orientation.[53] In a study of antibody immobilisation on polystyrene beads,[54] high pH was used to first generate a "preorientation" step before surface attachment. At high pH, mutual anionic repulsion retarded antibody adsorption, giving time for reorientation to occur near the surface; at this early stage the high anionic nature of the Fab' region led to its facing away from the surface (Figure 2.4). The more hydrophobic Fc region was then able to attach to the surface by hydrophobically driven attachment. The stability of the attached antibody layer was maintained by steric restriction at the surface using PEG as a co-immobilised packing layer.

2.4.2 Enzyme-Based Systems

The ability of substrate to permeate macroscale, reactive enzyme layers has enabled the use of thick-film techniques for their construction. The only real limitation has been the significant diffusive distances for product/reactant that limit the speed of response. However, even this is only of importance when a steady-state output is needed as in the case of continuous monitoring devices. For biosensors specifically, one of the most popular routes to immobilisation has been use of the homo-bifunctional reagent glutaraldehyde to crosslink the enzyme to give a matrix that also incorporates an inert protein, typically albumin. Glutaraldehyde itself has a somewhat protean structure, with thirteen different forms apparently present depending on solution conditions,[55] and though the exact reaction with a protein is not certain, linkage with amine groups *via* the aldehyde function looks to be the most important and, moreover, involves an oligomeric form of the reagent, rather than the monomer. The tendency has been to take an empirical approach to enzyme immobilisation based on trial and error, and to generate enzyme membranes of the requisite thickness and activity, but based on an unknown crosslinking density; this becomes a more challenging approach to design when the enzyme in question is unstable or vulnerable to the distorting effect of the crosslinking process. Unless a quencher is introduced, the crosslinking, moreover, may be difficult to make reproducible or controllable. One general outcome is that at low enzyme concentration, damaging intramolecular crosslinking is promoted, whereas the preferred intermolecular crosslinking is only promoted at high enzyme concentration. One way in which the latter is facilitated, of course, is through the inclusion of an inert protein.

Surface retention has been explored using glutaraldehyde bridging, but leads to limited surface layers and insufficient loading if only mono- or oligolayers are formed. A recent example with glutaraldehyde involved an aminated surface, where enzyme was preadsorbed, and subsequent crosslinking led to high loading as well as bridging of enzyme to the surface.[56] Enhanced stabilisation of enzyme after crosslinking has also been achieved by initial precipitation of enzyme.[57] By creating insoluble aggregates of enzyme through desolvation and then crosslinking, higher thermal and storage stability has been demonstrated. A further potential advantage of this crosslinked enzyme aggregate (CLEA) approach is that the source enzyme does not have to be of high purity; this is aspect conferred by the aggregation step.[58] The chemical crosslinking component is important for stabilisation, not just as a means of avoiding dissolution, but in modulating structural stability. Compared with solution enzyme, there are greater constraints to conformational fluctuations, and the ability to transition into nonactive conformers is thereby diminished; ultimately it is a balance between intrinsic stabilisation energies within the enzyme molecule *vs* the dynamics of thermodenaturation.[59]

Enzyme mobility and binding to different surfaces is affected both by the nature of the surface and by the bridging molecule used, as was shown in stability/loading studies of glucose oxidase, lactate oxidase and horseradish peroxidase, respectively, on aminated glass and cellulosic surfaces.[60]

Interestingly, bridging molecule structure, where an aromatic reagent was used, had a greater influence on stability as enzyme size increased, and in the case of glucose oxidase there was a substantially effect on effective K_m. Cellulose is an attractive matrix for dry reagent chemistries, and the potential for photopatterning on it was demonstrated by its functionalisation with an azidobenzene bridge; the latter, photoactivated to a reactive nitrene, enabled ready enzyme binding.[61] Cellulose can also act as a host for other polymeric supports, as shown with a preformed glycidyl methacrylate linkage that then allow for epoxy group mediated binding to enzyme (urease) *via* amino side groups.[62] The general use of paper-based supports and their bioactivation for diagnostics has been reviewed by Pelton.[63] Now, with the advent of microarrays for proteomics, high-density arrays are featured at a wide range of polymeric and glass substrates, the associated chemistries have diversified and are frequently based on photoactivatable bridging compounds.[64]

2.4.3 Aptamers as Biorecognition Elements

Aptamers are synthetic oligonucleotides selected for their specific target affinities and have shown a broad propensity to bind to ions, small organics and proteins.[65] Their development and tailoring is a burgeoning scientific field, likely to contribute eventually to clinical diagnostics in the way that antibodies have done. They are, however, more stable than antibodies, and can be chemically generated, rather than needing a biological system for their production. Their immobilisation for diagnostics essentially follows the route already used for genomic sensors and arrays.

Adsorption of aptamers is less effective than that of proteins; there are fewer functional groups for noncovalent interactions. An aptamer for ATP, for example, retained activity when adsorbed to microcrystalline cellulose, but also showed ready, ionic-strength-dependent, desorption.[66] DNA-loaded gold nanoparticles that had been interlinked by an aptamer able to bind to adenosine disaggregated in the presence of adenosine and furnished a semiquantitative paper colorimetric assay.[67] Here, whilst the aptamer itself served as a noncovalent bridge to DNA strands on the nanogold, the latter needed to be attached covalently.

Entrapment of aptamer within a sample-accessible phase can allow for its ready retention, while allowing a sufficient degree of conformational flexibility such that a binding event can trigger transduction. One such arrangement involved the entrapment of RNA aptamer within sol gels[68] and binding to a target led to structural change inducing fluorescence; the reporter mechanism involved induced separation of a fluorophore/quencher combination. The polarity and charge of the matrix had important effects on mobility and the access of the target molecule. The balance of these features provided for the necessary level of conformational freedom of RNA, whilst allowing it to be stabilised; ssRNA is less stable relative to ssDNA, and is also more susceptible to enzymic degradation. By adsorbing ssRNA to graphene oxide, resistance to nuclease digestion was demonstrated for a theophylline aptamer.[69] Mechanisms for this are unclear, but possibilities considered were altered local ionic composition, conformational remodelling and steric hindrance. In this regard this may be a general nanostructure effect as was suggested by the parallel effect of single-walled carbon nanotubes at DNA.[70]

Covalent attachment of aptamers has been by far the most popular route to diagnostic device design, and has been reviewed by Balamurugan *et al.*[71] The techniques established are anticipated by those used for nucleic acid detection. Thiol-derivatised DNA is a classic approach that readily allows for self-assembled monolayers on gold, but consideration needs to be given to both the mutual charge repulsion of anionic single-strand DNA and the complicating aspect of nonspecific adsorption. With these provisos, reproducible immobilisation on both planar and particulate gold is achievable under mild conditions. As with other immobilised biomolecules, the direct exposure of aptamer to the solution environment is critical to its function. Whilst the structural distortions that arise at multitethered proteins are not an issue with aptamers, there is nevertheless a need to design appropriate spacers and spacer lengths to the surface to optimise binding capacity and affinity. Ethylene glycol repeat units as spacers have had the beneficial effect of reducing nonspecific binding as well as improving binding effectiveness.[72] A variant of this has been to immobilise the ethylene glycol motif on to a gold surface after aptamer attachment; this made for higher aptamer loading then when simultaneous attachment was carried out.[73] Aromatic spacers can create more compact, stacked, arrays at a surface, and this could also assist with higher loading.

Conducting polymers such as polypyrrole offer an impedimetric means of registering affinity binding, but also provide a more general and easily deposited surface with useful functional side groups. Thus, with electropolymerised biotinylated pyrrole,[74] an avidin bridge to biotinylated aptamer was readily formed. The affinity system here was an aptamer for prion protein (Figure 2.5); the spacer DNA sequence used to immobilise the aptamer affected binding affinity,

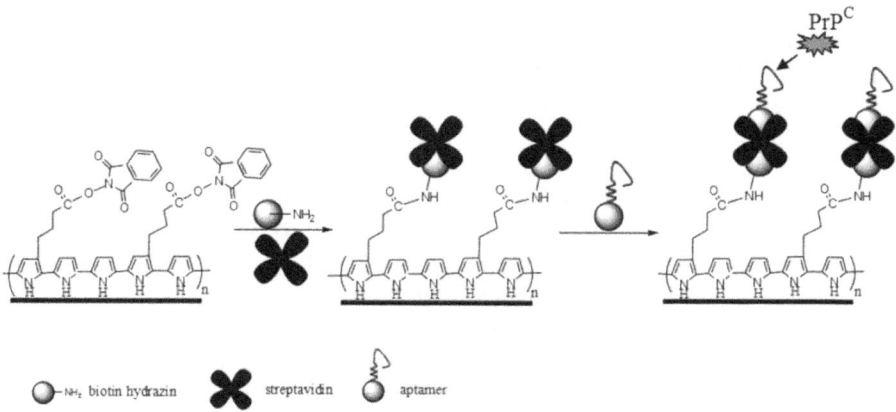

Figure 2.5 Synthetic steps in aptasensor functionalisation of a gold surface: (a) electropolymerisation of pyrrole with a hydroxyphthalimido derivative of pyrrole to create a composite layer, (b) functionalisation with biotin using biotin hydrazide, (c) surface attachment of biotinylated aptamer for human cellular prion protein (PrP^C) through a streptavidin-biotin linkage. (Reproduced from ref. 74 with permission.)

dsDNA, instead of ssDNA, with its more rigid spacer properties giving improved surface organisation and ultimately a higher binding affinity.

In another electropolymerisation procedure, pyrene with pendant lysines was deposited on Pt;[75] the latter allowed for copper (II) binding and through this linkage to an aptamer that had been functionalised with histidines for his-tag attachment. Inorganic surfaces may also be used for aptamer attachment. Thus, an aminated diamond surface functionalised with terephthalic acid enabled binding of an amino-terminated RNA aptamer;[76] the diamond here formed the sensing interface of a field effect device for HIV-Tat protein. One versatile inorganic phase is the quantum dot, usable for high-sensitivity colorimetric assay, and many of the traditional bridging methods applied to polymeric surfaces are now being reported for quantum dots.[77,78]

2.5 Electrochemical Sensors

2.5.1 Ion-Selective Electrodes

Ion-selective electrodes (ISEs), established with the glass pH electrode, have been a forerunner of many reagent-free, miniaturised, reversibly responsive systems advanced for extra-laboratory testing. The original principles enunciated, those of interfacial charge generation at an ion interactive interface, have not changed, nor has the perennial dependence on a stable reference electrode, needed to maintain a stable emf in the face of a variable solution electrolyte composition.[79] Special problems arise when using ISEs in biological fluids, and these remain an issue even with advanced, contemporary devices. So whilst fundamental chemistry has not changed greatly, what has advanced the field for clinical use has been the re-engineering of ISEs such that they are now miniaturised, robust, solid state and use integrated reference electrodes.

With new affinity ligands, some ISEs have achieved improved selectivity and sensitivity.[80] However, despite constant efforts towards new ligand synthesis, in the clinical context at least, it has been difficult to extend ISEs beyond the traditional repertoire of H^+, Na^+, K^+, Ca^{2+}, Cl^- and possibly Li^+, Mg^{2+} and NH_4^+. Certainly, trace metal ion detection and measurement of organic ions remains elusive. There, however, remains an unassailable market for ISEs for the core repertoire, key factors being their ability to operate in whole blood without sample separation or dilution, key advantages for POCT. Dilution is possible, and does permit better handling in a flow system as well as providing convergence with a more buffer-like matrix; the practical benefit of this is greater assay reproducibility and extension of the assay concentration range, *e.g.* as might be required for urine analysis. A drawback to dilution, as with any assay, is dilution error due to major variation in protein content, often seen in critical-care patients.[81]

Sodium measurement originally used modified glass, a variant on the glass pH electrode, but now PVC membranes loaded with ionophore, notably monensin, an antibiotic, or its derivatives, are used. In samples such as urine, the lipophilic PVC membrane can extract nonpolar solutes that then interfere with the response, leading to a loss of selectivity as well as to drift. However, urine assay is possible provided a nonpolar pre-extraction is carried out.[82] An interesting assay variant has been the use of sonication to create PVC microspheres loaded with a sodium ionophore. Within the same phase, a fluorescent chromionophore for H^+ was incorporated; uptake of Na^+ led to chromionophore deprotonation, simultaneous ejection of H^+ and thereby a change in microsphere fluorescence.[83] Such an arrangement was considered for Na^+ to be of use in high-throughput analysis. Newer ionophores continue to emerge, one example was based on a calyx[4]arene ester.[84] This sensor system was configured with a solid-state internal electrolyte comprising polypyrrole stabilised with a cobalt organic counteranion. The resulting electrode had enhanced sensitivity and represents a general solid-state ISE model. Crown ethers such as cyclic ionophores have been frequently studied, but adequate binding for clinical use has been a challenge. One variant where a silicon atom formed a part of the ether ring has given improved performance with low-pH effects and detection limits in the micromolar range for sodium.[85]

The microbial ionophore valinomycin has provided for high-selectivity measurement of potassium, with manipulation of internal electrolyte in traditional electrodes enabling extension of sensitivity limits.[86] An alternative solid internal contact electrode was based on poly(3-octylthophene).[87] This lipophilic electron conductor avoids the response instabilities associated with development of intervening aqueous films with the ionophore layer. Graphene has also been tried as a solid contact material.[88]

Various synthetic ionophores have been developed for ionised calcium, which give relatively trouble-free measurement, and commercial versions of magnesium sensors have also emerged, though these electrodes, based on diketone and amide antibiotics, have calcium ion interference and need a correction step for background calcium levels.[89] Chloride sensing has been achieved through the use of a quaternary ammonium ion loaded membrane, but there can be interference from other halogen ions and bicarbonate.[90] Lithium ion measurement was subject to Na^+ interference with earlier ionophores. An improvement has been possible with a redesigned crown ether chromionophore (Figure 2.6), admittedly for optical sensing, but design considerations for size selectivity are better understood.[91] Thus, bulky side groups

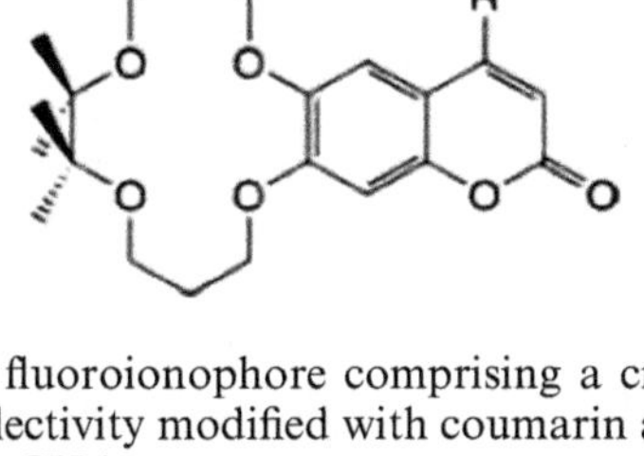

Figure 2.6 Li$^+$ selective fluoroionophore comprising a crown ether with bulky side groups for selectivity modified with coumarin as fluorophore. R is CH$_3$ or (CH$_2$)4-CH $=$ CH$_2$).
(Reproduced from ref. 91 with permission.)

on the crown ether prevented formation of sandwich complexes, avoiding the bivalent chelation of larger ions, whilst enhancement of ring rigidity prevented entry of smaller ions. pH electrodes using ionophores have, to a degree, been able to replace their glass counterpart; in one example tridodecylamine was used as part of a pH-responding ion-selective field effect transducer (ISFET).[92]

There remains a challenge to extrapolate from core measured ions in medicine to other diagnostically important ions, such as those that are present in trace amounts or that are organics. Design of affinity molecules based on modelling and simulation may secure the needed extension of the measurement span.

2.5.2 Amperometric Enzyme Biosensors

One reason for the difficulty in extending the analytical repertoire of ISEs, is the need for a new synthetic affinity molecule for each new target ion. Large numbers of ionophores have been reported, but relatively few have delivered on the key practical criteria of selectivity, sensitivity, rapid binding and the equally necessary rapid dissociation. By contrast, amperometric sensors do not need the invention of a new binding agent for each new analytical target. Not only can tuning of the polarising voltage provide for initial selectivity, but this can be augmented through generic barrier membranes to screen out interferents. Most importantly, redox enzymes and electron mediators are able to, respectively, bind to the solute of interest and facilitate the translation of this into an analytical signal.[93] The early days of amperometric biosensors, moreover, were considerably aided by a combination of a strong need for clinical blood glucose measurement (market pull) and the availability of an exceptionally robust enzyme (glucose oxidase) capable of providing stable biosensors (technology push). This favourable set of circumstances has, alas, not been repeated for any other biological target, though a high degree of commercial effort and innovation have underpinned developments in the last couple of decades.

An example of the conversion to practical realisation of a redox enzyme-based system has been the glucose and lactate sensors advanced by Roche Diagnostics.[94] Here, the oxidase enzymes for the respective metabolites were formulated in acrylate polymer to give stable, adherent thick films over a screen-printed working electrode on one face, the other face protected by a biocompatible membrane comprising a PVC copolymer with vinyl acetate/vinyl maleinate. Another practical adaptive step here was a MnO_2–carbon ink composite electrode that regenerated O_2 from the enzyme reaction to help replenish the co-substrate, so extending the linear range while also allowing for measurements at reducing polarising voltages, the latter minimising electrochemical interference.

A combination of enzyme, electron mediator and selectivity barrier was reported for a disposable glucose sensor that used SiO_2 nanoparticles as a means of controlling transmembrane diffusion.[95] While permeability is a key factor determining sensor performance, it may not be constant for a given membrane, and hydration and ageing effects warrant more detailed analysis. In an attempt to follow solution effects, Hsu *et al.*[96] monitored thickness and refractive-index changes at a thin glucose oxidase layer; refractive index altered in solution in a nonlinear relationship with pH, but thickness was unaffected. This confirms that quite subtle structural changes do occur in enzyme layers and that their tracking may identify better immobilisation strategies. Regardless of this, the continued emphasis on substrate/O_2 diffusion into the enzyme layer may give a one-sided picture of performance variables. In a modelling study using controlled layer-by-layer deposition of barrier films at a glucose electrode,[97] it was found that a greater barrier could actually lead to a bigger response. This was concluded to be due to a parallel reduced permeability to H_2O_2, and therefore its accumulation within the sensing layer.

Nanoscale electrodes, especially combined with new materials, provide a further means of modifying baseline sensor performance, even without use of separate barrier membranes. At a single-wall carbon-nanotube electrode, the entrapment of glucose oxidase within a polypyrrole layer considerably extended the upper linear range, while also allowing for detection limits down to the micromolar range.[98] Alternatively, carboxylated graphene enabled self-assembly of glucose oxidase, and with this a direct electron exchange with the enzyme prosthetic group, radically lowering the polarisation voltage needed for a response.[99] A more natural direct electron transfer is achieved using a PQQ-glucose dehydrogenase; here the PQQ (pyrroloquinline quinone) cofactor is able to relay electrons to a mediator, or directly to a polarised working electrode, and does not have an O_2 requirement; ruthenium hexamine and an osmium complex has been utilised as the electron relays to construct commercial glucose strips.[100] In an example of direct electron transfer, a nanoporous carbon electrode was used.[101] The PQQ-glucose dehydrogenase enzyme is not as selective as glucose oxidase, but the problem has been overcome using genetically engineered enzyme. In a similar way an FAD-glucose dehydrogenase has been modified to give greater glucose selectivity.[102] Even in the case of classical glucose oxidase, mutagenesis of the amino acid residues involved in the reaction with O_2 has allowed for a downregulation of oxidative activity with parallel upscaling of intrinsic dehydrogenase activity, so enabling use of artificial mediator for an oxygen-independent response.[103]

Lactate sensors are based on lactate oxidase, a low-stability enzyme with a high lactate affinity, leading, respectively, to low operational stability and a limited analytical range. Sensor design, as with any O_2-dependent system needs to promote a high transport rate for O_2 *vs* the substrate to maximise the linear range of operation. The intrinsic high diffusion coefficient of O_2 is advantageous in this regard, as is its regeneration from the H_2O_2 electrode reaction. One modelling study showed that enzyme layer concentrations of O_2 were minimally altered in the lactate oxidase layer compared with bulk solution $[O_2]$.[104] A taylored screen-printed lactate electrode with cobalt phthalocyanine as electron mediator for hydrogen peroxide enabled, the operating voltage was reduced.[105] Lactate oxidase can be stabilised by chemical crosslinking, provided the enzyme survives this initial binding step. Further improved storage stability is achievable, and may be effected *via* the resulting membrane matrix; an interesting enhancement of stability was seen when mucin was used in the crosslinking layer.[104] Electrode architecture and surface area affect sensitivity, as was shown using a 3D macroporous gold construct to immobilise the lactate oxidase.[106] The general construction approaches for lactate sensors has been reviewed by Nikolaus and Strehlitz.[107]

The advantage of surface area increase was demonstrated for a cholesterol oxidase-based cholesterol electrode where the enzyme was adsorbed onto a nanostructured ZnO film.[108] The practical use, challenges of sample delivery, electrode configuration, timing and temperature control have been described for a cholesterol strip sensor that used oxidase coupled with horseradish per-oxidase and a ferricyanide mediator.[109] Even direct electrochemical oxidation of cholesterol at platinum nanoparticle assemblies on carbon nanotubes has been demonstrated;[110] response was not affected by interferents, apparently; whether such a system can provide enzyme-free clinical assay remains to be seen.

Creatinine measurement requires a more complex, multienzyme, sequence. Despite the challenges of multienzyme functional compatibility under common reaction conditions, it has been possible to combine the classic sequence of creatinine amidohydrolase, creatine amidinohydrolase and sarcosine oxidase to achieve the respective conversion of creatinine to creatine and then to sarcosine and finally H_2O_2 end product.[111] Typically, an inner membrane barrier is used to reduce electrochemical interference, but in an alternative approach, Shin *et al.*[112] used a PbO_2 oxidising prelayer to inactivate such redox active compounds. In addition, a hydrophilic polyurethane barrier enabled transport of cationic creatinine, while limiting entry of creatine. Gel entrapment of the enzymes avoids denaturation, but retention of activity was also demonstrated by facile enzyme attachment to porous polypropylene functionalised with poly(acrylic acid),[113] and even chemical crosslinking seems possible.[114] A factor that has not received the level of attention it needs is the inhibition of enzyme activity by Ag^+ from Ag/AgCl reference electrodes. In the case of creatinine amidohydrolase, rapid loss of activity was seen with $1\,\mu M\ Ag^+$;[115] a covering barrier of cellulose acetate over the reference was able to protect against Ag^+ leaching.

2.5.3   Immunosensors

Direct electrochemical detection of an antigenic target, through binding to an immobilised antibody, will generate surface-charge effects, and these may be registered as impedimetric or potentiometric change.[116] This is an attractive, simple means of detection, but in reality the strong effects of biological sample solution variables and nonspecific binding of macromolecules make it difficult to consider such systems for clinical use. Practical progress has been achieved, therefore, mainly by the use of tried and tested electrochemical and redox enzyme labels, utilising traditional immunoassay formats classified, as with any other immunoassay, as competitive or noncompetitive.

A classical noncompetitive immunoassay is represented in an IgA assay[117] where horseradish peroxidase (HRP) was used as the label on a binding second antibody (Figure 2.7). An immobilised first antibody on a gold electrode provided the affinity surface for initial IgA attachment, and nonspecific protein binding was reduced by near-surface poly(ethylene glycol). HRP, in the presence of H_2O_2, oxidised o-dianisidine (ODS) to give a product that was detectable at the electrode. This arrangement shows both the feasibility of using amperometric formats and the difficulty of converting any such approach into a single-step assay.

One way of achieving assay simplicity is to automate reagent presentation to the electrode surface. A microfluidic arrangement was devised by Liang *et al.*[118] for brain natriuretic peptide (BNP) where Prussian blue as electron mediator was used to amplify the response from the second antibody label. A multiparameter (carcinoembryonic antigen, α-fetoprotein, β-human chor-iogonadotropin, carcinoma antigen 125) competitive immunoassay has been reported. This used limited antibody in solution for competitive binding between sample antigen and membrane immobilised antigen. The resulting surface binding of antibody provided a measure of target antigen level; HRP was the tag on the antibody, and membrane-entrapped mediator enhanced the enzyme signal.[119] A similar competitive assay was realised for D-dimer in serum, but here the immobilised antigen had been retained on magnetic beads, later captured at a working electrode surface.[120] The advantage of the

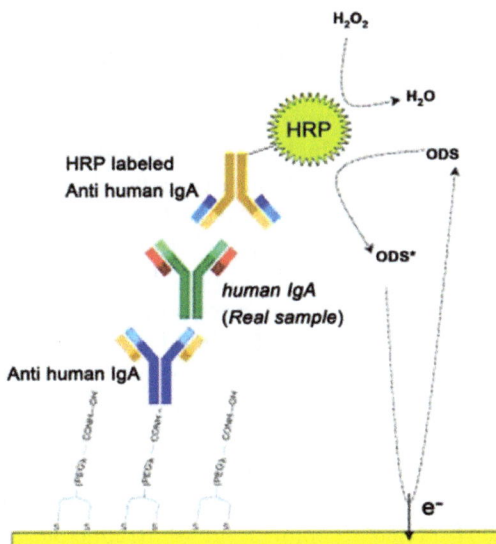

Figure 2.7   Schematic of an electrochemical immunoassay for IgA. Bipodal carboxy terminated alkyl thiols provide a surface function for immobilising capture antibody *via* carbodiimide/N-hydroxy-succinimide bridging. The second antibody is horseradish peroxidase (HRP) labelled to catalyse an indicator reaction using O-dianosidine as mediator.
(Reproduced from ref. 117 with permission.)

distributed, particulate phase was that it accelerated binding exchanges, while the facility for magnetic transfer meant both a simplified separation of free from bound antibody and a way of carrying out electrochemical detection under better controlled solution conditions. In an alternative format, a competitive immunoassay for *Staphylococcus aureus* involved the limited antibody being immobilised on magnetic particles, with labelled antigen in solution.[121] Because binding here involves only one antibody, two epitopes are not needed, and so small molecules can be estimated. Oestradiol was measured in this way using monoclonal antibody attached to gold nanoparticles;[122] competitive binding between HRP labelled and unlabelled oestrodiol was measured electrochemically *via* enzymatically generated benzoquinone.

2.5.4 DNA Sensors

Many types of DNA sensors have been reported, but they generally require PCR amplification to generate sufficient target DNA and depend on some form of labelling, *e.g.* electrochemical, for determining the hybridisation reaction. There are, therefore, parallels with immunosensors. Intrinsic electrochemical properties of DNA bases have been utilised for sensing DNA, as have DNA-induced impedimetric and potentiometric changes at surfaces, but these are invariably basic studies in highly controlled solution.

An unconventional means of generating an amplified signal at captured DNA was to use a gold nanoparticle labelled probe DNA to hybridise to cytomegalovirus DNA.[123] Here, oxidative dissolution of the nanoparticles released high quantities of Au(III) ions that were then determined by anode-stripping voltammetry. Duplex DNA intercalating agents are frequently used to identify hybridisation. In one example, methylene blue was used as the agent, and was able to reduce ferricyanide which served as a diffusible redox reporter.[124] A common intercalater is the electrochemically active dye Hoechst 33258, which can be oxidised at modest polarising voltages. This mode of reporting has been used to detect single base-pair mismatches,[125] though sensitivity is limited. The binder was used for detection of PCR-amplified DNA mixtures using an electrode array.[126] New forms of label continue to be developed; thus, noncovalent labelling was achieved using Zr(IV).[127] The highly charged ion has strong affinity for dsDNA phosphate groups and was able to bridge to an anionic ferrocene carboxylate reporter. Redox cycling of the latter led to electrochemical signal generation at hybridised DNA. The exploitation of labels has been extensively reported at DNA sensors, and optical methods will add to the effort on electrochemical approaches, but the challenge will be of how to simplify procedures sufficiently for POCT. There remain a set of irreducible steps, so the most plausible strategy would appear to be their automation.

2.6 Micromechanical Transduction

Microgravimetric and surface acoustic wave (SAW) sensors offer two distinct mass-based transduction modalities that have the ability to register

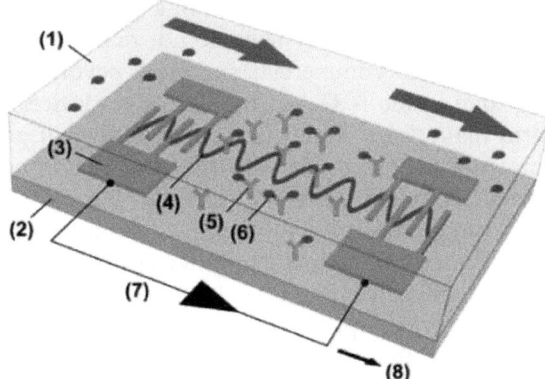

Figure 2.8 Schematic of surface acoustic wave (SAW) immunosensor in contact with (1): arrows indicate flow direction. Device components are: (2) a piezo-electric crystal (3) interdigitated transducers to set up the acoustic wave, (4) the surface acoustic wave, (5) capture antibodies (6) target molecules. The driving electronics (7) generate change in the output signal (8) when the target molecule binds to the sensor surface.
(Reproduced from ref. 130 with permission.)

protein/macromolecule binding. Variables such as deposited surface mass, layer thickness and post-binding surface microviscosity change all contribute to the response, but a simple quantitative link can, nevertheless, be made with the amount bound. Typically, antibodies are used to functionalise the surface for selectivity. Low molecular weight species such as drugs can be measured, but a label is needed to generate a high-mass product. High-sensitivity detection of protein breast cancer markers was demonstrated, for example, at a SAW device functionalised with antibody;[128] coupling chemistry that delivered an orientated antibody gave better sensitivity. In another study, precoating of the SAW surface with a glass coating reduced acoustic velocity, so increasing surface wave confinement and therefore sensitivity.[129] Additional use of a poly(vinylamine) phase reduced nonspecific binding, allowing assay in diluted whole blood. A schematic of an immunosensor based on a SAW device is shown in Figure 2.8. The general field and the potential for commercialisation have been reviewed.[130]

2.7 Commercialisation

POCT systems as consumer products demand considerable multidisciplinary input. Only in this way is the integration of multifunctional components possible. Final design goes well beyond the original chemistry concept, and the often perceived advantages of the original chemistry become submerged in quite unexpected problems not amenable to chemical manipulation. Overall, engineering and analytical principles require to converge in order to address the special fabrication and design needs of robust analysis. The associated development costs are not trivial, and are orders of magnitude greater than those of the original research. This ramp-up is as much of a barrier as the technical issues and a reason why many promising systems do not reach the market place.

There have been notable commercial successes beyond glucometers. One well-established system is the i–STAT; the cartridge devices of this deskilled analytical platform make for a highly attractive analytical package.[131] While not achieving a matched simplicity, the Vantix™ sandwich immunoassay system with its antigen-coated potentiometric sensor goes a long way to simplifying this type of assay;[132] immunoassay is inherently difficult to simplify. The Alpha DX, a STAT whole-blood analysis system alternatively approaches immuno-assay by fluorescent sandwich immunoassay *via* target molecule capture at a test disc.[133] In a further example, a radial geometry immunoassay (Stratus CS®) has been developed for cardiac markers[134] where a dendrimer bridging molecule improves the presentation of capture antibody. For the Biosite Triage® there is linear tracking of reagent and antigen, a similar platform has also been advanced for brain natriuretic peptide.[135]

A commercial ABL800 FLEX analyser used for measurements of pH, blood gas, electrolyte, oximetry and metabolite parameters has also been developed.[136]

2.8   Conclusions

Transformation to the solid state has been an important route to producing POCT devices, enabling incorporation of both biological and artificial functional components. The technology for immobilisation of delicate biological reagents, especially, has allowed the development of a diverse range of measurement platforms. In addition, the wider range of transduction and registration methods have extended the ways in which biorecognition can be followed. While there are still no real alternatives to biological receptor systems for the measurement of complex organics, their conversion to robust reagents, either through chemistry or genetic engineering, is likely to expand.

Parallel developments in miniaturisation and lamination technologies have enabled radical design changes. The increased use of micro- and nano-constructs, along with multilayer deposition techniques has allowed creation of ever more complex devices with greater dimensional compression. A parallel exists with the functional compression of electronic devices. There will be a time, also, when device incorporation into electronics will become seamless and the added value of data interpretation and transmission fully realised. This would not be just a technological goal for its own sake, but a means of data sharing to enhance clinical interpretation.

Recent miniaturisation effort has led on to newer materials such as CNTs and graphene. These may eventually deliver practical devices that allow the utilisation of new engineering principles. Tomorrow's POCT devices may well become less intrusive than those of today with the newer materials. If automated preanalytical processes and sample handling can be similarly re-engineered, then convergence to the "lab-on-a-chip" concept will be possible in the real world. Some of the subcomponents set to make up such platforms have been described in this chapter. What remains as an intriguing next phase is their integration into seamless systems.

Acknowledgement

The authors wish to thank EPSRC for generous funding through Project EP/H009744/1 (ESPRIT).

References

1. C. P. Price, *Brit. Med. J.*, 2001, **322**, 1285.
2. P. B. Luppa, C. Muller, A. Schlichtiger and H. Schlebusch, *Trac-Trends Analyt. Chem.*, 2011, **30**, 887.
3. J. Y. Wick, *The Consultant Pharmacist: J. Am. Soc. Consultant Pharmacists*, 2010, **25**, 416.
4. C. P. Price and L. J. Kricka, *Clin. Chem.*, 2007, **53**, 1665.
5. T. Vo-Dinh, B. M. Cullum and D. L. Stokes, *Sens. Actuators B-Chem.*, 2001, **74**, 2.
6. R. Pauliukaite, M. Schoenleber, P. Vadgama and C. M. A. Brett, *Anal. Bioanal. Chem.*, 2008, **390**, 1121.
7. S. M. Ng, P. Krishnaswamy, R. Morissey, P. Clopton, R. Fitzgerald and A. S. Maisel, *Am. J. Cardiol.*, 2001, **88**, 611.
8. J. McCord, R. M. Nowak, P. A. McCullough, C. Foreback, S. Borzak, G. Tokarski, M. C. Tomlanovich, G. Jacobsen and W. D. Weaver, *Circulation*, 2001, **104**, 1483.
9. J. E. Dechant, W. A. Symm and J. E. Nieto, *Vet. Surg.*, 2011, **40**, 811.
10. Y. C. Manabe, Y. P. Wang, A. Elbireer, B. Auerbach and B. Castelnuovo, *Plos One*, 2012, **7**, e34319.
11. J. Porter, N. O'Loan, B. Bell, J. Mahoney, M. McGarrity, R. I. McConnell and S. P. Fitzgerald, *Anal. Bioanal. Chem.*, 2012, **403**, 3051.
12. M. A. Qasaimeh, S. G. Ricoult and D. Juncker, *Lab on a Chip*, 2013, **13**, 40.
13. T. H. Schulte, R. L. Bardell and B. H. Weigl, *Clin. Chim. Acta*, 2002, **321**, 1.
14. P. Yager, T. Edwards, E. Fu, K. Helton, K. Nelson, M. R. Tam and B. H. Weigl, *Nature*, 2006, **442**, 412.
15. A. E. Herr, A. V. Hatch, D. J. Throckmorton, H. M. Tran, J. S. Brennan, W. V. Giannobile and A. K. Singh, *Proc. Nat. Acad. Sci. USA*, 2007, **104**, 5268.
16. H. Chang, R. Khan, Z. Rong, A. Sapelkin and P. Vadgama, *Biofabrication*, 2010, **2**, Art No 035002.
17. M. K. Ramasubramanian and S. P. Alexander, *Biomed. Microdevices*, 2009, **11**, 217.
18. L. Gervais and E. Delamarche, *Lab on a Chip*, 2009, **9**, 3330.
19. S. Hsiung, C. Chen and G. Lee, *Lab on a chip*, 2009, **9**, 3330.

20. D. S. Millington, R. Sista, A. Eckhardt, J. Rouse, D. Bali, R. Goldberg, M. Cotten, R. Buckley and V. Pamula, *Sem. Perinatol.*, 2010, **34**, 163.
21. A. Bhattacharyya and C. M. Klapperich, *Biomed. Microdevices*, 2007, **9**, 245.
22. M. H. Zweig and A. Jackson, *Clin. Chem.*, 1986, **32**, 674.
23. B. Walter, *Method Enzymol.*, 1998, **137**, 794.
24. R. Sommer, C. Nelson and A. Greenquist, *Clin. Chem.*, 1986, **32**, 1770.
25. R. H. Ng, K. M. Sparks and B. E. Statland, *Clin. Chem.*, 1992, **38**, 1371.
26. M. J. Pugia, J. A. Lott, J. A. Profitt and T. K. Cast, *J. Clin. Lab. Anal.*, 1999, **13**, 180.
27. A. Tripodi, V. Chantarangkul, M. Primignani, A. Dell'Era, M. Clerici, F. Iannuzzi, A. Aghemo, M. Cazzaniga, F. Salerno and P. M. Mannucci, *J. Hepatol.*, 2009, **51**, 288.
28. A. Warsinke, *Anal. Bioanal. Chem.*, 2009, **393**, 1393.
29. S. Puertas, M. Moros, R. Fernandez-Pacheco, M. R. Ibarra, V. Grazu and J. M. de la Fuente, *J. Phys. D – Appl. Phys.*, 2010, **43**.
30. W. U. Dittmer, T. H. Evers, W. M. Hardeman, W. Huijnen, R. Kamps, P. de Kievit, J. H. M. Neijzen, J. H. Nieuwenhuis, M. J. J. Sijbers, D. W. C. Dekkers, M. H. Hefti and M. F. W. C. Martens, *Clin. Chim. Acta*, 2010, **411**, 868.
31. G. A. Posthuma-Trumpie, J. Korf and A. van Amerongen, *Anal. Bioanal. Chem.*, 2009, **393**, 569.
32. S. Choi, E. Y. Choi, H. S. Kim and S. W. Oh, *Clin. Chem.*, 2004, **50**, 1052.
33. W. E. Doering, M. E. Piotti, M. J. Natan and R. G. Freeman, *Adv. Mater.*, 2007, **19**, 3100.
34. G. A. Posthuma-Trumpie, J. H. Wichers, M. Koets, L. B. J. M. Berendsen and A. van Amerongen, *Anal. Bioanal. Chem.*, 2012, **402**, 593.
35. M. Rayev and K. Shmagel, *J. Immun. Methods*, 2008, **336**, 9.
36. P. F. Mens, A. van Amerongen, P. Sawa, P. A. Kager and H. D. F. H. Schallig, *Diagnos. Microbiol. Infect. Dis.*, 2008, **61**, 421.
37. K. A. Edwards and A. J. Baeumner, *Anal. Bioanal. Chem.*, 2006, **386**, 1613.
38. N. V. Zaytseva, R. A. Montagna, E. M. Lee and A. J. Baeumner, *Anal. Bioanal. Chem.*, 2004, **380**, 46.
39. N. V. Kulagina, K. M. Shaffer, F. S. Ligler and C. R. Taitt, *Sens. Actuators B-Chem.*, 2007, **121**, 150.
40. A. W. Martinez, S. T. Phillips and G. M. Whitesides, *Proc. Natl. Acad. Sci. USA*, 2008, **105**, 19606.
41. F. Kong and Y. F. Hu, *Anal. Bioanal. Chem.*, 2012, **403**, 7.
42. J. Lehtio, J. Sugiyama, M. Gustavsson, L. Fransson, M. Linder and T. T. Teeri, *Proc. Natl. Acad. Sci. USA*, 2003, **100**, 484.
43. S. Di Risio and N. Yan, *Colloids Surf. B-Biointerfaces*, 2010, **79**, 397.
44. W. Zhang, W. T. Ang, C.-Y. Xue and K.-L. Yang, *ACS Appl. Mater. Interfaces*, 2011, **3**, 3496.

45. M. Khan, X. Li, W. Shen and G. Garnier, *Colloids and Surfaces B-Biointerfaces*, 2010, **75**, 239.
46. L. Pardo, W. C. Wilson and T. J. Boland, *Langmuir*, 2003, **19**, 1462.
47. J. Wang, B. Yiu, J. Obermeyer, C. D. M. Filipe, J. D. Brennan and R. Pelton, *Biomacromolecules*, 2012, **13**, 559.
48. H. Yan, A. Ahmad-Tajudin, M. Bengtsson, S. Xiao, T. Laurell and S. Ekstrom, *Analyt. Chem.*, 2011, **83**, 4942.
49. Q. Ma, T.-Y. Song, P. Yuan, C. Wang and X.-G. Su, *Colloids Surf. B-Biointerfaces*, 2008, **64**, 248.
50. H. Wang, Y. L. Liu, Y. H. Yang, T. Deng, G. L. Shen and R. Q. Yu, *Anal. Biochem.*, 2004, **324**, 219.
51. P. Peluso, D. S. Wilson, D. Do, H. Tran, M. Venkatasubbaiah, D. Quincy, B. Heidecker, K. Poindexter, N. Tolani, M. Phelan, K. Witte, L. S. Jung, P. Wagner and S. Nock, *Anal. Biochem.*, 2003, **312**, 113.
52. K. Bonroy, F. Frederix, G. Reekmans, E. Dewolf, R. De Palma, G. Borghs, P. Declerck and B. Goddeeris, *J. Immun. Methods*, 2006, **312**, 167.
53. J. Buijs, J. W. T. Lichtenbelt, W. Norde and J. Lyklema, *Colloids Surf. B-Biointerfaces*, 1995, **5**, 11.
54. X. Yuan, D. Fabregat, K. Yoshimoto and Y. Nagasaki, *Colloids Surf. B-Biointerfaces*, 2012, **99**, 45.
55. I. Migneault, C. Dartiguenave, M. J. Bertrand and K. C. Waldron, *Biotechniques*, 2004, **37**, 790.
56. F. Lopez-Gallego, L. Betancor, C. Mateo, A. Hidalgo, N. Alonso-Morales, G. Dellamora-Ortiz, J. M. Guisan and R. Fernandez-Lafuente, *J. Biotech.*, 2005, **119**, 70.
57. M. N. Gupta and S. Raghava, *Methods Mol. Biol. (Clifton, N.J.)*, 2011, **679**, 133.
58. R. A. Sheldon, *Biochem. Soc. Trans.*, 2007, **35**, 1583.
59. A. Haouz, J. M. Glandieres and B. Alpert, *Febs Lett.*, 2001, **506**, 216.
60. J. C. Tiller, R. Rieseler, P. Berlin and D. Klemm, *Biomacromolecules*, 2002, **3**, 1021.
61. U. Bora, P. Sharma, K. Kannan and P. Nahar, *J. Biotech.*, 2006, **126**, 220.
62. C. Tyagi, L. K. Tomar and H. Singh, *J. Appl. Polym. Sci.*, 2009, **111**, 1381.
63. R. Pelton, *Trac-Trends Analyt. Chem.*, 2009, **28**, 925.
64. P. Jonkheijm, D. Weinrich, M. Koehn, H. Engelkamp, P. C. M. Christianen, J. Kuhlmann, J. C. Maan, D. Nuesse, H. Schroeder, R. Wacker, R. Breinbauer, C. M. Niemeyer and H. Waldmann, *Angew. Chem.-Int. Ed.*, 2008, **47**, 4421.
65. D. H. J. Bunka and P. G. Stockley, *Nature Rev. Microbiol.*, 2006, **4**, 588.
66. Su, R. Nutiu, C. D. M. Filipe, Y. Li and R. Pelton, *Langmuir*, 2007, **23**, 1300.
67. W. Zhao, M. M. Ali, S. D. Aguirre, M. A. Brook and Y. Li, *Anal. Chem.*, 2008, **80**, 8431.

68. C. Carrasquilla, P. S. Lau, Y. Li and J. D. Brennan, *J. Am. Chem. Soc.*, 2012, **134**, 10998.
69. L. Cui, Z. Chen, Z. Zhu, X. Lin, X. Chen and C. J. Yang, *Analyt. Chem.*, 2013, **85**, 2269.
70. Y. Wu, J. A. Phillips, H. Liu, R. Yang and W. Tan, *ACS Nano*, 2008, **2**, 2023.
71. S. Balamurugan, A. Obubuafo, S. A. Soper and D. A. Spivak, *Anal. Bioanal. Chem.*, 2008, **390**, 1009.
72. S. J. Hurst, A. K. R. Lytton-Jean and C. A. Mirkin, *Analyt. Chem.*, 2006, **78**, 8313.
73. X. Zhang and V. K. Yadavalli, *Biosens. Bioelectron.*, 2011, **26**, 3142.
74. A. Miodek, A. Poturnayova, M. Snejdarkova, T. Hianik and H. Korri-Youssoufi, *Anal. Bioanal. Chem.*, 2013, **405**, 2505.
75. Y. Wenjuan, A. Le Goff, N. Spinelli, M. Holzinger, G.-W. Diao, D. Shan, E. Defrancq and S. Cosnier, *Biosens. Bioelectron.*, 2013, **42**, 556.
76. M. Levy, S. F. Cater and A. D. Ellington, *Chem. Biochem.*, 2005, **6**, 2163.
77. X.-C. Chen, Y.-L. Deng, Y. Lin, D.-W. Pang, H. Qing, F. Qu and H.-Y. Xie, *Nanotechnology*, 2008, **19**.
78. G. Mayer, *Angew. Chem.-Int. Ed.*, 2009, **48**, 2672.
79. R. P. Buck and E. Lindner, *Acc. Chem. Res.*, 1998, **31**, 257.
80. E. Bakker, P. Buhlmann and E. Pretsch, *Electroanalysis*, 1999, **11**, 915.
81. G. Dimeski, T. J. Morgan, J. J. Presneill and B. Venkatesh, *J. Crit. Care*, 2012, **27**.
82. F. Phillips, K. Kaczor, N. Gandhi, B. D. Pendley, R. K. Danish, M. R. Neuman, B. Toth, V. Horvath and E. Lindner, *Talanta*, 2007, **74**, 255.
83. R. Retter, S. Peper, M. Bell, I. Tsagkatakis and E. Bakker, *Analyt. Chem.*, 2002, **74**, 5420.
84. I. A. M. de Oliveira, D. Risco, F. Vocanson, E. Crespo, F. Teixidor, N. Zine, J. Bausells, J. Samitier and A. Errachid, *Sens. Actuators B-Chem.*, 2008, **130**, 295.
85. S. Chandra and H. Lang, *Sens. Actuators B-Chem.*, 2006, **114**, 849.
86. W. Qin, T. Zwickl and E. Pretsch, *Analyt. Chem.*, 2000, **72**, 3236.
87. K. Y. Chumbimuni-Torres, N. Rubinova, A. Radu, L. T. Kubota and E. Bakker, *Analyt. Chem.*, 2006, **78**, 1318.
88. F. Li, J. Ye, M. Zhou, S. Gan, Q. Zhang, D. Han, and Li Niu, *Analyst*, 2012, **137**, 618.
89. G. Dimeski, T. Badrick and A. St John, *Clinica Chim. Acta*, 2010, **411**, 309.
90. G. Dimeski and A. E. Clague, *Clin. Chem.*, 2004, **50**, 1106.
91. D. Citterio, J. Takeda, M. Kosugi, H. Hisamoto, S.-i. Sasaki, H. Komatsu and K. Suzuki, *Analyt. Chem.*, 2007, **79**, 1237.
92. N. Abramova and A. Bratov, *Talanta*, 2010, **81**, 208.
93. P. Vadgama and P. W. Crump, *Analyst*, 1992, **117**, 1657.
94. B. P. H. Schaffar, *Anal. Bioanal. Chem.*, 2002, **372**, 254.

95. Z. Chen, C. Fang, H. Wang and J. He, *Electrochimica Acta*, 2009, **55**, 544.
96. C.-C. Hsu, W.-L. Lan, Y.-C. Chen and Y.-Y. Sung, *Opt. Lasers Eng.*, 2013, **51**, 388.
97. R. A. Croce, Jr., S. Vaddiraju, F. Papadimitrakopoulos and F. C. Jain, *Sensors*, 2012, **12**, 13402.
98. F. Valentini, L. G. Fernandez, E. Tamburri and G. Palleschi, *Biosens. Bioelectron.*, 2013, **43**, 75.
99. B. Liang, L. Fang, G. Yang, Y. Hu, X. Guo and X. Ye, *Biosens. Bioelectron.*, 2013, **43**, 131.
100. A. Heller and B. Feldman, *Chem. Rev.*, 2008, **108**, 2482.
101. V. Flexer, F. Durand, S. Tsujimura and N. Mano, *Analyt. Chem.*, 2011, **83**, 5721.
102. Y. Yamashita, S. Ferri, M. L. Huynh, H. Shimizu, H. Yamaoka and K. Sode, *Enzyme Microbial. Technol.*, 2013, **52**, 123.
103. Y. Horaguchi, S. Saito, K. Kojima, W. Tsugawa, S. Ferri and K. Sode, *Int. J. Mol. Sci.*, 2012, **13**, 14149.
104. M. R. Romero, F. Ahumada, F. Garay and A. M. Baruzzi, *Analyt. Chem.*, 2010, **82**, 5568.
105. T. Shimomura, T. Sumiya, M. Ono, T. Ito and T.-a. Hanaoka, *Anal. Chim. Acta*, 2012, **714**, 114.
106. M. Gamero, J. L. G. Fierro, E. Lorenzo and C. Alonso, *Electroanalysis*, 2013, **25**, 179.
107. N. Nikolaus and B. Strehlitz, *Microchimica Acta*, 2008, **160**, 15.
108. N. Batra, M. Tomar and V. Gupta, *Analyst*, 2012, **137**, 5854.
109. R. Foster, J. Cassidy and E. O'Donoghue, *Electroanalysis*, 2000, **12**, 716.
110. J. Yang, H. Lee, M. Cho, J. Nam and Y. Lee, *Sens. Actuators B-Chem.*, 2012, **171**, 374.
111. T. Tsuchida and K. Yoda, *Clin. Chem.*, 1983, **29**, 51.
112. J. H. Shin, Y. S. Choi, H. J. Lee, S. H. Choi, J. Ha, I. J. Yoon, H. Nam and G. S. Cha, *Analyt. Chem.*, 2001, **73**, 5965.
113. G. H. Hsiue, P. L. Lu and J. C. Chen, *J. Appl. Polym. Sci.*, 2004, **92**, 3126.
114. M. B. Madaras, I. C. Popescu, S. Ufer and R. P. Buck, *Anal. Chim. Acta*, 1996, **319**, 335.
115. J. A. Berberich, A. Chan, M. Boden and A. J. Russell, *Acta Biomater.*, 2005, **1**, 193.
116. A. Warsinke, A. Benkert and F. W. Scheller, *Fresenius J. Analyt. Chem.*, 2000, **366**, 622.
117. L. Carlos Rosales-Rivera, J. Lluis Acero-Sanchez, P. Lozano-Sanchez, I. Katakis and C. K. O'Sullivan, *Biosens. Bioelectron.*, 2012, **33**, 134.
118. W. Liang, Y. Li, B. Zhang, Z. Zhang, A. Chen, D. Qi, W. Yi and C. Hu, *Biosens. Bioelectron.*, 2012, **31**, 480.
119. J. Wu, F. Yan, J. Tang, C. Zhai and H. Ju, *Clin. Chem.*, 2007, **53**, 1495.
120. M. Gamella, S. Campuzano, F. Conzuelo, A. Julio Reviejo and J. M. Pingarron, *Electroanalysis*, 2012, **24**, 2235.

121. B. Esteban-Fernandez de Avila, M. Pedrero, S. Campuzano, V. Escamilla-Gomez and J. M. Pingarron, *Anal. Bioanal. Chem.*, 2012, **403**, 917.
122. X. Liu and D. K. Y. Wong, *Talanta*, 2009, **77**, 1437.
123. L. Authier, C. Grossiord, P. Brossier and B. Limoges, *Analyt. Chem.*, 2001, **73**, 4450.
124. E. M. Boon, D. M. Ceres, T. G. Drummond, M. G. Hill and J. K. Barton, *Nature Biotechnol.*, 2000, **18**, 1096.
125. Y. S. Choi and D. H. Park, *J. Korean Phys. Soc.*, 2004, **44**, 1556.
126. M. Takahashi, J. Okada, K. Ito, M. Hashimoto, K. Hashimoto, Y. Yoshida, Y. Furuichi, Y. Ohta, S. Mishiro and N Gemma, *Clin. Chem.*, 2004, **50**, 658.
127. P. Wipawakarn, H. Ju and D. K. Y. Wong, *Anal. Bioanal. Chem.*, 2012, **402**, 2817.
128. F. J. Gruhl and K. Laenge, *Anal. Biochem.*, 2012, **420**, 188.
129. S. Rupp, M. von Schickfus, S. Hunklinger, H. Eipel, A. Priebe, D. Enders and A. Pucci, *Sens. Actuators B-Chem.*, 2008, **134**, 225.
130. K. Laenge, B. E. Rapp and M. Rapp, *Anal. Bioanal. Chem.*, 2008, **391**, 1509.
131. I. R. Lauks, *Acc. Chem. Res.*, 1998, **31**, 317.
132. J. Cork, R. M. Jones and J. Sawyer, *J. Immun. Methods*, 2013, **387**, 140.
133. F. S. Apple, F. P. Anderson, P. Collinson, R. L. Jesse, M. C. Kontos, M. A. Levitt, E. A. Miller and M. M. Murakami, *Clin. Chem.*, 2000, **46**, 1604.
134. S. Altinier, M. Mion, A. Cappelletti, M. Zaninotto and M. Plebani, *Clin. Chem.*, 2000, **46**, 991.
135. Y. Fischer, K. Filzmaier, H. Stiegler, J. Graf, S. Fuhs, A. Franke, U. Janssens and A. M. Gressner, *Clin. Chem.*, 2001, **47**, 591.
136. http://www.radiometer.com.

CHAPTER 3

Blood-Glucose Biosensors, Development and Challenges

YUAN WANG*[a] AND MADELEINE HU[b]

[a] Siemens HealthCare Diagnostics, 511 Benedix Ave. Tarrytown, NY 10591, United States; [b] The College of New Jersey, 2000 Pennington Road, Ewing, NJ 08628-0718, United States
*Email: yuan_wang@siemens.com

3.1 Introduction

Diabetes mellitus is the most common endocrine disorder of carbohydrate metabolism. It is one of the major causes of premature illness and death worldwide. The World Health Organization (WHO) estimated that 285 million people, corresponding to 6.4% of the world's adult population, lived with diabetes in 2010.[1] The number is expected to increase to 439 million by 2030, corresponding to 7.8% of the world's adult population. A sedentary lifestyle combined with changes in eating habits and increasing obesity is thought to be the main cause of such an increase. The National Health and Nutrition Examination Survey III shows that two-thirds of adults in the US diagnosed with Type 2 diabetes have a body mass index (BMI) of 27 or greater, which is classified as overweight and unhealthy.

With an increasingly diabetic population, the Blood Glucose Monitoring System (BGMS) is becoming an ever-important tool for diabetes management.[2–8] The goal of BGMS is to help the patients achieve and maintain normal blood glucose concentrations in order to delay or even prevent the progression of microvascular (retinopathy, nephropathy and neuropathy) and macrovascular (stroke and coronary artery disease) complications. The findings of the

RSC Detection Science Series No. 2
Detection Challenges in Clinical Diagnostics
Edited by Pankaj Vadgama and Serban Peteu
© The Royal Society of Chemistry 2013
Published by the Royal Society of Chemistry, www.rsc.org

Diabetes Control and Complications Trial (DCCT) and the United Kingdom Prospective Diabetes Study (UKPDS) clearly showed that intensive control of elevated levels of blood glucose in patients with diabetes decreases the occurrence of complications such as nephropathy, neuropathy and retinopathy and may also reduce the occurrence and severity of large blood vessel disease.[9–11] In addition, the system can also be useful for detecting hypoglycemia and providing real-time information for immediate adjustment in medications, dietary, and physical activities.[6,12] Regular and frequent measurement of blood glucose can provide valuable information for optimizing and/or changing patient treatment strategies.

Due to the aforementioned rapid increase in diabetes population and the growing awareness for good diabetes management, the growth of BGMS market is accelerating. In 2004, BGMS accounted for approximately 85% of the world market for biosensors, which had been estimated to be around $5 billion USD.[13] According to a recent report by Global Industry Analysts, Inc., the global market for glucose biosensors and strips would reach $11.5 billion USD by 2012.

In this chapter, we will review the development history of the BGMS, explain the principal chemical mechanisms on which various systems are based, discuss issues relating to accurate and robust detection, including minimizing of interference, and, finally, look at challenges and future trends.

3.2   History of Blood-Glucose Biosensor

In 1932 Warburg and Christian reported the "yellow enzyme" from yeast changed to colorless upon oxidizing its substrate and reassumed the yellow color after reoxidation by oxygen.[14] In 1939, Franke and Deffner isolated glucose oxidase (GOx) from *Aspergillus niger* and revealed that the GOx reaction center containing riboflavin that was involved in its reduction/oxidation reaction.[15] The mechanism of the glucose and glucose oxidase (GOx) reaction has been established and is well understood. Glucose reacts with yellow FAD-GOx to generate FADH$_2$-GOx which is then oxidized by oxygen, producing FAD-GOx and hydrogen peroxide.

$$\text{Glucose} + \text{FAD-GOx} \longrightarrow \text{gluconolactone} + \text{FADH}_2\text{-GOx} \qquad (3.1)$$

$$\text{FADH}_2\text{-GOx} + \text{O}_2 \longrightarrow \text{FAD-GOx} + \text{H}_2\text{O}_2 \qquad (3.2)$$

The first glucose assay using electrochemical methods was based on hydrogen peroxide oxidation of I^- to I_2 (dark blue), where the concentration of H$_2$O$_2$ was determined by I^-/I_2 potentiometry.[16,17]

$$\text{H}_2\text{O}_2 + 2\text{H}^+ + 2\text{I}^- \longrightarrow \text{H}_2\text{O} + \text{I}_2. \qquad (3.3)$$

The first portable blood glucose meter, the Ames Reflectance Meter (ARM), was introduced by Miles Laboratories (Elkhart, IN, USA) in 1969. A drop of blood was added onto a Dextrostix and left for one minute before being washed off. The Dextrostix was then placed in the ARM and a beam of light was shined on it.[18–21] A darker blue strip would be generated by a higher H_2O_2 content and would reflect back less light, therefore indicating higher blood glucose content. However, this is not how a typical glucose biosensor works today.

3.2.1 First Generation of Biosensors

Considerable advances in glucose biosensors were made after the introduction of the oxygen electrode by Clark and Lyons in 1962.[22] However, problems were encountered with various background oxygen levels that led to irreproducible results. Updike and Hicks stabilized GOx in a polyacrylamide gel on an oxygen electrode that corrected for background oxygen levels and significantly simplified the electrochemical glucose assay.[23,24] The first amperometric enzyme glucose sensor was developed in 1973, in which hydrogen peroxide levels were analyzed instead of the highly variable oxygen reduction current.[25] Nevertheless, the necessary oxidation of hydrogen peroxide required a relatively higher potential, at which common endogenous reducing substances, such as ascorbic acid and uric acid, *etc.*, would be oxidized, causing a falsely elevated glucose reading.

The first generation of biosensors was dependent on the presence of oxygen as a cofactor to ensure the catalytic regeneration of the FAD center (as in reaction 3.2). This method of sensor generation faced two major problems: interference from electroactive species of endogenous reducing substances and common drugs such as acetaminophen, and a dependence on free oxygen as a catalytic mediator.

Although GOx is generally a very specific and selective enzyme, the potential range at which the hydrogen peroxide is oxidized coincides with the oxidation potential of a number of compounds found in blood. These compounds include both endogenous and exogenous substances (Table 3.1). In addition, blood protein and chloride ions can be absorbed onto the electrode's surface and leave the sensor susceptible to fouling.[26]

Oxygen dependence is a problem specific to this generation. The oxygen content in various blood samples can be considerably different. In arterial blood samples, the normal partial pressure of oxygen (pO_2) is in a range of

Table 3.1 Common endogenous and exogenous interference substances.

Endogenous substances	Exogenous substances
Ascorbic acid	Acetaminophen
Bilirubin	Alcohol
Chloride	L-Dopa
L-Cystine	Salicylic acid
Protein	Tolazamide
Uric acid	

75–100 mmHg, whereas the pO_2 level in venous blood is about 40 mmHg. Since the regeneration of FAD-GOx requires one oxygen molecule per glucose molecule (reaction 3.2), depending on the dissolved oxygen level in the blood sample, the errors surrounding this high oxygen dependence is quite significant.

To overcome the problems encountered in the first generation glucose biosensors, a number of approaches have been developed. A comprehensive review paper describing different methods used for elimination of electrochemical interference in glucose biosensors was published by Jia *et al.*[27] These methods include coating the sensor with a preoxidizing layer to convert the interference substances into electrochemically inert forms before reaching the electrode surface;[28–34] the use of a permselective membrane to block electroactive substances reaching electrode surface; utilizing a modified layer, a mediator, or nanomaterial to allow a reduced potential where the effects of electroactive substances are negligible; and use of scanning electrochemical microscopy (SECM) to detect depletion of electroactive species in the diffusion layer.

Many methods were also proposed to minimize the errors caused by various background oxygen levels. These include the use of a polymer membrane that limits glucose diffusion, allowing more oxygen input relative to glucose which eases the oxygen dependence issue;[35–38] the use of oxygen-rich electrode materials that act as an internal source of oxygen;[39,40] and replacement of GOx with an oxygen insensitive enzyme, glucose dehydrogenase (this will be discussed further later).

3.2.2   Second Generation of Glucose Biosensors

The limitations of the first generation of blood glucose biosensors were overcome by the second generation of biosensors, for which mediated blood glucose biosensors were developed. Varieties of nonphysiological electron acceptors have been synthesized and can be utilized to facilitate electron transfer from the enzyme reaction center to the electrode surface. These mediators include ferrocene derivatives,[41,42] ferrocyanide,[43] quinones,[44,45] methylene blue,[14] thionine, pyocyanine, safranine T[15] and transition-metal complexes.[46] A comprehensive review paper by Heller and Feldman was published in 2008.[47] In the review paper, mediators are categorized into organic, inorganic, and metal mediators. Each of these mediators possess its own attributes such as low molecular weight, low redox potential to avoid interfering substances, high stability, resistance to forming side compounds, and low toxicity, which is especially important for *in vivo* glucose biosensors.

In addition to the mediator, redox hydrogels were also used for shuttling the electron from the enzyme reaction center to the electrode. In this case, enzymes are immobilized in the redox hydrogels and the electron transfer is realized by self-exchange between rapidly reduced and rapidly oxidized redox functions tethered to backbones of a crosslinked polymer network.[48] Because the redox hydrogels envelop the redox enzymes, they electrically connect the enzyme reaction centers to electrodes regardless the spatial orientation of the enzymes at the electrode surface. They also connect multiple enzyme layers that lead to a

10-fold higher, or in some cases, even 100-fold higher current densities, than an enzyme monolayer on the electrode surface.[49–52] The electron conductivity is at its highest when the redox hydrogel is poised at its redox potential.

The first research on the amperometric determination of blood glucose using a redox couple-mediated and GOx-catalyzed reaction was reported in 1970.[53] However, this study did not lead to a rapid adaptation of the technology in home testing.[54] Although the mediator reacts with the enzyme at a considerably faster rate than oxygen, the dissolved oxygen may also compete with the mediator when FAD-GOx enzymes are present. In recent years, GDH, an oxygen-insensitive enzyme, has been widely used in the BGMS industry.

The first electrochemical blood glucose monitor for diabetic patients home testing was launched in 1987 as ExacTech by Medisense Inc. It was a pen-sized monitor and used PQQ-GDH and a ferrocene derivative as a mediator.[55] Its success led to a revolution in diabetes management. Various BGMSs use redox mediators coupled with either GOx or GDH enzymes. This strategy has been widely commercialized.

3.2.3 Third Generation of Glucose Biosensor

The third-generation glucose biosensors are based on direct electron transfer between the enzyme and the electrode without the need for toxic mediators. Because the FADH2 is buried at a deep center, about 13–15 Å below the electrode-contacting periphery of its glycoprotein,[56] direct electron transfer from FADH2 of GOx to an electrode surface is much slower. In recent years, rapid advances in the development of nano and porous materials have greatly increased electrode surface area and dynamics.[57–59] In 2003, Xiao *et al.*[60] bound the FAD site of GOx to 1.4 nm gold nanocrystals, showing that it is possible to electro-oxidize glucose when the electrons are relayed to an electrode through a gold nanoplug in the GOx. However, the contact to the electrodes was poor and a high potential was required.

Other approaches used for direct electron transfer between enzyme and electrode include the use of an overoxidized boron-doped diamond electrode (BDDE)[61] in which the carboxyl group on the electrode surface is chemically crosslinked to the enzyme through glutaraldehyde.

Although the complications of tailored mediators are avoided in this third generation of glucose sensors, the efficiency of the electron conductivity, selectivity and sensitivity of the sensors remain as challenges. Therefore, the third-generation glucose biosensors are not yet adopted widely in the current BGMS market.

3.3 Sensors for BGMS

Though many techniques are available for BGMS, including electrochemical, optical, thermometric, piezoelectric and magnetic,[62] the majority of current glucose biosensors are electrochemistry based, using second generation blood glucose biosensor technology. Enzymatic amperometric glucose biosensors are

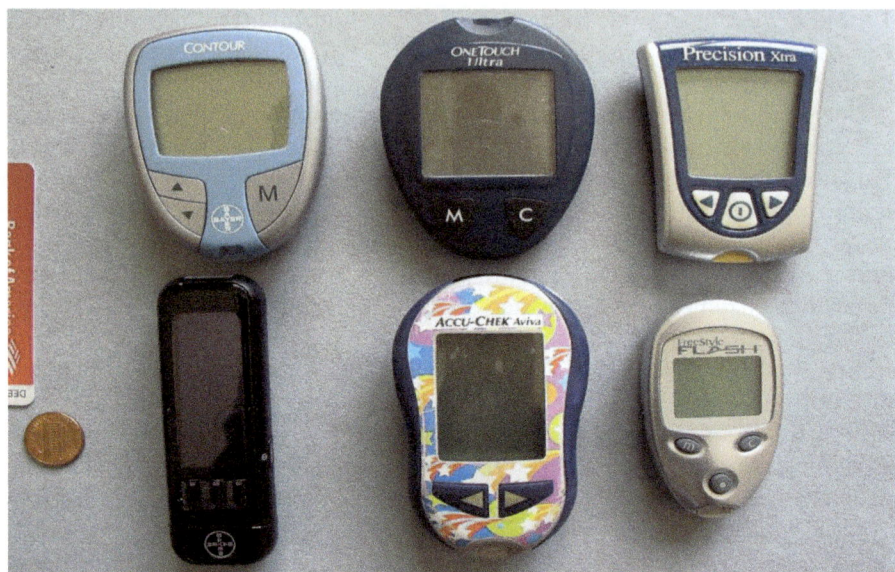

Figure 3.1   Commercially available enzymatic amperometric blood glucose monitoring Systems from major manufacturers (Bayer, Johnson and Johnson, Roche and Abbott, left to right, top to bottom).

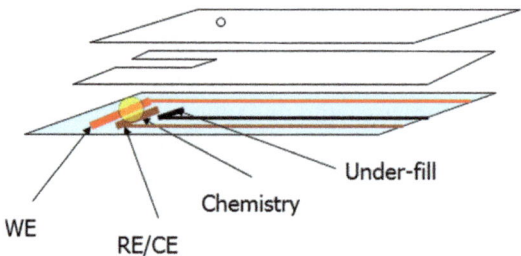

Figure 3.2   The structure of a basic amperometric BGMS sensor.

the most common devices commercially available, and have been widely studied over the last few decades (Figure 3.1). As a result, we will focus on this type of sensors in this section, and discuss in details of the construction of modern BGMS's.

3.3.1   Electrode

Electrode substrate, working electrode, reference electrode and/or counter-electrode can be used as base electrodes for BGMSs. The structure of a BGMS sensor is demonstrated in Figure 3.2.

3.3.1.1   Electrode Substrate

The base electrode for a BGMS is normally constructed on a thin piece of plastic. This plastic material has to be thermally and chemically stable. It serves

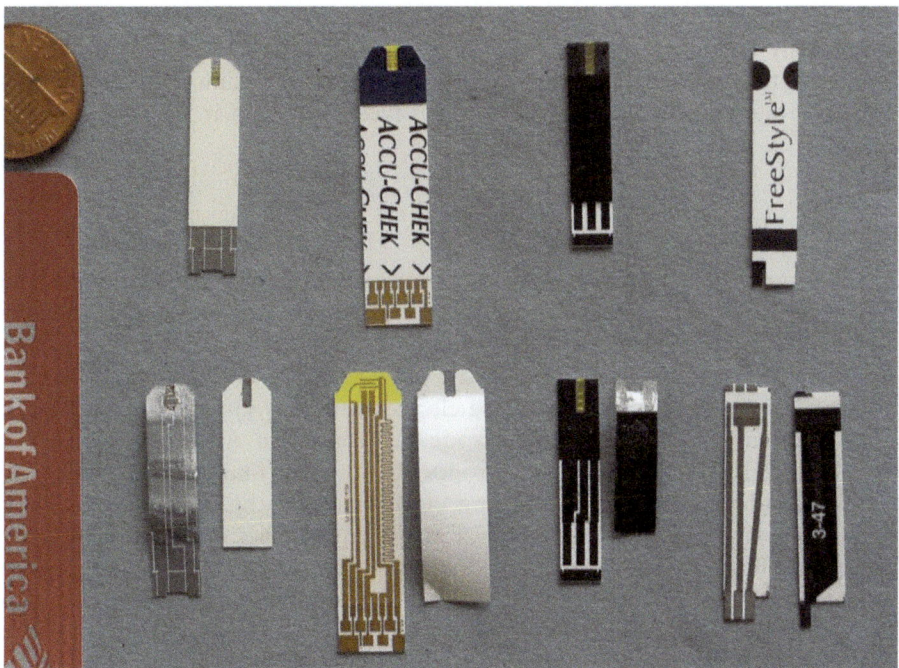

Figure 3.3 Commercialized amperometric BGMS sensors (left to right: Bayer, Roche, Johnson & Johnson and Abbott. Top: sensor strips; bottom: after sensors are opened to view the inside.

as a foundation for the electrode on which the conductive materials will be either screen printed or vapor deposited. As these processes may involve high temperature and organic solvents, the substrate materials should have a high glass-transition temperature and be inert to common organic solvents. The common electrode substrates are polyethylene terephthalate (PET), polycarbonate, polyimide, polyethylenenapthalate (PEN). polyethersulfone (PES) and cyclic polyolefin. The BGMS commercially available are illustrated in Figure 3.3.[63]

3.3.1.2 Working Electrode

The working electrode is the place on which the reaction of glucose oxidation occurs. The glucose-derived electrons exit the sample and transfer to the meter. Common working electrodes can consist of precious metals such as gold (Au) and palladium (Pd), or carbon. While Au and Pt are normally coated onto a plastic substrate by vapor deposition, a carbon working electrode is screen printed from a carbon paste – a mixture of carbon particles and a polyester binder. Because the numbers of electrons generated from a working electrode are directly related to its area, the working electrode area must be known and kept constant during the manufacturing process. This area is generally defined by an insulating dielectric layer or by laser abrasion (for metal electrodes). The distance between the working electrode and the reference/counterelectrode is normally kept to a minimum in achieving for a small sample volume and reduced interelectrode electrolytic resistance. Another design is to have working and reference electrodes positioned face to face for reduced sample volume.[64]

3.3.1.3   Reference/Counterelectrode

Most commercially available BGMS test strips are two-electrode systems. The reference and counterelectrodes are combined into a single electrode. The reference electrode ensures a constant potential applied onto the working electrode and the counterelectrode provides an electrical path for glucose-derived electrons generated from the working electrode to be transferred to the meter. The common reference/counterelectrode can consist of Au, Pd or Ag/AgCl. The Au or Pd reference/counterelectrodes are normally made of using the same material and process as the working electrode, and the area is defined by removing a fine line of metal material at the edge of desired area using laser abrasion. This practice minimizes the manufacturing step and thus reduces the cost, as the manufacturer can make metal-coated plastic sheets in one step and define the working and reference/counterelectrode area by laser abrasion. Ag/AgCl reference/counterelectrodes are usually formed by screen printing an Ag/AgCl ink, which is made of Ag and AgCl particles with polyester binding material.

3.3.1.4   Fluid-Detection Electrode

It is important that the same amount of blood sample reacts with the chemistry within the reaction chamber. This is because the quantity of the enzyme inside the reaction chamber is fixed. If the sample volume was different every time, a variable glucose reading would be expected. To monitor the sample volume inside the reaction chamber, fluid detector electrodes are often placed at the end of the chamber.[65,66] When these electrodes do not detect any fluid, an error code is typically displayed on the device to inform the end user that there is insufficient sample and there is a need to repeat the test with a new sensor.

3.3.2   Reaction Chamber

The reaction chamber is a miniature electrochemical cell. It is formed on the top of the electrode with a spacer defining the edges and a cover plate facing the electrode surface. The spacer used is normally a double-sided sticky material that keeps the cover plate and the electrode together. The cover plate is normally transparent and can be treated with a wetting agent to ensure a fast capillary reaction by which the blood sample can be drawn into the reaction chamber quickly. The size of the reaction chamber corresponds to the product sample volume.

3.3.3 Chemistry

Chemistry is at the heart of a glucose biosensor – glucose is oxidized and the signal generated is proportional to its concentration in a blood sample. In general, the chemistry consists of electrolyte, polymer, surfactant, mediator and enzyme. Buffer salt is normally used as an electrolyte to provide conductivity between electrodes. Hydrophilic polymer is present in the chemistry as a thickening agent to adjust the viscosity of the chemistry so that it can be easily and reproducibly dispensed. The polymer also prevents the chemistry from being washed away from its deposition position when the blood sample is pulled into the reaction chamber by capillary reaction. Surfactant as a wet agent is often used to ensure a rapid chemical hydration. All commercialized blood glucose sensors contain a mediator in the chemistry, which shuttles the electron from the reaction center to the electrode surface. As a result, these types of sensors can work at a relatively low potential that minimizes the effect from common interference substances. The common redox mediators are potassium ferricynide and ruthenium complexes.

Enzymes are the main component of the chemistry. There are two types of enzymes that have been used for the commercialized BGMS: glucose oxidase and glucose dehydrogenase. Glucose oxidase is a very stable, specific and cost-effective enzyme. However, as mentioned in Section 3.2, the response of a glucose sensor made of glucose oxidase is subject to influence from the dissolved oxygen concentration in the sample. Because the BGMS is designed for patient home testing and the oxygen-level difference in a normal person's capillary blood is in a relatively narrow range, glucose oxidase is still a common enzyme for some commercialized BGMSs. Nevertheless, this type of sensor normally uses capillary or oxygenized venous blood to calibrate the system during the product development. It needs extra precautions if the system is used in a hospital or a clinical lab where venous or arterial blood samples are available and could be tested.

The mechanism of the reaction using glucose oxidase is plotted in Figure 3.4. Glucose is oxidized to gluconolactone, and glucose oxidase is reduced to FADH-GOx. Then ferricyanide is reduced to ferrocyanide, and reduced FADH-GOx is oxidized to FAD-GOx. Oxygen as an oxidizing agent competes

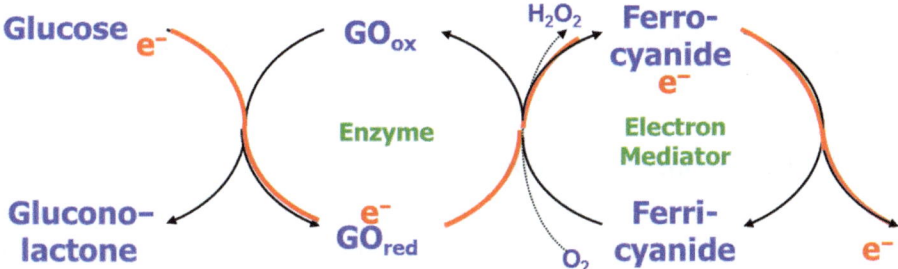

Figure 3.4 Mechanism of the reaction using glucose oxidase.

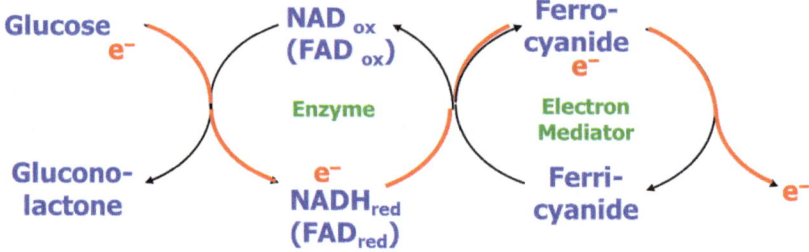

Figure 3.5   Mechanism of the reaction using GDH.

for the electron, thus artificially reducing the glucose signal and providing a false low glucose reading.

To reduce the oxygen effect, oxygen-insensitive enzymes such as pyrrolo-quinoline quinine – glucose dehydrogenase (PQQ-GDH),[67–76] nicotinamide adenine dinucleotide – GDH (NAD-GDH)[77–84] and flavin adenine dinucleotide – GDH (FAD-GDH)[85–95] have been widely used for glucose biosensors. However, PQQ-GDH has been found to lack specificity. It not only reacts with glucose but also with maltose, galactose and xylose, and produces falsely high results. It is therefore advised that patients need extra caution to check the medicine they take to make absolutely sure it will not interfere with their blood-glucose measurement. Dialysis patients should not use the blood-glucose sensor that uses PQQ-GDH technology. This is because isosmotic peritoneal dialysis solution contains glucose polymer (icodextrin) as the primary osmotic agent, icodextrin is metabolized by alpha-amylase into oligosaccharides with a lower degree of polymerization, including maltose, maltotriose and maltotetraose, *etc*. In addition to the problem with the specificity of PQQ-GDH enzyme, GDH-type enzyme is overall more expensive than GOx enzyme. The mechanism of the reaction using GDH is shown in Figure 3.5.

3.3.4   Detection Method

The most common electrochemical detection methods used in BGMSs are amperometry and coulometry. The amperometric method measures the current generated from the reaction at a potential, this current is proportional to the glucose concentration in the sample. The coulometric method measures the amount of coulombs produced from the reaction; the quantity of the coulombs is related to the glucose concentration in the sample.

3.3.5   Correction Algorithm

The BGMS product is designed to test capillary blood samples. Depending on the patient's age, gender and health condition, there are variations in hematocrit level, dissolved oxygen content and electrochemical interferents, *etc*. Each of these variations can have a significant effect on test accuracy and lead to a biased result. To improve the accuracy, manufacturers normally employ a series of algorithms to minimize these variations.

3.3.5.1 Hematocrit Correction

The hematocrit levels for male and female adults are in the range of 42–54% and 38–46%, respectively. Because the BGMs measure the glucose that is only in the liquid phase, and the higher the hematocrit the lower the liquid content, samples with a higher hematocrit level will have a lower glucose reading than those with a lower hematocrit level, even when the actual glucose concentrations in these samples are the same.[96–114]

To correct the hematocrit variation, some BGMs systems have hematocrit electrodes on board to simultaneously measure the glucose and hematocrit levels of a testing sample; they apply correction algorithms to correct the hematocrit effect before displaying a final glucose result.[115–117] Other BGMs systems rely on electrochemical methods[118] and polymers in the chemistry[119] to correct or reduce hematocrit effect. However, the majority of commercial available BGMs do not have hematocrit correction algorithm, instead, they calibrate to a middle hematocrit level (*i.e.* 45%). In this case, samples with 45% hematocrit give very accurate test results. Those with either higher or lower hematocrit level give less accurate results. In general, samples with hematocrit levels greater than 45% will generate a low glucose response and have a negative bias, while samples with hematocrit levels less than 45% will have a high glucose response and have a positive bias. Manufacturers usually optimize the chemistry formulation to make the BGMs less sensitive to the hematocrit level and maintain the assay accuracy within regulatory requirement.

3.3.5.2 Temperature Correction

In addition to hematocrit-correction algorithms, temperature-correction algorithms are also widely used to correct the temperature effect on the glucose measurement as generically illustrated in Figure 3.6. The glucose biosensor is an enzyme-based assay. The enzymatic reaction is very dependent on temperature. As a result, BGMs are normally equipped with temperature-detection capacities. One common approach is to have one or two thermometer(s) embedded inside the handheld meter. When the meter is in thermal equilibrium with the atmosphere where blood testing is performed, the temperature reading from the thermometer is very close to the temperature of the glucose test strip. However, when the meter is not equilibrated to the testing environment, temperature offset can cause inaccurate readings.[120–122]

To overcome this problem, multiple approaches have been proposed. One approach is to have an IR sensor near the glucose strip insertion port so that the temperature of the glucose test strip can be monitored instantaneously, and there will be no equilibrium time needed.[123,124] However, the accuracy of the IR sensor is subject to environmental interference such as breezes and vibration. Another approach is to have a temperature sensor printed on the glucose strip

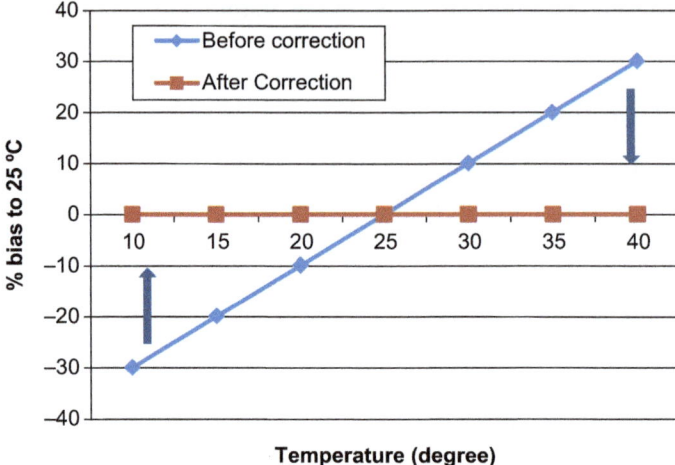

Figure 3.6   Percentage bias before and after temperature correction.

and measure the temperature at the same time as the glucose. The final blood glucose results are calculated based on the temperature measurement at the time.[125,126] In addition, a detachable temperature measurement unit,[127] AC/DC (alternative current/direct current) measurement[128] and measuring current difference between the first and second voltage[129] have also been proposed.

3.3.6   Calibration of BGMS

Millions and millions of glucose biosensors are manufactured every year. These glucose test strips are randomly sampled and tested by the manufacturer before they enter the market to characterize their analytical performance, including linearity, coefficient of variation, hematocrit and oxygen dependence, response to electrochemical interferents and other factors. The manufacturing release test is normally performed with spiked whole blood of low, medium and high glucose concentrations covering the manufacturer claimed detection range. The performance of the glucose test strips is compared with a reference method and the difference between the two is plotted. The Yellow Spring Instrument (YSI) glucose analyzer is a commonly used reference method in the BGMS industry. The glucose test strips will be released only if their performance meets the acceptance criteria. Each glucose test strip batch has an assigned value that is associated with its calibration information, such as slope and intercept. This information is programmed into a chip that is shipped with the test strips. Users are advised to check and make sure the number (codes) on a test strip bottle matches with that of the calibration chip and always apply a new chip into the meter when a new glucose test strip bottle is used.

Some BGMS systems have "no coding" technology. In this case, the manufacturing process and incoming raw materials need to be extremely well controlled. Manufacturing quality control is a key element in ensuring production of reproducible glucose test strips.

There are two types of BGMS systems: plasma calibrated and whole-blood calibrated. This is because some countries measure glucose concentration in whole blood while others measure glucose concentration in plasma. In theory, plasma glucose readings are higher than whole-blood glucose readings for a specific sample. The difference coefficient is normally in a range of 1.12 to 1.15 (*i.e.* plasma glucose = 1.12× whole blood glucose), depending on the chemistry of a glucose test strip. Because plasma glucose and whole-blood glucose readings are interchangeable, the manufacturers usually perform glucose testing in either plasma or whole-blood samples and convert to other readings when needed.

3.3.7 Performance Validation of BGMS

The BGMS is a user-friendly device. It is easy to operate, requires a very small blood sample and provides test results in a few seconds. Even though the BGMS is a home-monitoring device, the quality of its measurement is very important, because many patients, especially Type 1 diabetic patients, rely on the glucose reading to adjust their insulin intake. Therefore, there are a number of regulatory recommendations, and analytical performance criteria are available to guide the manufacturers during the development of their products. The International Organization for Standardization (ISO) 15197:2003 recommends an allowable error of ± 15 mg/dL for blood glucose levels <75 and ± 20% for blood glucose levels ≥75 mg/dl.[130] In 1987, the American Diabetes Association (ADA) recommended a goal of total error of <10% at glucose concentrations between 30 and 400 mg/dL 100% of the time when compared to the reference laboratory measurements.[131] On March 16–17 2010 the FDA and the toxicology devices unit of the Center for Devices and Radiological Health (CDRH) held a public meeting to discuss the issues with the current BGMS manufacturers and health care professionals regarding whether the accuracy standard for the current systems was acceptable. At the end of a two-day meeting, FDA and other stakeholders appeared to be coming together; the consensus was that the overall accuracy of BGMS should be at least ± 15% total error at the upper range of glucose results, tighter than the current ± 20% standard presented in the 10-year-old ISO standard used by FDA to clear the BGMS products for market.

The ISO 15197:2003 has been widely used to evaluate BGMS analytical performances including repeatability, precision, reliability and accuracy. A detailed evaluation plan is listed below.

3.3.7.1 Repeatability

According to ISO 15197:2003, repeatability of a BGMS is evaluated at five blood glucose levels: 30–50, 51–110, 111–150, 151–250, and 251–400 mg/dL. The evaluation requires ten glucose meters to be used and equilibrated to 25 ± 1°C and 50 ± 10% RH (relative humidity). On each meter, 10 measurements are taken. The mean value, the standard deviation (SD) and the coefficient of variation (CV) for each meter from the ten measurements are calculated. The grand mean, the pooled variance, the pooled SD (with 95% confidence interval) and the pooled CV are also calculated.

3.3.7.2 Precision

The precision of a BGMS is evaluated using control materials at three glucose concentrations with 10 glucose meters. It requires one measurement per day per sample over a 10-day period for each of 10 meters. The mean, SD and CV for each meter from the ten measurements are calculated. The grand mean, the pooled variance, the pooled SD (with 95% confidence interval) and the pooled CV are calculated. The pooled SD and CV are measures of the intermediate precision of a single system over multiple days.

3.3.7.3 Reliability

The reliability evaluation is mainly to check the robustness of the BGMS. The glucose meters are exposed to extreme conditions that the system claims to operate under, such as lowest and highest temperatures and humidity. The analytical performances before and after harsh conditions are compared, and the differences should be statistically insignificant. The glucose meters are also evaluated for their mechanical strength using drop testing and vibration.

3.3.7.4 Accuracy

Based on the ISO 15197:2003 requirement, the accuracy of a BGMs is evaluated in seven different glucose categories using fresh-blood samples from diabetic patients, see Table 3.2. Because there is a strict requirement for the number of samples per glucose category, it is very hard to obtain fresh-blood samples at extreme glucose levels. Blood samples at very high and very low glucose concentrations were created by either glycolyzing a blood sample to lower the glucose concentration or supplementing a blood sample with a concentrated glucose solution to increase the glucose level. All glucose measurements are compared with a reference value and the difference between the two is calculated.

Table 3.2   ISO 15197 required glucose distribution for performance evaluation.

Glucose concentration (mg/dL)	Number of samples
< 50	5
51–80	15
81–120	20
121–200	30
201–300	15
301–400	10
>400	5

3.3.7.5 Hematocrit Independence

A BGMS normally has its hematocrit (HCT) claim range stated in the product insert. Depending on the chemistry and correction algorithm, hematocrit effects can be significantly different among different systems. In general, the higher the HCT of a blood sample, the lower its glucose reading. To evaluate the HCT independence, blood samples are altered by either removing the plasma to increase the HCT level or adding the plasma to decrease the HCT level to reach the target HCT levels. Within the claimed HCT levels, all results are expected to be within ±15% of the reference value.

3.3.7.6 Interference

There are a number of variables that can interfere with the performance of a BGMS. These variables can be categorized as related to the environment, patient and endogenous and exogenous substances. The environment-related variables include temperature, humidity and altitude, while patient-related variables include hypertension, hypoxemia, HCT, hydration, icterus, lipemia and other health conditions. Endogenous and exogenous substance-related variables are normally electrochemically active substances such as acetaminophen, ascorbic acid, uric acid, salicylic acid, bilirubin, tetracycline, dopamine, ephedrine, ibuprofen, L-DOPA, methy-DOPA, tolazamide and other frequently administered diabetes drugs.

In addition to the aforementioned interference, enzymes used in the glucose biosensor can induce inaccuracy as well. PQQ-GDH is known to be a non-specific enzyme. It oxidizes not only glucose but also maltose, maltotriose, maltotetraose, and icodextrin. Icodextrin is an osmotic agent, commonly used for patients undergoing peritoneal dialysis.[132,133] It is metabolized during systemic circulation into different glucose polymers, mainly maltose, which leads to a significant overestimation of glucose results.

3.3.8 Manufacturing Process

Millions of test strips are manufactured each year. Many different manufacturing methods have been explored and adopted to produce high-performance, cost-effective glucose test strips. Screen printing is the most common manufacturing method. The conductive raw material is printed on the substrate *via* a patterned screen.[134–137] The screen is made of either nylon or polyester[134,135] or silk[136] material having certain mesh count with patterns from light-sensitive emulsion. Metal sputtering is another method to produce a conductive layer on the substrate.[138–145] Conductive metals such as gold and platinum, *etc.* are sputtered onto the substrate to cover a target area or the entire sheet. A laser beam is used to create the electrodes as desired. There are alternative methods of constructing testing strips. For example, one method uses injection molding to make the substrate with holes, and conductive raw materials are inserted into the holes to create electrodes.[146,147]

Once the electrodes are formed, an insulation layer is printed on to define the sample reaction area. An enzyme solution is then deposited onto the surface of the electrodes. Different methods have been reported for enzyme deposition, including screen printing,[147–153] slot-die coating,[154] spraying,[155] dispensing[156,157] and inkjet printing.[158,159] Among these methods, screen printing and dispensing are the most widely used methods in the BGMS industry. A transparent lid with a reclosable adhesive backing is normally used to make a complete BGMS glucose test strip. This reclosable adhesive backing is patterned so that adhesives only exist in regions not corresponding to the reaction area.[160]

It is of the essence that the performance across glucose test strips is consistent between inter- and intramanufacturing batches. Since a raw material lot can only produce certain quantities of glucose test strips, it is inevitable that there will be raw-material changes during production. Therefore, inspection of incoming materials is very important. Manufacturers are expected to set up tight specifications for each raw material prior to mass production. An inline automatic inspection is normally conducted during production and defective products are crossed out and removed from the good ones before the products are packaged. During the production, if the response slope and intercept are found to be outside of acceptable ranges, a quick adjustment can be made by changing the working electrode area and reagent mediator amount.[150]

3.4 Clinical Utility and Potential Problems

Although the BGMS is designed only to monitor the patients' blood glucose at home and it is not best suited for the diagnosis of the diabetes or to make a clinical decision, millions of diabetic patients constantly rely on the results of the BGMS to adjust their insulin dosage, or to treat hypoglycemia. An error in the BGMS reading can sometimes cause dreadful outcomes. It is the manufacturer's responsibility to develop the most accurate BGMS that meets all regulatory requirements. It is the healthcare personnel's responsibility to recommend a BGMS system and to provide proper training to patients based on individual circumstances. It is patients' responsibility to operate the BGMS system following the manufacturer's use guide.

The performance of a BGMS system is extensively evaluated by the manufacturer before its commercialization. Clinical trials are also conducted to ensure the system is easy to use and provide reliable blood glucose results in the hands of both healthcare professionals and patients. However, controversial results were reported on the real-world accuracy of the BGMS.[161–166] The studies conducted by the manufacturers or healthcare professionals are normally carried out in a controlled environment representing the best situation, while extreme environmental conditions or abnormal situations could occur in the real world.

As we know, the embedded temperature sensor in the BGMS only measures the temperature inside the meter, not the testing environment. A biased result can occur when the BGMS is not equilibrated to the ambient temperature. For example, some patients like to keep their BGMS in their pocket, or leave it in the car during hot summer or cold winter days and then conduct a test right away without waiting a required 15–20 min for the meter to be equilibrated to the ambient temperature, as suggested by the manufacturer.

The BGMS glucose test strips have a finite lifetime when stored at manufacturer-specified temperature and humidity. Their shelf life can be shortened if the vial is left open for a certain period of time or the test strips are exposed to extreme temperature and humidity. A biased result can be obtained when expired or environmentally stressed test strips are used.

Most BGMSs have alternative site-testing options, in which blood testing can be done in areas such as the palm, forearm and upper arm. Because the skin of these areas contains fewer nerves than the fingertip, alternative site testing could be less painful and may be more acceptable to patients. However, the fingers contain a more concentrated network of blood vessels as compared to the forearm, which allows the blood in the finger to adjust more quickly to rapid changes in glucose. This means that readings can be delayed at alternate sites during times of rapidly changing blood glucose, making it more difficult to identify hyperglycemia or hypoglycemia. A 15–30-min lag is often observed at these sites when the glucose level is not in a stable status.

In addition, miscoding the meter, testing without washing one's hands, leaving the lid open after taking the test strip out and using expired test strips are all common mistakes that could happen in the real world. Certain substances in patients' blood specimen such as triglycerides, uric acid, oxygen, ascorbic acid (vitamin C) and hematocrit as well as drug residues and the products of their metabolism will also interfere with the accuracy of the BGMS.

3.5   Continuous Glucose Monitoring System (CGMS)

The BGMS has been a very useful tool in diabetes management. However, it can only provide discreet glucose information at the time of the test; the glucose trend in the next few minutes is still unknown. As a result, effective actions cannot be taken to correct an upcoming hyperglycemia or treat an impending hypoglycemia. To compensate the shortage of BGMSs, continuous monitoring of blood glucose was proposed in 1974.[167] Shichiri *et al.*[168] demonstrated the first *in vivo* needle-type enzymatic Continuous Glucose Monitoring System (CGMS) in 1982. Due to biofouling issues from the protein and coagulation factors and the risk of thromboembolism, subcutaneous glucose monitor approaches that reflect the blood glucose level have been adapted over the continuous blood glucose monitor.[169–173] The first commercial implantable needle-type biosensor that measured the glucose concentration in interstitial fluid was marketed by Minimed. However, it did not provide real-time data; the results could only be viewed retrospectively at a physician's office after downloading.[174] Today, there are three FDA-approved needle-type real-time CGMS devices available, including the Minimed Guardian by Medtronic, SEVEN by Dexcom and Freestyle Navigator by Abbott (although Abbott discontinued Freestyle Navigator in 2011). These devices display the real-time glucose concentration every one to five minutes and use the disposable sensors that can be changed every three to seven days.[175]

Although the CGMS displays real-time glucose concentrations and provides a glucose trend, the FDA has not approved use of the CGMS in place of the BGMS. Many improvements are required, including accuracy, biocompatibility, calibration, long-term stability, specificity, linearity and miniaturization. Despite the improvements needed for the CGMS, researchers are looking into closed-loop systems where a CGMS and an insulin pump are connected together to form an artificial pancreas. The insulin pump will deliver extra insulin when the patient's blood glucose level is elevated, and switch off and send an alert when the patient experiences hypoglycemia. Though it may take a great deal of effort and many years to realize the artificial pancreas, future prospects for diabetes management are very exciting.

References

1. J. E. Shaw, R. A. Sicree and P. Z. Zimmet, *Diabet. Res. Clin. Pract.*, 2010, **87**, 4.
2. N. Poolsup, N. Suksomboon and S. Rattanasookchit, *Diabet. Technol. Ther.*, 2009, **11**, 775.
3. G. H. Murata, J. H. Shah, R. M. Hoffman, C. S. Wendel, K. D. Adam, P. A. Solvas, S. U. Bokhari and W. C. Duckworth, *Diabet. Care*, 2003, **26**, 1759.
4. S. Skeie, G. B. Kristensen, S. Carlsen and S. Sandberg, *J. Diabet. Sci. Technol.*, 2009, **3**, 83.
5. S. L. Tunis and M. E. Minshall, *Curr. Med. Res. Opin.*, 2010, **26**, 151.
6. E. I. Boutati and S. A. Raptis, *Diabet. Care*, 2009, **32**(Suppl. 2), S205.
7. L. G. Jovanovic, *Endocr. Pract.*, 2008, **14**, 239.
8. M. J. O'Kane and J. Pickup, *Ann. Clin. Biochem.*, 2009, **46**, 273.
9. The Diabetes Control and Complications Trial Research Group, *N. Engl. J. Med.*, 1993, **329**, 977.
10. R. R. Holman, S. K. Paul, M. A. Bethel, D. R. Matthews and H. A. Neil, *N. Engl. J. Med.*, 2008, **359**, 1577.
11. I. M. Stratton, A. I. Adler, H. A. Neil, D. R. Matthews, S. E. Manley, C. A. Cull, D. Hadden, R. C. Turner and R. R. Holman, *BMJ*, 2000, **321**, 405.
12. American Diabetes Association, *Diabet. Care*, 2010, **33**, S11.
13. J. D. Newman and A. P. Turner, *Biosens. Bioelectron.*, 2005, **20**, 2435.
14. O. Warburg and W. Christian, *Biochem. Z.*, 1932, **254**, 438.
15. W. Franke and M. Deffner, *Anal. Chem.*, 1939, **541**, 117.
16. H. V. Malmstadt and H. L. Pardue, *Analyt. Chem.*, 1961, **33**, 1040.
17. H. L. Pardue, R. K. Simon and H. V. Malmstadt, *Analyt. Chem.*, 1964, **36**, 735.
18. M. I. Drury, F. J. Timoney and P. Delaney, *J. Ir. Med. Assoc.*, 1965, **56**, 52.
19. M. S. Jensen, *Ugeskr-Laeger*, 1965, **127**, 709.

20. W. Korp, *Wien. Med. Wochenschr.*, 1965, **115**, 435.
21. V. Schmidt, *Ugesk-Laeger*, 1965, **127**, 706.
22. B. C. Clark and C. Lyons, *Ann. New York Acad. Sci.*, 1962, **556**, 46.
23. S. J. Updike and G. P. Hicks, *Nature*, 1967, **214**, 986.
24. S. J. Updike and G. P. Hicks, *Science*, 1967, **158**, 270.
25. G. G. Guilbault and G. J. Lubrano, *Anal. Chim. Acta*, 1973, **64**, 439.
26. K. E. Toghill and R. G. Compton, *Int. J. Electrochem. Sci.*, 2010, **5**, 1246.
27. W. Z. Jia, K. Wang and X. H. Xia, *Trends Analyt. Chem.*, 2010, **29**, 306.
28. J. Anzai, H. Takeshita, Y. Kobayashi, T. Osa and T. Hoshi, *Analyt. Chem.*, 1998, **70**, 811.
29. I. Mao, P. G. Osborne, K. Yamamoto and T. Kato, *Analyt. Chem.*, 2002, **74**, 3684.
30. A. Collins, E. Mikeladze, M. Bengtsson, M. Kokaia, T. Laurell and E. Csoeregi, *Electroanalysis (NY)*, 2001, **13**, 425.
31. G. Cui, S. J. Kim, S. H. Choi, H. Nam and G. S. Cha, *Analyt. Chem.*, 2000, **72**, 1925.
32. S. H. Choi, S. D. Lee, J. H. Shin, J. Ha, H. Nam and G. S. Cha, *Anal. Chim Acta*, 2002, **461**, 251.
33. J. H. Shin, Y. S. Choi, H. J. Lee, S. H. Choi, J. Ha, I. J. Yoon, H. Nam and G. S. Cha, *Analyt. Chem.*, 2001, **73**, 5965.
34. I. J. Xu, X. L. Luo, Y. Du and H. Y. Chen, *Electrochem. Commun.*, 2004, **6**, 1169.
35. J. C. Armour, J. Y. Lucisano, B. D. McKean and D. A. Gough, *Diabetes*, 1990, **39**, 1519.
36. R. Pauliukaite, M. Schoenleber, P. Vadgama and C. Brett, *Anal. Bioanal. Chem.*, 2008, **390**, 1121.
37. P. Vadgama, I. Christie, Y. Benmakroha and S. Reddy, WO 9402585 A1 1994.
38. P. Vadgama, S. Reddy and M. Kyrolainen, WO 9717607 A1 1997.
39. J. Wang and F. Lu, *J. Am. Chem. Soc.*, 1998, **120**, 1048.
40. J. Wang, J. W. Mo, S. Li and J. Poter, *Anal. Chim. Acta*, 2001, **441**, 183.
41. A. E. G. Cass, G. Davis, G. D. Francis, H. A. O. Hill, W. J. Aston, I. J. Higgins, E. V. Plotkin, L. D. L. Scott and A. P. F. Turner, *Analyt. Chem.*, 1984, **56**, 667.
42. A. Mulchandani and S. Pan, *Anal. Biochem.*, 1999, **267**, 141.
43. S. Tsujimura, S. Kojima, K. Kano, T. Ikeda, M. Sato, H. Sanada and H. Omura, *Biosci., Biotechnol., Biochem.*, 2006, **70**, 654.
44. M. G. Loughran, J. M. Hall and A. P. F. Turner, *Electroanalysis*, 1996, **8**, 870.
45. K. Lau, S. A. L. de Fortescu, L. J. Murphy and J. M. Slater, *Electroanalysis*, 2003, **15**, 975.
46. C. Taylor, G. Kenausis, I. Katakis and A. Heller, *J. Electroanal. Chem.*, 1995, **396**, 511.
47. A. Heller and B. Feldman, *Chem. Rev.*, 2008, **108**, 2482.
48. A. Aoki and A. Heller, *J. Phys. Chem.*, 1993, **97**, 11014.

49. L. Ye, M. Haemmerle, A. J. J. Olsthoorn, W. Schuhmann, H. L. Schmidt, J. A. Duine and A. Heller, *Analyt. Chem. Soc.*, 1993, **65**, 238.
50. F. Mao, N. Mano and A. Heller, *J. Am. Chem. Soc.*, 2003, **125**, 4951.
51. N. Mano, F. Mao and A. Heller, *Chem. Commun.*, 2004, **18**, 2116.
52. N. Mano, F. Mao and A. Heller, *J. Electroanal. Chem.*, 2005, **574**, 347.
53. D. L. Williams, A. R. Doig and A. Korosi, *Analyt. Chem.*, 1970, **42**, 118.
54. J. Hu, *Biosens. Bioelectron.*, 2009, **24**, 1083.
55. D. R. Matthews, R. R. Holman, E. Bown, J. Steemson, A. Watson, S. Hughes and D. Scott, *Lancet*, 1987, **1**, 778.
56. H. J. Hecht, H. M. Kalisz, J. Hendle, R. D. Schmid and D. Schomburg, *J. Mol. Biol.*, 1993, **229**, 153.
57. Y. Wang and F. Caruso, *Chem. Commun. (Cambridge, UK)*, 2004, **1528**.
58. S. Wu, H. Ju and Y. Liu, *Adv. Funct. Mater.*, 2007, **17**, 585.
59. S. Bao, C. M. Li, J. Zang, X. Cui, Y. Qiao and J. Guo, *Adv. Funct. Mater.*, 2008, **18**, 591.
60. Y. Xiao, F. Patolsky, E. Katz, J. F. Hainfeld and I. Willner, *Science*, 2003, **299**, 1877.
61. J. Wu and Y. Qu, *Anal. Bioanal. Chem.*, 2006, **385**, 1330.
62. J. D. Newman and A. P. Turner, *Biosensors: Principles and Practice*; Portland Press: London, UK, 1992, **27**, 147.
63. R. Heydarpour, M. K. Shi and K. Kian and K. Tsai, US 20020110673, 2002.
64. J. F. Benjamin, A. Heller, E. Heller, F. Mao, J. A. Vivolo, J. V. Funderburk, F. C. Colman and R. Krishnan, US 6299757 B1, 2001.
65. I. Harding, S. Diamond, R. Williams and S. G. Iyengar, US 7547382 B2 2009.
66. B. W. Bae, B. S. Kang, S. G. Park, S. D. Lee and M. J. Kwon, US 7297248 B2, 2007.
67. W. Jin, U. Wollenberger and F. W. Scheller, *Biol. Chem.*, 1998, **379**, 1207.
68. M. Zayats, E. Katz, R. Baron and J. Willner, *Am. Chem. Soc.*, 2005, **127**, 12400.
69. O. A. Raitman, F. Patolsky, E. Katz and I. Willner, *Chem. Commun. (Camb.)*, 2002, **17**, 1936.
70. M. Alkasrawi, I. C. Popescu, B. Mattiasson, E. Csoregi and V. Laurinavicius, *Anal. Commun.*, 1999, **36**, 395.
71. A. Malinauskas, J. Kuzmarskyte, R. Meskys and A. Ramanavicius, *Sens. Actuators, B: Chem.*, 2004, **B100**, 387.
72. A. Suzumura, T. Koyama and S. Maeda, JP 2005333872 A, 2005.
73. C. Zhao and G. Wittstock, *Biosens. Bioelectron.*, 2005, **20**, 1277.
74. J. Razumiene, J. Barkauskas, V. Kubilius, R. Meskys and V. Laurinavicius, *Talanta*, 2005, **67**, 783.
75. G. Li, H. Xu, W. Huang, Y. Wang, Y. Wu and R. Parajuli, *Measure. Sci. Technol.*, 2008, **19**, 065203/1.
76. T. Yoshioka, *Chem. Sens.*, 2009, **25**(Suppl. B), 25.
77. J. Kulys, L. Tetianec and I. Bratkovskaja, *Biotech. J.*, 2010, **5**, 822.
78. P. N. Bartlett and R. G. Whitaker, *Biosensors*, 1987, **3**, 359.

79. P. N. Bartlett, E. Simon and C. S. Toh, *Bioelectrochemistry*, 2002, **56**, 117.
80. L. Gorton and E. Dominguez, *J. Biotechnol.*, 2002, **82**, 371.
81. E. J. D'Costa, I. J. Higgins and A. P. Turner, *Biosensors*, 1986, **2**, 71.
82. M. Zhang, C. Mullens and W. Gorski, *Analyt. Chem.*, 2007, **79**, 2446.
83. L. I. Boguslavsky, L. Geng, I. P. Kovalev, S. K. Sahni, Z. Xu, T. A. Skotheim, V. Laurinavicius, B. Persson and B and L. Gorton, *Biosens. Bioelectron.*, 1995, **10**, 693.
84. S. Kojima, S. Tsujimura, K. Kano, T. Ikeda, M. Sato, H. Sanada and H. Omura, *Chem. Sens.*, 2004, **20**(Suppl. B), 768.
85. A. Hiratsuka, K. Fujisawa and H. Muguruma, *Anal. Sci.*, 2008, **24**, 483.
86. H. Kawaminami, M. Kitabayashi and Y. Nishiya, US 20120171708 A1, 2012.
87. H. Shimizu and W. Tsugawa, *Electrochemistry (Tokyo, Japan)*, 2012, **80**, 375.
88. H. Inoue, Y. Akasaka and K. Shirosaki, WO 2011118758 A1, 2011.
89. K. Nishio and H. Ido, WO 2011007792 A1, 2011.
90. K. Yuuki and Y. Nakanishi, PCT Int. Appl., WO 2009069381 A1, 2009.
91. W. Tsugawa and K. Sode, PCT Int. Appl., WO 2009037838 A1, 2009.
92. M. Kitabayashi, Y. Tsuji, Y. Nishiya and T. Kishimoto, US 20080090278 A1, 2008.
93. N. Tanaka and K. Yuuki, WO 2007139013 A1, 2007.
94. M. Kitabayashi, Y. Tsuji, H. Aiba, H. Kawaminami, T. Kishimoto and Y. Nishiya, WO 2007116710 A1, 2007.
95. S. Tsujimura, S. Kojima, K. Kano, T. Ikeda, M. Sato, H. Sanada and H. Omura, *Biosci. Biotech. Biochem.*, 2006, **70**, 654.
96. S. Kos, A. van Meerkerk, J. van der Linden, T. Stiphout and R. Wulkan, *Clin. Chem. Lab. Medic.*, 2012, **50**, 1573.
97. R. W. A. Peake, M. S. Devgun and I. A. D. O'Brien, *Brit. J. Biomed. Sci.*, 2012, **69**, 41.
98. W. J. Cheng, C. W. Lin, T. G. Wu, C. S. Su and M. S. Hsieh, *Clin. Chim. Acta*, 2013, **415**, 152.
99. K. Wiener, *Diabet. Med. and J. Brit., Diabet. Assoc.*, 1991, **8**, 172.
100. M. Daves, R. Cemin, B. Fattor, G. Cosio, G. L. Salvagno, F. Rizza and G. Lippi, *Biochem. Med.*, 2011, **21**, 306.
101. I. Jday-Daly, C. Augereau-Vacher, C. De Curraize, M. Fonfrede, G. Lefevre, B. Lacour and C. Hennequin-Le Meur, *Annal. Biol. Clin.*, 2011, **69**, 55.
102. S. Sano, T. Kishi and M. Gotoh, *Tonyobyo (Tokyo, Japan)*, 2009, **52**, 895.
103. K. Hirose, S. Inoue, S. Miyahara, K. Fujisawa, K. Misawa, S. Kubota, T. Sakuma, W. Yumita and K. Nakajima, *Iryo to Kensa Kiki-Shiyaku*, 2009, **32**, 423.
104. F. Yakushiji, H. Fujita, H. Suzuki, R. Joukyu, M. Yasuda, Y. Terayama, K. Nagasawa, A. Ohwada, K. Taniguchi, K. Fujiki, M. Shimojo and H. Kinoshita, *Diabet. Tech. Therapeut.*, 2009, **11**, 369.

105. R. J. Slingerland, W. Muller, J. T. Meeues, I. van Blerk, C. Gouka-Tseng, R. Dollahmoersid, C. Witteveen and K. Vroonhof, *Nederlands Tijdschrift voor Klinische Chemie en Laboratoriumgeneeskunde*, 2007, **32**, 202.

106. Y. Mitsuyoshi, M. Noguchi, K. Ohtake, A. Shibata, H. Kikuchi and M. Murata, *Iryo to Kensa Kiki-Shiyaku*, 2008, **31**, 411.

107. M. J. Kwon and S. Y. Lee, *Daehan Jindan Geomsa Ui Haghoeji*, 2008, **28**, 8.

108. E. S. Kilpatrick, A. G. Rumley, H. Myint, M. H. Dominiczak and M. Small, *Annals Clin. Biochem.*, 1993, **30**, 485.

109. W. J. Cheng, C. W. Lin, T. G. Wu, C. S. Su and M. S. Hsieh, *Clin. Chim. Acta; Int. J. Clin. Chem.*, 2013, **415**, 152.

110. B. Solnica, J. Skupien, B. Kusnierz-Cabala, K. Slowinska-Solnica, P. Witek, A. Cempa and M. T. Malecki, *Clin. Chem. Lab. Med.: CCLM/FESCC*, 2012, **50**, 361.

111. M. Daves, R. Cemin, B. Fattor, G. Cosio, G. L. Salvagno, F. Rizza and G. Lippi, *Biochemia medica: casopis Hrvatskoga drustva medicinskih biokemicara/HDMB*, 2011, **21**, 306.

112. B. H. Ginsberg, *J. Diabet. Sci. Tech.*, 2009, **3**, 903.

113. Z. Tang, J. H. Lee, R. F. Louie and G. J. Kost, *Arch. Path. Lab. Med.*, 2000, **124**, 1135.

114. P. B. Barreau and J. E. Buttery, *Diabet. Care*, 1988, **11**, 116.

115. L. V. Rao, F. Jakubiak, J. S. Sidwell, J. W. Winkelman and M. L. Snyder, *Clin. Chim. Acta*, 2005, **356**, 178.

116. E. Yamanishi and H. Tokunaga, JP 2005147990 A, 2005.

117. M. Teodorczyk, M. Cardosi and S. Setford, *J. Diabet. Sci. Tech.*, 2012, **6**, 648.

118. P. B. Musholt, C. Schipper, N. Thome, S. Ramljak, M. Schmidt, T. Forst and A. Pfutzner, *J. Diabet. Sci.*, 2011, **5**, 1167.

119. Y. S. Yu, US 5789255 A, 1998.

120. K. Nerhus, P. Rustad and S. Sandberg, *Diabet. Tech. Therapeut.*, 2011, **13**, 883.

121. M. J. Haller, J. J. Shuster, D. Schatz and R. J. Melker, *Diabet. Tech. Therapeut.*, 2007, **9**, 1.

122. Y. Takahashi, M. Tomioka, H. Ohkawa, M. Sanaka and Y. Iwamoto, *Igaku to Yakugaku*, 2000, **44**, 109.

123. M. Z. Kermani, S. Noujaim and T. Jetter, US 20100130838 A1, 2010.

124. H. Haar, WO 2001033214 A2, 2001.

125. S. Chinnayelka, J. Fei, Y. Wang, N. Parasnis, H. C. S. Sun, R. Gifford and S. Charlton, US 20110312004 A1, 2011.

126. J. H. Eom, KR 980316 B1, 2010.

127. M. S. Park, M. S. Cha, Y. C. Ahn and Y. B. Choi, WO 2010032911 A1, 2010.

128. D. W. Burke, L. S. Kuhn, T. A. Beaty and V. Svetnik, US 20040157338 A1, 2004.

129. M. Z. Kermani and M. Teodorczyk, US 2012012341 A1, 2012.

130. *In vitro diagnostic test systems – Requirements for blood-glucose monitoring systems for self-testing in managing diabetes mellitus*, International Organization for Standardization: Geneva, Switzerland, 2003.
131. American Diabetes Association. Consensus statement on self-monitoring of blood glucose. *Diabet. Care*, 1987, **10**, 93.
132. W. Janssen, G. Harff, M. Caers and A. Schellekens, *Clin. Chem.*, 1998, **44**, 2379.
133. S. O. Oyibo, G. M. Pritchard, L. McLay, E. James, I. Laing, R. Gokal and A. J. Boulton, *Diabet. Med.*, 2002, **19**, 693.
134. Yasuo Nippon Tokushu Fabric Corp. Emori Ltd; Kenji Shin-Etsu Chemical Co. Fushimi Ltd; KeiChi Akebono Sangyo Co. Ishikawa Ltd; Toru Shin-Etsu Chemical Co. Nakanishi Ltd and Susumu Shin-Etsu Chemical CO., EP 0208616 A1,1986.
135. U. Susumu and N. Toru, EP 0220121 A2, 2012.
136. Y. S. Choi, US 2003/0024811 A1, 2003.
137. A. Joseph, WO 2012025208 A1, 2012.
138. M. Chira, N. Jiko, H. Matsuzaki and K. Takagi, EP 2479312 A1, 2012.
139. T. Hariu, M. Inoue and K. Watanabe, EP 0376709 A3, 1990.
140. K. Inoue and S. Matsuzaki, EP 1431414 A1, 2004.
141. R. A. Emigh and W. B. Willett US 5687600, 1994.
142. T. A. Lee, J. S. Chen, C. H. Yeh, L. H. Chen, C. H. Chao and H. H. Chen, US 2002/0083571 A1, 2002.
143. K. Hasegawa and N. Ishii, US 8252127 B2, 2011.
144. A. Inoue, H. Kimura, A. Nakamura, K. Sasamori, H. Takahashi and M. Yahagi, EP 1652960 A1, 2004.
145. C. M. Huang, US 20040256228 A1, 2004.
146. C. T. Hsu, H. H. Chung, D. M. Tsai, M. Y. Fang, H. C. Hsiao and J. M. Zen, *Analyt. Chem.*, 2009, **81**, 515.
147. M. F. Cardosi and M. O'Connell, EP 2243841 A1 2010.
148. W. Tuan, EP 2365329 A1, 2011.
149. Z. C. Chen, C. Fang, H. Y. Wang and J. S. He, *Electrochim. Acta*, 2009, **55**, 544.
150. G. Macfie, R. Marshall, M. Alvarez-Icaza, K. Delaney, K. Duffus, T. Jarvis, M. Lamacka, N. Phippen and R. Russell, WO 2009090392 A1, 2009.
151. J. Marcinkeviciene and J. Kulys, *Biosens. Bioelectron.*, 1993, **8**, 209.
152. C. A. Galan-Vidal, J. Munoz, C. Dominguez and S. Alegret, *Sens. Actuators, B: Chem.*, 1998, **B52**, 257.
153. O. W. H. Davies, J. F. McAleer and R. M. Yeudall, WO 2001073109 A2, 2001.
154. C. D. Wilsey and D. Mosoiu, WO 2004113917 A2, 2004.
155. H. Yoshida, T. Baba and K. Shirasaki, JP2011099693 A, 2011.
156. J. M. Gleisner, CA 2142026 A1, 1996.
157. H. Sakamoto, Y. Takahashi, Y. Higuchi and T. Yamaguchi, JP 09101297 A, 1997.

158. O. Frey, P. D. van der Wal, N. F. de Rooij and M. Koudelka-Hep, *Procedia Chem.*, 2009, **1**, 281.
159. Y. S. Chao, TW 291900 B, 2008.
160. S. Wooldridge and C. Bolam, US 20100051455 A1, 2010.
161. N. Lalic, T. Tankova, M. Nourredine, C. Parkin, U. Schweppe and I. Amann-Zalan, *Diabet. Tech. Therap.*, 2012, **14**, 338.
162. G. E. Ritchie, A. P. Kengne, R. Joshi, C. Chow, B. Neal, A. Patel and S. Zoungas, *Diabet. Care*, 2011, **34**, 44.
163. G. V. McGarraugh, W. L. Clarke and B. P. Kovatchev, *Diabet. Tech. Therap.*, 2010, **12**, 365.
164. A. Pfützner, M. Mitri, P. B. Musholt, D. Sachsenheimer, M. Borchert, A. Yap and T. Forst, *Curr. Med. Res. Opin.*, 2012, **28**, 525.
165. B. Hunsley and W. Ryan, *J. Diabet. Sci. Tech.*, 2007, **1**(2), 173.
166. B. H. Ginsberg, *J. Diabet. Sci. Tech.*, 2009, **3**, 903.
167. A. M. Albisser, B. S. Leibel, T. G. Ewart, Z. Davidovac, C. K. Botz, W. Zingg, H. Schipper and R. Gander, *Diabetes*, 1974, **23**, 397.
168. M. Shichiri, R. Kawamori, Y. Yamasaki, N. Hakui and H. Abe, *Lancet*, 1982, **2**, 1129.
169. D. S. Bindra, Y. Zhang, G. S. Wilson, R. Sternberg, D. R. Thevenot, D. Moatti and G. Reach, *Analyt. Chem.*, 1991, **63**, 1692.
170. E. Csoregi, D. W. Schmidtke and A. Heller, *Analyt. Chem.*, 1995, **67**, 1240.
171. C. Henry, *Analyt. Chem.*, 1998, **70**, 594A.
172. D. W. Schmidtke, A. C. Freeland and A. Heller, *Proc. Natl. Acad. Sci. USA*, 1998, **95**, 294.
173. K. Rebrin and G. M. Steil, *Diabet. Tech. Ther.*, 2000, **2**, 461.
174. T. M. Gross, B. W. Bode, D. Einhorn, D. M. Kayne, J. H. Reed, N. H. White and J. J. Mastrototaro, *Diabet. Techn. Ther.*, 2000, **2**, 49.
175. M. Cox, *J. Pediatr. Health Care*, 2009, **23**, 344.

CHAPTER 4

Recent Progress in the Electrochemical Detection of Disease-Related Diagnostic Biomarkers

ALINA VASILESCU,[a] WOLFGANG SCHUHMANN[b] AND
SZILVESZTER GÁSPÁR*[a]

[a] International Centre of Biodynamics, 1B Intrarea Portocalelor,
060101 – Bucharest, Romania; [b] Analytische Chemie-Elektroanalytik & Sensorik,
Ruhr-Universität Bochum, Universitätsstr. 150, Bochum D-44780, Germany
*Email: sgaspar@biodyn.ro

4.1 Introduction

The European market of *in vitro* diagnostics has grown 2–4% annually between 2006 and 2010, and 0.9% from 2010 to 2011. The market revenues generated by the European *in vitro* diagnostics industries were of € 10847 million in 2011 (see European In Vitro Diagnostic Market Statistics Report 2011 by the European Diagnostic Manufacturers Association). Although a decline was anticipated for 2012 (due to the economic recession), it is clear that an increasing number of *in vitro* diagnostic tests are available and used in order to correctly diagnose and manage diseases. Many of these tests are carried out in specialized laboratories, on a wide range of complex instruments, and with high costs (in terms of both money and time). Great effort is currently directed towards making diagnostic tests faster and cheaper and also towards making some of

RSC Detection Science Series No. 2
Detection Challenges in Clinical Diagnostics
Edited by Pankaj Vadgama and Serban Peteu
© The Royal Society of Chemistry 2013
Published by the Royal Society of Chemistry, www.rsc.org

them directly usable by doctors (for point-of-care diagnostics) or even patients (at home). Electrochemical sensors are usually characterized by high sensitivities, they can easily be miniaturized and then integrated into microfluidic systems, and they require no complex instrumentation. Therefore, there is great enthusiasm concerning the use of electrochemical sensors in disease diagnostics (at least when judged by the large number of papers published on this matter).

Electrochemical sensors are solid/liquid interfaces characterized by electrical currents or potentials that are dependent on the presence and the concentration of an analyte of interest. Simple interfaces (such as those between bare electrodes and solutions) have a current or potential output that is dependent on many factors. Therefore, in order to develop an electrochemical sensor, the solid/liquid interface must be carefully modified to maximize the effect of the analyte of interest and minimize the effect of other factors. One elegant way of achieving the desired selectivity for electrochemical sensors is based on proteins that are able to selectively bind and/or convert their substrate. As we will see in the next sections, the selectivity of antibodies and/or enzymes can be added to electrodes by immobilizing these biomolecules onto the electrodes. Even though immobilized on the electrode, these biomolecules perform almost as in their natural environment, that is, they recognize their substrate even from mixtures with other compounds. Having a good and selective electrochemical biosensor is thus possible if the physicochemical changes associated with the recognition event are sensitively converted into currents or potentials.

In order to detect the minute (physical or chemical) changes associated with the recognition event, several electrochemical methods are in use. These methods are identified by names that might not say a lot for nonelectrochemists (*e.g.* square-wave voltammetry, SWV) but in the end all of them consist of applying a potential or current program (such as sweeps, pulses, *etc.*) and measuring the resulting current or potential response (continuously or intermittently). Fundamentals of electrochemical methods will not be very much detailed in the coming paragraphs. Details can be found elsewhere.[1] Details about electrochemical impedance spectroscopy (EIS, a method that will be often mentioned in the coming sections), are found also in a very recent review together with details about its use in bioanalysis.[2]

Often, the sensitivity of the electrochemical method alone is not sufficient to observe the interaction between the analyte and the selected biorecognition element. In these situations different amplification methods are used. The most common of these methods is the use of an enzyme label in the affinity interaction between an antibody and an antigen (exactly as in enzyme linked immuno sorbent assay, ELISA). By using the enzyme label, because each immunocomplex will generate hundreds of thousands of detectable molecules, the detection of really low concentrations of analyte is possible.

The nature of the sensor material also significantly impacts the sensitivity of electrochemical detection. Advantages brought in recent years by nanomaterials were rationalized in terms of high surface/volume ratio and very good electrocatalytic properties (*e.g.* carbon nanotubes[3,4]) or high adsorption capacity of biomolecules with preservation of their activity (*e.g.* Au nanoparticles[4]).

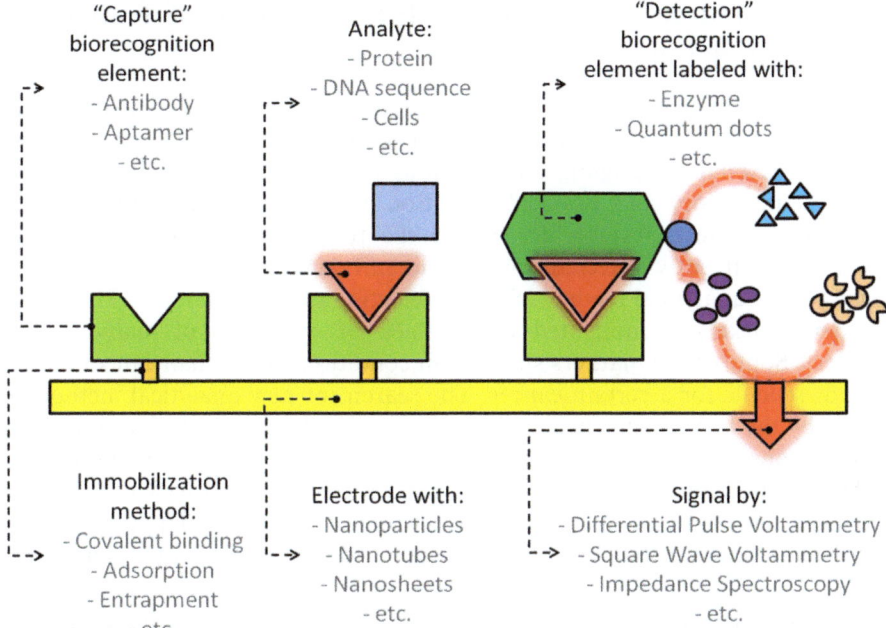

Figure 4.1 Schematic representation of the main elements and working principle of electrochemical biosensors for disease diagnostics. Such sensors are often quite similar to a sandwich-type ELISA. However, the "capture" recognition element is immobilized onto electrodes instead of the plastic plates used in ELISA. Moreover, the affinity binding between the analyte and the recognition element is electrochemically observed with or without the use of a second, labelled, "detection" biorecognition element. Higher sensitivities and shorter analysis times are usually obtained as compared to ELISA.

The schematic representation of the elements of the most common electrochemical biosensors developed for disease diagnostics is shown in Figure 4.1.

The present chapter will detail representative examples of electrochemical sensors for the diagnosis of seven, often-targeted diseases: cancer, heart diseases, acquired immunodeficiency syndrome (AIDS), hepatitis, rheumatoid arthritis (RA), celiac disease, and urinary tract infection (UTI). It will also highlight the challenges still faced by such sensors. A relatively new and thorough review on electrochemical sensors and biosensors, but not necessarily for disease diagnostics, can be found elsewhere.[5] Biosensors in clinical chemistry, but not necessarily electrochemical, were also reviewed elsewhere.[6,7] Electrochemical sensors that make use of nanomaterials and target biomedical applications (including diagnostics) were also recently reviewed elsewhere.[8–10] Some other reviews put the emphasis on practical aspects of the research on electrochemical biosensors towards point-of-care diagnostics, including mainly data from published clinical trials.[11–13]

4.2 Electrochemical Sensors for Detection of Cancer Biomarkers

There is no single test that can be used to undoubtedly diagnose cancer. Examination of tissue and cell samples (*e.g.* collected by biopsy), scans (X-ray, magnetic resonance imaging, *etc.*), various blood tests (*e.g.* complete blood count), and history and physical condition examination are combined with tumor marker tests instead. Regarding the tumor marker tests, significant efforts are currently directed towards identifying new markers as well as towards new analytical methods for the detection of these markers. The search for new markers is motivated by the lack of specificity of many current markers. Some of the markers show up in conditions other than cancer and/or are not specific for a certain cancer. The search for new analytical methods is driven by the lack of sensitivity, the high cost, the high complexity and/or the long response time of the currently existing methods to detect markers. The present section offers an overview of cancer biomarkers for which electrochemical sensors were proposed in the recent literature. There are other excellent reviews on the electrochemical detection of cancer biomarkers.[14–16]

4.2.1 Electrochemical Sensors for Carcinoembryonic Antigen

Carcinoembryonic antigen (CEA) is a 180–200 kDa glycoprotein that is found in concentrations of less than 2.5 ng/mL (5–10 ng/mL) in the blood of healthy nonsmoker (smoker) adults.[17] However, high concentrations of CEA have been reported in patients with different cancers (*e.g.* gastrointestinal cancer). Serum CEA values larger than 25 ng/mL are highly suggestive of metastatic cancer.[17]

In order to electrochemically detect CEA, anti-CEA antibodies were immobilized onto glutathione-modified Au nanoparticles, and then the so-obtained particles were entrapped into a film of poly(ortho-aminophenol) electrodeposited onto Au electrodes.[18] When incubated with samples containing CEA, the Faradaic impedance of such modified electrodes increased proportionally with the concentration of CEA. The approach is characterized by an important advantage: it is among the very few approaches that use no label and thus it is simple. This label-free electrochemical sensor for CEA detection is characterized by a detection limit of 0.1 ng/mL and a linear range of 0.5–20 ng/mL. However, unspecific binding to this sensor requires additional investigations. It was shown that the poly(ortho-aminophenol) film resists well to unspecific binding when tested with lysozyme (1 mg/mL) or with serum (diluted 50 times). However, lysozyme is far from being the most abundant protein in serum. Moreover, if serum samples are diluted 50 times (as in the tests of unspecific binding), normal CEA levels will be not measured with this sensor because they fall below its detection limit.

A lower CEA detection limit (0.01 ng/mL) was achieved using comb-like, interdigitated carbon electrodes (as working electrodes), a glassy carbon electrode (GCE, as counterelectrode), polystyrene microparticles and a set of three antibodies.[19] Mouse anti-CEA antibody (1st antibody) was immobilized onto the polystyrene microparticles and used to capture CEA from samples. Subsequently, the microparticles were incubated with rabbit polyclonal anti-CEA antibody (2nd antibody) and goat polyclonal anti-IgG antibody β-galactosidase conjugate (3rd antibody). β-galactosidase is an enzyme that produces p-aminophenol when p-aminophenyl-β-D-galactopyranoside is available. Therefore, the concentration of CEA in the sample can be revealed by electrochemically measuring how much p-aminophenol is produced. The novelty of the approach is represented by the way p-aminophenol is quantified by using a "dual amplification system". P-aminophenol is redox cycled in between the interdigitated electrodes (*i.e.* each p-aminophenol molecule is several times oxidized on the anode and then immediately reduced on the cathode). This redox cycling represents the first amplification system. Each time p-aminophenol is oxidized to quinoneimine on the anode of the interdigitated electrodes, silver ions are reduced onto the glassy carbon counterelectrode (that communicates with the anode through a salt bridge). The higher the concentration of p-aminophenol, and the longer the redox cycling is allowed to proceed, the more silver is deposited onto the counterelectrode. This silver accumulation is thus the second amplification system. The problems associated with nonspecific binding are eliminated by incubation of microparticles with bovine serum albumin, BSA, (immediately after immobilization of the mouse anti-CEA antibody) and by the fact that the counterelectrode (where the silver deposit was formed) is never exposed to the sample.

Yet another possibility is to carry out the sandwich assay of CEA in a microtiter plate (*i.e.* not on the surface of the electrode) and use an electrochemical sensor as the detector to quantify the activity of the enzyme label (after it is eluted from the immunocomplex).[20] The activity of the horseradish peroxidase (HRP) label linked to the detection anti-CEA antibody was quantified using a thionine-modified Au electrode and differential pulse voltammetry. A detection limit of 0.2 ng/mL CEA was achieved based on this approach. Good agreement was also observed between results obtained with this electrochemical approach and immunoradiometric assay while analyzing (a relatively low number) of human serum samples. However, the approach retains some of the drawbacks of ELISA (*e.g.* the quite long incubation times).

4.2.2 Electrochemical Sensors for Prostate-Specific Antigen

Prostate-specific antigen (PSA) is a glycoprotein produced in the prostate gland with a molecular mass of 26 kDa (without its carbohydrate moieties).[21] The more elevated the blood PSA concentration of a man is, the more likely it is that he has prostate cancer. However, there are benign conditions that cause PSA levels to increase as well as prostate cancers that do not induce a PSA level increase.[22] Therefore, diagnosis of prostate cancer involves also investigations other than tumor marker tests.[23]

The electrochemical detection of PSA was achieved with a sandwich-type immunoassay carried out with i) a graphite electrode modified first with Au nanoparticles and then with capture anti-PSA antibodies, and ii) magnetic beads modified with both detection anti-PSA antibodies and about 7500 HRP molecules.[24] Nonspecific binding to the sensor was eliminated by blocking the sensor surface with 0.4% Casein and 0.05% Tween-20 in phosphate buffer. Immunocomplexes formed on the sensor (after incubation with sample) were quantified by measuring the current resulting from the bioelectrochemical (*i.e.* HRP-mediated) reduction of hydrogen peroxide at -0.3 V. The approach allowed a detection limit of 0.5 pg/mL in 10 µL undiluted serum sample to be obtained. This detection limit represents a 2000-fold increase compared to the detection limit achieved by the same approach carried out with single HRP-labeled antibodies. Measurements of PSA in real samples gave excellent correlations with standard ELISA.

In a very similar approach the electrochemical sandwich-type immunoassay was carried out using i) a GCE modified first with iron-oxide nanoparticles and then with capture anti-PSA antibodies, and ii) iron-oxide nanoparticles decorated with detection anti-PSA antibodies and HRP.[25] Nonspecific binding to the sensor was this time eliminated by blocking the sensor surface with 1% BSA solution. Immunocomplexes formed on the sensor (after incubation with sample) were quantified by measuring the current resulting from the bioelectrochemical (*i.e.* HRP-mediated) reduction of hydrogen peroxide at -0.2 V. The immunosensor was characterized by a linear range from 0.005 to 50 ng/mL and a detection limit of 4 pg/mL. Experiments made with spiked full-blood serum allowed recoveries from 92.1 to 112.5% to be observed. There was a good agreement between results obtained with the electrochemical sensor and ELISA in the same serum samples.

GCEs were sequentially modified with silver-coated mesoporous silica nanoparticles (to increase the active surface area and improve the electrochemical properties of the electrode), with anti-PSA antibodies (to capture PSA from samples), and BSA (to eliminate nonspecific binding).[26] The resulting sensors were interrogated with cyclic voltammetry and hydroquinone as indicator (the more PSA present in the sample the smaller were the current peaks for hydroquinone). The sensor detected PSA with a linear range from 0.05 to 50.0 ng/mL, and a detection limit of 15 pg/mL. When used to detect PSA in spiked human serum samples, recoveries between 98% and 101% were obtained. When results obtained with human serum samples were compared to ELISA, a maximum relative error of 6.4% was observed. However, only a small number of tests were carried out with real samples, a shortcoming that characterizes many studies on electrochemical sensors for diagnostics.

4.2.3 Electrochemical Sensors for Other Protein Cancer Biomarkers

A smaller number of different sensors were developed to detect other biomarkers than CEA and PSA. These sensors are shortly described in this section.

The 62-kDa c-Myc protein, the product of the c-myc proto-oncogene, is a nuclear phosphoprotein with DNA binding properties.[27] Elevated expression of the c-Myc correlates with poor prognosis in head and neck squamous cell carcinoma.[28] The electrochemical detection of c-Myc was only recently reported.[29] The capture antibody was immobilized onto cysteamine-modified Au electrodes, while the detection anti-c-Myc antibody was immobilized onto Au nanoparticles. The formation of the immunocomplex (between the capture antibody, the antigen, and the detection antibody) was detected using EIS. The sensor shows a linear range from 4.3 pmol/L to 43 nmol/L and an estimated detection limit of 1.5 pmol/L. Recovery between 96.2 and 109% was achieved in 1% serum samples.

Mucin 1 is a 122-kDa molecular weight protein that is expressed on the apical surface of epithelial cells of many different tissues. Abnormalities in mucin 1 expression are associated with many cancer types (*e.g.* breast, ovarian and lung).[30,31] Mucin 1 was electrochemically detected with an aptamer-based sensor.[32] Using aptamers as a biorecognition element instead of the more commonly used antibodies has several advantages.[33] Aptamers are small, chemically stable, and their production is cost effective. In this case (as in many other cases) the aptamer used as the biorecognition element had a methylene blue label and a "hairpin" conformation. When the aptamer is immobilized on the electrode surface, the methylene blue label is positioned very close to the electrode, leading to a high current signal in cyclic voltammetry (CV) or square-wave voltammetry (SWV) due to the short electron-transfer distance. Once the aptamer binds mucin 1 its conformation changes and the electron-transfer distance between the methylene blue label and the electrode increases. The higher the concentration of mucin 1 is the smaller the current signal of the methylene label becomes. The aptamer-based mucin 1 sensor displayed a detection limit of 50 nM and a dynamic response range up to 1.5 µM.

Squamous cell carcinoma antigen (SCCA) belongs to the serine protease inhibitor family of proteins and is a useful tumor marker for diagnosis and management of squamous cell carcinoma.[34,35] This antigen was electrochemically detected based on a sandwich-type immunoassay.[36] The GCE to be used in the assay was first decorated with nitrogen-doped graphene sheets. These sheets are characterized by high active surface areas and good conductivities that are two important properties for the development of electrochemical sensors. The first anti-SCCA antibody was immobilized onto these sheets with glutaraldehyde. The second antibody was adsorbed onto Pt and Fe_2O_3 nanoparticles. High amounts of antibodies can be loaded onto such particles due to their large surface area. Moreover, the two components of the nanoparticles (Pt and Fe_2O_3) show a synergistic effect in catalyzing the reduction of hydrogen peroxide. Therefore, such nanoparticles can replace peroxidase that is commonly used in immunoassays and can improve the robustness of the assay due to the better stability as compared with the commonly used enzymes. The immunosensor displayed a linear range from 0.05 to 18 ng/mL and a detection limit of 15.3 pg/mL.

α-enolase (ENOA) is a metabolic enzyme that also acts as a plasminogen receptor and by this mediates activation of plasmin and extracellular matrix degradation. In tumor cells, ENOA is upregulated and supports anaerobic proliferation, promotes cancer invasion, and is subjected to specific post-translational modifications.[37] ENOA was detected using two antibodies, an anti-ENOA monoclonal antibody that was adsorbed onto the electrode surface and an anti-ENOA polyclonal antibody that was labelled with Au nanoparticles.[38] After the usual incubation and washing steps, the Au nanoparticles of the formed immunocomplexes were oxidized in 0.1 M HCl at 1.2 V for 120 s, and the resulting gold ions were quantified using SWV. The immunosensor was characterized by a linear range from 10^{-8} to 10^{-12} g/mL and a detection limit of 11.9 fg (equivalent to 5 µL of a 2.38 pg/mL solution). Compared to the more commonly used enzyme labels, the Au nanoparticles used in this approach are more stable and do not require substrates or long incubation times to accumulate significant amounts of reaction products.

Murine double minute 2 (MDM2) is a protein that is encoded by the MDM2 gene, and an important negative regulator of the p53 tumor suppressor. It mediates the ubiquitination of p53 leading to its degradation by the proteasome. Elevated expression of MDM2 is observed in a significant proportion of different types of cancer.[39] A label-free electrochemical sensor for the detection of MDM2 was built by immobilizing anti-MDM2 antibodies onto Au electrodes.[40] Such modified electrodes are able to bind MDM2 and binding of MDM2 increases the faradaic impedance of the modified electrodes in a concentration-dependent manner. The selectivity of the sensors was tested by challenging them with recombinant human epidermal growth factor receptor (HEGFR). However, 1 ng/mL was the highest HEGFR concentration tested while higher concentrations of proteins can be expected in real samples. The storage stability of the sensors was also tested. After 2 weeks at 4 °C the sensor retained 81.8% of its initial signal to 1 ng/mL MDM2. The sensors were also used to detect MDM2 in normal mouse tissue homogenates and cancerous mouse tissue homogenates. In samples of 0.1% normal mouse tissue homogenate spiked with different concentrations of MDM2, a linear range from 1 to 10 ng/mL and a detection limit of 1.3 pg/mL were found. When analyzing the same cancerous tissue homogenate the sensor reported a MDM2 concentration of 7.8 ng/ml while the ELISA kit reported a MDM2 concentration of 10.9 ng/mL. While the ELISA required 5 h to provide results, the electrochemical sensor reported on the MDM2 concentration in less than 1 h.

4.2.4   Electrochemical Sensors for Simultaneous Detection of Several Protein Biomarkers

Over the years many of the cancer biomarkers have been discovered to be not very specific to a certain cancer and/or to have concentrations affected also by benign disorders. One very convenient approach to achieve a better disease diagnosis in such conditions is based on analytical tools simultaneously reporting on several cancer-related biomarkers in a single run. Such tools could also simplify the working procedure, increase throughput, and reduce the cost per test. It is relatively easy to make electrodes very small due to the readily available technology from the microelectronics industry, and to have several electrode systems integrated into the same device. Such multielectrode systems can then be used for the simultaneous electrochemical detection of several biomarkers. The analysis of signals from such devices might overcome the mentioned lack of specificity, and thus reliability of individual tests.

The simultaneous electrochemical detection of cancer antigen 125 (CA 125), CA 15-3, and CA 19-9 was recently reported.[41] Magnetic beads were decorated with monoclonal antibodies for all three antigens and used to capture the antigens from samples. Polymer dendrimers were decorated with metal sulfide quantum dots and polyclonal rabbit antibodies (detection antibodies) for the three antigens. For detection of CA 125, CA 15-3, and CA 19-9 the polymer dendrimers were decorated with Zn sulfide, Cd sulfide, and Pb sulfide quantum dots, respectively. After carrying out a sandwich-type immunoassay with the modified magnetic beads and dendrimers, and after dissolving the quantum dots, anodic stripping voltammetry (ASV) was used to quantify each of the three released metal ions and thus to quantify the three antigens from the sample. This multiplexed immunoassay enabled the simultaneous detection of the three cancer biomarkers in a single run with dynamic ranges of 0.01–50 U/ mL and detection limits of 0.005 U/mL.

The simultaneous electrochemical detection of interleukin 6 (IL-6), interleukin 8 (IL-8), vascular endothelial growth factor (VEGF), and vascular endothelial growth factor-C (VEGF-C) was also reported.[42] Magnetic beads were decorated with antibodies against the target proteins ($\sim 100\,000$ antibodies per bead) as well as with horseradish peroxidase (HRP, $\sim 400\,000$ enzyme molecules per bead). These beads were then used to carry out immunomagnetic separation of the target proteins from samples before being injected into a microfluidic device with an array of microelectrodes as detectors. Each microelectrode was modified with antibodies against one of the four target proteins and thus could bind only those magnetic beads that came out from the separation step bearing the respective protein. The concentration of such beads (and thus the concentration of the analyte) was finally determined by electrochemically detecting the activity of the HRP label with hydroquionone. The assay duration was 50 min. The analysis of 78 oral cancer patient serum samples and 49 controls gave clinical sensitivity of 89% and specificity of 98% for oral-cancer detection.

In another publication by the same authors the simultaneous electrochemical detection of PSA, prostate specific membrane antigen (PSMA), platelet factor-4 (PF-4), and IL-6 in the same sample is described.[43] Antibodies against these four biomarkers for prostate cancer were immobilized onto four different electrodes of the same array. Before being modified with the antibodies, the four electrodes were modified with a "forest" of single-wall carbon nanotubes in order to increase their active surface area and improve their electrochemical behavior. The antigen–antibody complexes were electrochemically detected and quantified using antibodies labelled with HRP, hydrogen peroxide and an electrochemically active cosubstrate of HRP (a quinone). The electrochemical detection presented excellent correlation with results obtained by ELISA.

4.2.5 Electrochemical Sensors for Genetic Markers of Cancer

Detection of DNA (or RNA) sequences with electrochemical sensors requires approaches quite similar to those described above for detecting protein biomarkers. A single-strand DNA (or a peptide nucleic acid, PNA) sequence is immobilized onto the electrode (as the biorecognition element). If its complementary DNA (or RNA) sequence is present in the sample, binding between the two strands occurs. The event is then electrochemically observed.

TP53 gene encodes p53, a 53-kDa tumor-suppressor protein. Mutations in this gene are associated with a variety of human cancers.[44,45] The discrimination between intact and G→A single nucleotide polymorphism (SNP)-containing TP53 cancer biomarker sequences was demonstrated with an electrochemical sensor using a short DNA hairpin molecular beacon (5′-GT TGT GCA GCG CCT CAC AAC-3′) as biorecognition element.[46] The DNA hairpin molecular beacon was labelled with methylene blue before being immobilized onto the electrode. When complementary DNA from the sample binds to the DNA hairpin molecular beacon, it significantly changes its conformation and thus alters the distance between the methylene blue label and the electrode surface. Currents due to oxidation/reduction of methylene blue are very sensitive to the distance between methylene blue and electrode. Therefore, such sensors can be used to observe differences between hybridization with intact and SNP-containing sequences. Using a truncated DNA hairpin molecular beacon (only 20 nucleotides, instead of the previously used 26 nucleotides) proved to be advantageous in several ways. When using the truncated DNA hairpin molecular beacon, the sensor was characterized by improved selectivity for SNP, and the way it works switched from having a high current signal in the absence of any hybridization to having a high current signal exclusively in the presence of hybridization (*i.e.* switched from ON-OFF to an OFF-ON mode). This switch is advantageous because it avoids false positives due to desorption of the biorecognition element from the electrode surface.

A more complex electrode was also used to detect TP53 gene-based DNA sequences.[47] Aiming for electrodes with higher surface area that allows immobilization of larger amounts of the biorecognition element and thus achieving higher sensitivities, Au electrodes were modified with a network of electrospun nylon 6 fibres entrapping carboxylated carbon nanotubes and covered with polypyrrole. ssDNA was adsorbed onto such electrodes and used as recognition element just as in the above example. The ability of methylene blue to bind to the adsorbed ssDNA decreases after hybridization with the target DNA. SWV can then be used to quantify how much methylene blue binds to the interface and thus to quantify the extent of hybridization. Such sensors were able to detect 50 fM of the wild type TP53 sequence and to distinguish between the wild type and the three base mismatched sequence of TP53. However, the precise role of each nanomaterial was not investigated, a shortcoming that characterizes several recent papers mixing together many different nanomaterials.

Up to this point, blocking the sensor surface with an inert protein (before being used to make measurements in biological samples) was the only method mentioned to eliminate nonspecific binding. This method is the most often used one to eliminate nonspecific binding. However, there are also other options to minimize nonspecific binding. Carefully designing the solid/liquid interface of the sensors is another option. Three polyethylene glycol (PEG)-based thiol derivatives were recently used to build sensors for the electrochemical detection of oestrogen receptor-α (ESR1, a genetic marker involved in breast cancer).[48] The sensors were based on PEG alkanethiol (structure no. 1), a mixture of PEG alkanethiol and mercaptohexanol (structure no. 2) or a bipodal aromatic PEG alkanethiol (structure no. 3, see Figure 4.2). Best results were obtained with sensors built by the coassembly of ESR1 capture probes and bipodal aromatic PEG alkanethiol in a ratio of 1 : 100. These sensors were used for the analysis of a PCR product resulting from the amplification of the genetic material extracted from 20 MCF7 cells.

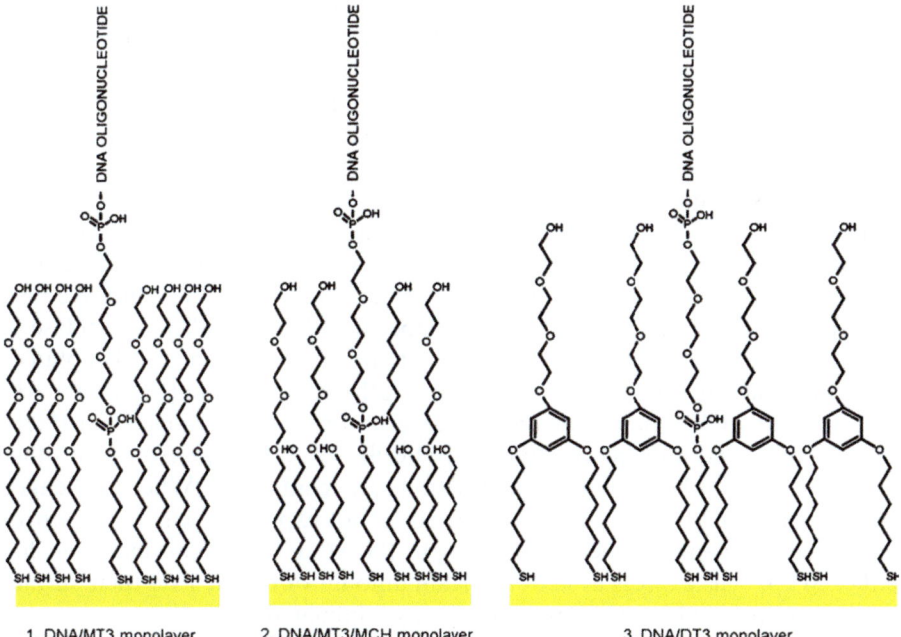

1. DNA/MT3 monolayer   2. DNA/MT3/MCH monolayer   3. DNA/DT3 monolayer

Figure 4.2 Polyethylene glycol and alkanethiol-based surfaces used in the electro-chemical detection of a genetic marker of breast cancer. Structure no. 3 proved to perform best by providing both sensitivity and ability to eliminate nonspecific binding.

Reprinted from Biosensors and Bioelectronics, Vol. 26, Henry, O.Y.F., Lluis Acero Sanchez, J., O'Sullivan, C.K., Bipodal PEGylated alkanethiol for the enhanced electrochemical detection of genetic markers involved in breast cancer, 1500–1506, (2010) with permission from Elsevier.

4.2.6   Electrochemical Sensors for Detection of Cancer Cells

Approaches to detect cancer cells with electrochemical sensors are not very much different than the above-described approaches to detect proteins and DNA sequences. A biorecognition element is again immobilized onto the electrode surface. However, this element now recognizes features that are on/in the cell membrane and not freely diffusing in the solution. Among the used recognition elements one finds aptamers, small molecules (with corresponding receptors on cancer cells), and of course antibodies. Representative examples of sensors based on each type of biorecognition element will be given below.

Aptamers are especially interesting biorecognition elements when it comes to detection of cells because their selection procedure can nowadays be carried out with whole cells.[49,50] Detection limits as low as 1000 cells were reported for a sensor combining graphene-modified electrodes with a 26-mer DNA aptamer that binds nucleolin that is overexpressed on the cancer-cell surface.[51] The sensor is interrogated using impedance spectroscopy and regenerated through DNA hybridization that releases the aptamer-bound cells (Figure 4.3).

Another aptamer-based electrochemical sensor involves the use of the anticancer drug daunomycin.[52] This drug interacts with the cell membrane of cancer cells. Daunomycin-decorated cancer cells are then selectively captured onto aptamer-modified electrodes. Impedance spectroscopy is used to detect the presence of the cells on the electrode surface. In order to increase the active surface area, the number of immobilized aptamers and finally the sensitivity of the sensor, the aptamers were immobilized on a conducting polymer–Au nanoparticle composite film.

The detection of MCF7 cancer cells with electrochemical sensors based on an aptamer–cell–aptamer sandwich architecture was also recently demonstrated.[53] Mucin 1-binding aptamer was both immobilized onto 2-mm diameter Au disk electrodes to act as capture aptamer and labeled with horseradish peroxidase to act as detection aptamer. The enzyme activity of the aptamer–cell–aptamer structures that are only formed when cancer cells are present in the sample, was detected using CV. The sensor was characterized by a linear range from 100 to 1×10^7 cells, and a detection limit of 100 cells (Obs.: Comparison with other sensors is difficult because no concentrations are mentioned). When challenged with 1×10^7 SP2/0 cells that are expected to not express mucin 1, the sensor generated a current signal corresponding to 100 MCF7 cells. This indicates that either SP2/0 cells express at least a certain amount of mucin 1 on their surface or the nonspecific binding is not sufficiently eliminated.

Immobilization of folic acid onto the electrode surface was also reported to assure the selective capture of cancer cells (*e.g.* HeLa or KB cells).[54,55] This was possible due to the fact that cancer cells overexpress folate receptors, while normal cells do not. The captured cancer cells were then detected using differential pulse voltammetry (DPV) carried out with surface-immobilized ferrocene. The concept assured a detection limit of 10 HeLa cells per mL.[54] Yet another possibility is to "interrogate" the sensor surface with impedance spectroscopy. This concept assured a detection limit of 20 KB cells per mL.[55]

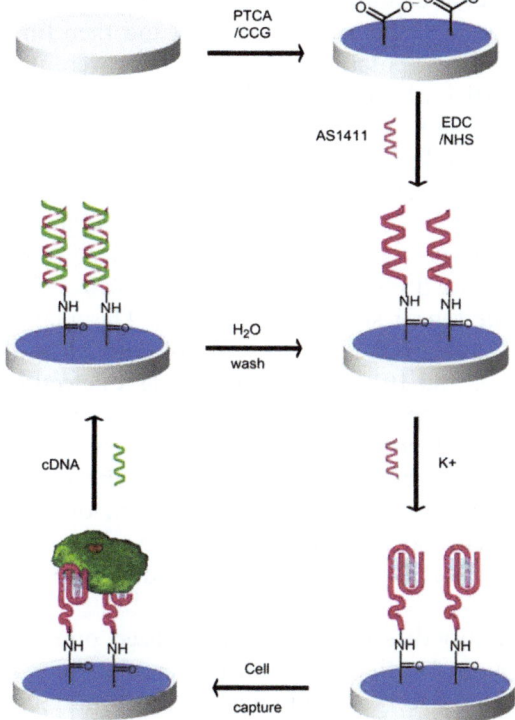

Figure 4.3 Cancer-cell detection based on graphene-modified electrodes and nucleolin-binding aptamer (AS1411) as biorecognition element. Graphene is modified with parylene tetra-carboxylic acid (PTCA) before being immobilized onto graphite electrodes using Nafion. AS1411 is then covalently attached to the carboxyl groups of the PTCA. The affinity binding between cancer cells and AS1411 (having G-quadruplex structure) was observed using EIS, CV, and fluorescence microscopy. Regeneration of the electrode surface using a second aptamer, complementary to AS1411, was possible.
Reprinted from Biomaterials, Vol. 32, Feng, L.Y., Chen, Y., Ren, J.S., and Qu, X.G. A graphene functionalized electrochemical aptasensor for selective label-free detection of cancer cells, 2930–2937, (2011) with permission from Elsevier.

No interferences from HEK 293 cells up to a concentration of 10^4 cells/mL were observed for the detection of the HeLa cells[54] but binding of HFL-I cells to the surface of the sensor for the KB cells was reported.[55]

The detection of low-abundance tumor cells (HepB3) using more traditional recognition elements, *i.e.* antibodies, was also demonstrated.[56] The authors of the study make use of graphene, which is immobilized onto the electrode surface using chitosan in order to assure larger active surface areas and improved electrochemical behavior of the electrode. However, the electrochemical properties of the graphene were not exploited because detection was carried out with a bismuth-film-modified GCE. Anti-EpCAM antibodies were immobilized onto graphene to act as capture antibodies. The electrodes modified in this way were incubated first with the sample (30 min) and then further treated in two different ways. As a first possibility, the electrodes were incubated for 50 min with SiO_2 nanoparticles bearing both CdTe quantum dots and anti-EpCAM antibodies and with SiO_2 nanoparticles bearing both ZnSe quantum dots and anti-GPC3 antibodies. EpCAM and GPC3 are both antigens (proteoglycans) from the surface of HepB3 cells. The formation of the immune complexes between the surface-immobilized antibodies, cells, and finally the nanoparticle-immobilized antibodies was detected and quantified using SWV. SWV was detecting the Zn and Cd ions resulting from dissolving the quantum dots accumulated onto the sensor surface with HNO_3. The quantification of HepB3 cells was possible based on the current peaks of both Zn and Cd. The detection limit was 5 cells/mL and 10 cells/mL when using a detection scheme based on Cd ions and Zn ions, respectively. The linear range of the sensor was up to 10^6 cells/mL. As a second possibility, the electrodes were incubated with SiO_2 nanoparticles bearing the same antibodies as in the first approach but also quantum dots that allow the fluorescent detection of the two immunocomplexes at two different wavelengths. Fluorescence microscopy was then used to visualize the HepB3 cells captured on the electrode surface and labelled with the nanoprobes.

Anti-EpCAM antibodies were also immobilized onto platinum microelectrodes. The so-obtained antibody-modified microelectrodes were used to detect ovarian cancer cells, SKOV3.[57] Binding of the cancer cells to the surface-immobilized anti-EpCAM antibodies was observed as a change in the impedance of the sensor. A detection limit of 4 SKOV3 cells/mL was obtained. Considering this and the previous example of electrochemical sensors for cancer-cell detection, it becomes obvious that anti-EpCAM antibody-modified electrodes will not be able to distinguish between different cancer cells.

The electrochemical detection of HEGFR 2-overexpressing breast cancer cells was also recently reported.[58] Unlike in the above approaches, which make use of either aptamers or antibodies, in this study the cancer cells were detected by a sandwich-type assay using both an antibody (anti-HEGFR 2, immobilized onto the electrode surface) and an aptamer (anti-HEGFR 2, immobilized onto hydrazine-modified Au nanoparticles). The role of the aptamer- and hydrazine-modified Au nanoparticles was to allow detection of the formed immuno-complexes through silver reduction (*i.e.* through silver staining). Silver staining could be observed both with an optical microscope and stripping voltammetry. When used to detect SK-BR-3 breast cancer cells in human serum samples, the method was characterized by an excellent detection limit of 26 cells/mL.

In addition to the above-described detection schemes employing aptamers, small molecules, or antibodies as biorecognition elements, cancer cells can also be detected based on the quantification of specific mRNA. Detection of circulating tumor cells in prostate cancer patient blood samples based on the electrochemical detection of mRNA from such cells was recently reported.[59] First, Au fractal microelectrodes with a nanostructured coating of palladium were fabricated by thin-film technology and electrodeposition. These microelectrodes were then modified with the following PNAs as capture probes: NH_2-C-G-D-gtc-att-gga-aat-aac-atg-gag-D-$CONH_2$ allowing detection of PSA mRNA, and NH_2-C-G-ata-agg-ctt-cct-gcc-gcg-ct-$CONH_2$ allowing detection of TMP/ERG Type III mRNA. The first probe enables unambiguous confirmation that the collected cells originate from a prostate tumor, while the second probe offers information for the classification of different types of prostate tumors because the TMPRSS2-ERG Type III gene fusion is a marker found in a subset of prostate tumors. Binding between the capture probe and target mRNA was observed using voltammetry and a system of two redox species, $Ru(NH_3)_6^{3+}$ and $Fe(CN)_6^{3-}$. Statistically significant changes in the current signal of the sensor were observed when it was exposed to mRNA at concentrations as low as 1 pg/μL or extracts from 100 cells/40 μL of cultured prostate cancer cells (VCaP). Finally, when analyzing blood samples from prostate cancer patients as well as controls the electrochemical sensor produced results that were consistent with established PCR-based techniques.

4.3 Electrochemical Sensors for Detection of Cardiac Biomarkers

Time delay has a major impact on the outcome of patients with cardiac problems. Unfortunately, heart diseases share symptoms with other diseases (*e.g.* acute myocardial infarction and pulmonary embolism were found to share more than chest pain[60]), and thus their diagnosis requires several investigations and time. Therefore, there is a great interest in improving both the accuracy and speed with which chest pain patients are diagnosed. A fast and reliable point-of-care device reporting on several cardiac biomarkers is seen as an appropriate tool to obtain such an improvement in diagnosis accuracy and speed.[61] This section will present a couple of electrochemical sensors reported in the literature for the detection of cardiac biomarkers. Most of these sensors are suitable to be integrated in point-of-care devices. Biosensors (not necessarily based on electrochemical principles) for the detection of cardiac biomarkers have also been reviewed elsewhere.[62]

Cardiac troponin T (35 kDa) and troponin I (29 kDa) are cardiac regulatory proteins that control the calcium-mediated interaction between actin and myosin. Raised serum concentrations of these proteins (> 0.1 μg/L for troponin T, and from 0.1 to 2 μg/L for troponin I) are now accepted as the standard biochemical marker for the diagnosis of myocardial infarction.[63,64] The electrochemical, label-free detection of cardiac troponin I was recently achieved using a carbon nanofibre electrode array modified with anticardiac troponin I antibodies.[65] Binding of the antigen to the antibody-modified electrode was observed with EIS. The sensor was characterized by a detection limit of 0.2 ng/mL, which is 25 times lower than that obtained with conventional methods.

About the same detection limit (148 pg/mL) was achieved for human cardiac troponin I detection using a sandwich-type immunoassay carried out with i) mouse antitroponin I monoclonal antibody as capture antibody immobilized

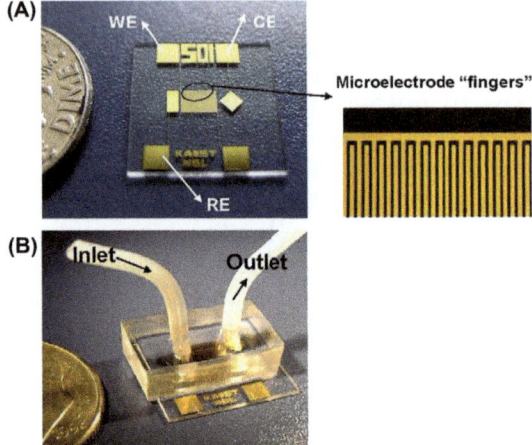

Figure 4.4   System of on-chip, planar electrodes (A) and electrochemical detector
obtained by fixing a PDMS microchannel onto electrodes (B). On-chip,
planar electrode systems are usually fabricated by means of thin-film
processing and consist of counterelectrode (CE), reference electrode (RE),
and working electrode (WE). The PDMS part of the detector also hosts
the inlet and outlet for a convenient loading of samples. The depicted
system was used to detect human cardiac troponin I.
Reprinted from Biosensors and Bioelectronics, Vol. 23, Ko, S., Kim, B.,
Jo, S.-S., Oh, S. Y., and Park, J.-K. Electrochemical detection of cardiac
troponin I using a microchip with the surface-functionalized poly(di-
methylsiloxane) channel, 51–59, (2007) with permission from Elsevier.

onto the walls of a polydimethylsiloxane (PDMS) microchannel, ii) alkaline
phosphatase-labeled mouse antitroponin I monoclonal antibody as detection
antibody, and iii) an array of interdigitated microelectrodes that form the
bottom of the PDMS microchannel (see Figure 4.4) and allow the electrochemical detection of the enzyme's reaction product.[66] It is worth noting that in
this approach the interface for the immobilization of the capture antibody is
advantageously separated from the interface where the electrochemical detection occurs. Thus, both the immobilization and the detection can be separately optimized without affecting each other. The PDMS wall, where the
capture antibodies are immobilized, is still kept close to the detection electrode
(only 50 μm from each other) in order to efficiently detect the product of the
enzyme reaction. In order to eliminate unspecific binding to the PDMS walls of
the microchannel, a 10-min incubation with 1 mg/mL BSA was performed.
Troponin I-proportional signals were obtained for concentrations from
0.2 ng/mL to 10 μg/mL (*i.e.* for the clinically relevant concentration domain) in
about 8 min from the injection of the analyte.
    The simultaneous detection of troponin I and human-heart-type fatty-acid-
binding protein (another important cardiac biomarker), by using an electrochemical sandwich-type immunoassay was also recently demonstrated.[67]
Excellent detection limits (1–3 fg/mL) were obtained mainly due to the use of

several nanomaterials. GCEs used in the assay were modified with both graphene nanoribbons (in order to increase active surface area and improve electrochemical behavior) and antibodies against the two biomarkers. The sandwich immunoassay was completed with titanium phosphate nanospheres modified in two different ways. Titanium phosphate nanospheres were modified with either antitroponin I antibodies and Cd or antifatty-acid-binding protein antibodies and Zn. Cd and Zn show different potentials in SWV. Therefore, the presence and the quantity of the markers could be determined by detecting and quantifying the two metals with SWV. Results obtained with this electrochemical immunoassay in the analysis of human serum samples correlated well with results obtained by ELISA.

A special category of electrochemical sensors is represented by sensors using electrochemiluminescence as the detection principle. Related sensors are special because, although they involve electrochemistry, electrochemiluminescence delivers an optical signal that is proportional to the concentration of the analyte of interest. Electrochemiluminescence-based detection of cardiac troponin I (from 10 pg/mL to 5.0 ng/mL, and with a detection limit of 4.5 pg/mL) was recently reported.[68] The approach makes use of a short linear peptide (FYSHSF-HENWPSK) that selectively binds troponin I. This peptide is immobilized both onto magnetic particles and onto liposomes containing bis(2,2'-bipyridine)-4,4'-dicarboxybipyridine ruthenium-di(N-succinimidyl ester) bis(hexafluorophosphate) as an electrochemiluminescent reagent. The modified magnetic particles and liposomes are then used to capture troponin I from samples and to label it, respectively. The particle–troponin–liposome complexes are subsequently separated from the sample, the liposomes disrupted, and the amount of released electrochemiluminescent reagent quantified by using a GCE poised to +1.15 V *vs.* Ag/AgCl. The sensor was successfully used to detect troponin I in four human serum samples. Another electrochemiluminescence-based sensor for troponin I detection made use of Au nanoparticles modified with both luminol and antitroponin I antibodies.[69] The modified particles were immobilized onto Au electrodes, which thus exhibited stable and strong electrochemiluminescence. In the presence of troponin I the intensity of this electrochemiluminescence decreased in a concentration-dependent manner. The sensor was characterized by a dynamic range from 0.1 to 1000 ng/mL and was used to detect troponin I in plasma samples.

The detection of troponin T was also demonstrated using an electrochemical sandwich-type immunoassay carried out with i) screen-printed electrodes obtained by printing a mixture of graphite powder, silver epoxy, and tetracyanoquinodimethane as redox compound for mediating the electron transfer between HRP and electrode, ii) antitroponin T capture antibodies that were immobilized onto polystyrene beads, and iii) HRP-labeled antitroponin T detection antibodies.[70] The antibody-modified beads were immobilized onto the screen-printed electrodes with glutaraldehyde. The so-modified electrodes were incubated with sample and then with HRP-labelled detection antibodies. The HRP of the formed immunocomplexes was determined by CV. The sensor was characterized by a linear range from 0.1 and 10 ng/mL and a detection limit of

0.2 ng/mL. The sensor was also calibrated with undiluted human-serum samples with known troponin T content as determined using ELISA. A linear dependence of the current signal on the troponin T concentration was again observed. An improved detection limit for troponin T detection (0.017 ng/mL) was obtained when using i) magnetic beads modified with antitroponin T antibodies to capture troponin T from samples, ii) HRP-labeled antitroponin T antibodies as detection antibodies, iii) a magnet to collect the magnetic beads onto the detection electrode, and iv) a screen-printed electrode to detect the 3,3′,5,5′-tetramethylbenzidine oxidized by the HRP label in the presence of H_2O_2.[71] This sensor was successfully tested also with serum samples.

Myoglobin is an oxygen-binding protein that is released into serum as early as 1 h after acute myocardial infarction. However, because it is found in a high concentration in both cardiac and skeletal muscle, myoglobin has poor clinical specificity and is not used by most hospitals to evaluate chest pain.[72] It was also documented that adding myoglobin measurements to troponin I measurements has a limited value for exclusion of myocardial infarction.[73] In spite of this limited diagnostic value of myoglobin, some electrochemical sensors for myoglobin detection have been reported. In a recent example, screen-printed electrodes were modified with didodecyldimethylammonium bromide-stabilized Au nanoparticles and with mouse antihuman myoglobin antibodies and then used to detect myoglobin in standard solutions and human plasma.[74] The presence of a heme prosthetic group in myoglobin opened up the possibility to detect and quantify myoglobin using SWV and direct electron transfer between the surface-immobilized heme and the electrode. The electrochemical immunosensor detected myoglobin in 30 min up to concentrations of 1780 ng/mL, and with a detection limit of 10 ng/mL. The current signal was found to linearly increase with the myoglobin content of human plasma samples from healthy donors and patients that were previously analyzed with a commercially available immunoassay. It is not clear whether the sensor is affected by nonspecific binding or not and whether the sensor can be used to determine the quantity of myoglobin in plasma samples or not. In a quite different approach, single polyaniline nanowires were modified with antimyoglobin monoclonal antibodies and used to detect myoglobin.[75] Nonspecific binding was eliminated by soaking the nanowires into 2 mg/mL BSA immediately after immobilization of the antibodies. Binding of myoglobin to the modified nanowires was detected in a few seconds as an increase in nanowire conductivity. The single nanowire biosensor detected myoglobin down to 1.4 ng/mL and showed no significant signal when challenged with 500 ng/mL BSA. However, if one takes into account the concentration of albumin in serum (~ 40 mg/mL) as well as the concentration of myoglobin in patients with documented acute myocardial infarction (~ 500 ng/mL[76]), it becomes obvious that the tested BSA concentration was just too small.

An increased concentration of serum c-reactive protein is an indicator of an inflammation in the body. Although this inflammation can be due to several reasons, c-reactive protein levels seem to be also correlated with levels of heart-disease risk. Concentrations higher than 3 mg/mL are indicators of a high risk of cardiovascular diseases.[77] A very simple, label-free electrochemical sensor for detection of c-reactive protein consisted of a macroporous Au electrode modified with anti-c-reactive protein antibodies.[78] The sensor detected c-reactive protein in concentrations from 0.1 to 20 ng/mL when interrogated with EIS. It was also used to detect c-reactive protein in three serum samples with acceptable accuracy.

4.4 Electrochemical Sensors for Detection of Acquired Immunodeficiency Syndrome

Acquired Immuno Deficiency Syndrome (AIDS) is the disease caused by human immune deficiency virus (HIV).[79] Recently developed electrochemical sensors, with potentials to detect an infection with HIV, fall into two categories. The first category of sensors detects the presence of HIV by detecting its proteins (*e.g.* p24 HIV capsid protein, HIV enzymes, *etc.*). The second category of sensors detects virus DNA (after reverse transcription PCR) based on hybridization between capture and complementary DNA sequences. These categories add to the classic possibility to detect infection with HIV by detecting the antibodies the body is producing as a response to the infection. While these classic tests are useful 2–8 weeks after infection, *i.e.* when the body produces sufficient antibodies to be detected, the new tests including their electrochemical versions are able to detect infection already in the acute phase of the infection.

The electrochemical detection of p24 antigen was achieved with capture antibodies immobilized onto Au nanoparticle-modified GCEs, detection antibodies labelled with HRP, and hydroquinone as redox mediator that is converted by HRP and then electrochemically detected.[80] Before use the sensor was immersed into 3% BSA in order to block the surface and reduce unspecific binding. DPV was used as a sensitive method to detect the oxidized redox mediator. The sensor displayed a linear range from 0.01 to 100 ng/mL and a sensitivity twice the sensitivity of the conventional ELISA.[80] p24 antigen was detected at much lower concentrations (subattogram per millilitre) using a label-free approach in which anti-p24 antibodies were immobilized onto Au electrodes and the binding between antigen and immobilized antibody was observed as a change in the capacitance of the resulting sensor.[81] In addition to the impressive detection limit this capacitive immunosensor is characterized by a short assay time ($\sim$20 min) and a selectivity good enough to allow detection in p24-spiked human serum samples.

The electrochemical detection of HIV-1 integrase, HIV-1 reverse transcriptase, and HIV-1 protease was achieved using Au electrodes modified with ferrocene-conjugated peptides (see Figure 4.5).[82] The sensors detected the HIV enzymes in nanomolar concentrations based on the fact that the electrochemical behavior of the ferrocene moiety is strongly altered following binding of the HIV enzyme to the surface-immobilized peptide. Binding affects both the potential and the amplitude of the current peaks recorded for the oxidation and

Figure 4.5   HIV enzyme detection using a Au electrode modified with a ferrocene-
conjugated peptide as biorecognition element. The peptides used to con-
struct the sensors were VVStaASta (for detection of HIV-1 protease),
VEAIIRILQQLLFIH (for detection of HIV-1 reverse transcriptase), or
YQLLIRMIYKNI (for detection of HIV-1 integrase).
Reproduced from ref. 84.

reduction of the ferrocene moiety in CV scans. The nonspecific binding to the
sensors was also investigated. However, the tests were carried out at relatively
high concentrations of the target analyte (100 nM) and low concentrations of
nontarget proteins of 300 nM, which is a low concentration if one considers the
protein content of human serum.

When using this sensor design in combination with impedance spectroscopy,
HIV-1 protease was detected in picomolar concentrations.[83] Sensors for HIV-1
protease detection were also obtained by immobilizing the ferrocene-
conjugated peptide (pepstatin) onto a composite material obtained from single-
walled carbon nanotubes and Au nanoparticles.[84] The detection limit for
HIV-1 protease detection with the composite material was 0.8 pM.

HIV-1 protease was also detected using its peptide substrate (GSGSARV-
LAEAGSGSC or GGGSGSGSARVLAEAGGGSGSGSC) that was first
conjugated to a magnetic bead (30 nm in diameter) and then immobilized onto
a Au electrode.[85] The impedance of such peptide-modified electrodes decreased
proportionally with the concentration of active HIV-1 protease in the sample.
The detection limit of the approach was 10 pg/mL. Selectivity was tested by
injection of 1 ng/mL of BSA that did not produce significant changes in the
impedance of the peptide-modified electrodes. However, this concentration is
very small compared to the protein concentrations that might be present in real
samples.

A HIV DNA biosensor was developed by covalently immobilizing HIV ssDNA fragments (5′-ACT GCT AGA GAT TTT CCA CAT-3′) to a GCE.[86] The hybridization of the immobilized HIV ssDNA fragment with its complementary DNA fragment (5′-ATG TCG AAA ATC TCT AGC AGT-3′) was evidenced by using DPV and $[Co(phen)_2IP]^{2+}$ as a novel electrochemical indicator/intercalator (where phen = 1,10-phenantroline and IP = imidazo[*f*] [1,10] phenanthroline). A detection limit of 27 pmol and a linear range from 1.6×10^{-10} to 6.2×10^{-9} mol was obtained. This sensor was not tested with real samples.

A much lower detection limit (2 aM) was obtained when a HIV DNA fragment was detected after making two changes to this approach.[87] The first change was to use two "auxiliary" DNA probes in addition to the capture and target DNAs. These "auxiliary" probes initiated a cascade of hybridizations once the initial hybridization between the target probe and the surface-bound capture probe occurred. The cascade eventually generated micrometer-long one-dimensional DNA structures on the sensor surface. The second change was to use $[Ru(NH_3)_6]^{3+}$ instead of $[Co(phen)_2IP]^{2+}$ as redox indicator for quantifying the formed DNA structures. When used in cell lysates or human serum (diluted 1 : 1), the sensor was still characterized by a good detection limit (5 aM).

HIV has two types, HIV-1 and HIV-2, but electrochemical sensors were mainly developed for HIV-1, because HIV-2 is relatively rare. However, an array of planar sensors suitable for detection of both HIV-1 and HIV-2 oligonucleotides was already reported.[88] Hairpin DNA probes (GCGAGCCTGGGAT-TAAATAAAATAGTAAGAATGTATAGCGCTCGC-$(CH_2)_6$-SH for HIV-1 and GCGAGCAAAGGACCAGGCGCAACTAAATTCAGCTCGC-$(CH_2)_6$-SH for HIV-2) were immobilized onto separate Au electrodes having dimensions of 1 mm × 2 mm and used as capture probes. These probes bind a lot more methylene blue in the single strand state than in the hybridized state. This property allows detection of hybridization by electrochemically (*e.g.* by SWV) observing the amount of methylene blue bound to the sensor surface before and after incubation with the investigated sample. The sequences GCTATACATTCTTACTATTTTATTTAATCCCAG and TGAATTTAG-TTGCGCCTGGTCCTTT were used as target probes specific for HIV-1 and HIV-2, respectively. Selectivity tests were also carried out with two sequences displaying single mismatch. The sensor array allowing the detection of oligonucleotides for both HIV-1 and HIV-2 was characterized by a linear range from 20 to 100 nM, a detection limit of 0.1 nM, and by the ability to discriminate between the complementary and single-base mutation oligonucleotides.

The detection of a polypurine tract sequence conserved in all HIV-1 strains with an electrochemical sensor was also demonstrated.[89] Unlike most of the mentioned DNA sensors that use ssDNA both as capture probe and as target probe, this sensor uses triplex-forming oligonucleotides as capture probe and target double-stranded (ds) DNA. The sensor employs the already described methylene blue- and SWV-based "ON-OFF" mode. The sensor detects dsDNA at concentrations as low as 10 nM and is selective enough to be used in complex sample matrices such as blood serum.

4.5 Electrochemical Sensors for the Detection of Hepatitis Biomarkers

One of the major causes of mortality worldwide is due to infection with hepatitis. There are five known hepatitis viruses: A (HAV), B (HBV), C (HCV), D (HDV) and E (HEV). Diagnosis of hepatitis viremia is based on the detection of the virus antigen or of the nucleic material of the virus. Recent progress in the field was pushed by the importance of early detection in asymptomatic patients suffering from chronic disease and by the need for personalized treatment. This has prompted the appearance of a new generation of commercial kits and new technologies. As more nanomaterials and technologies became available, electrochemical biosensors have emerged as viable alternatives to current standard tests, promising to surpass them in terms of cost, ease of operation and sensitivity. The vast majority of scientific literature is centred on Hepatitis B, while a smaller number of reports is focused on Hepatitis C.

4.5.1 Electrochemical Sensors for Hepatitis B

A variety of electrochemical sensors for diagnosis of hepatitis B have been reported based on amperometric,[90] voltammetric,[91] potentiometric,[92] conductometric,[93] and impedimetric methods.[3,94]

Hepatitis B surface antigen (HBsAg) is the earliest indicator of acute hepatitis B and it is frequently used for identifying infected people before apparition of the actual symptoms.[95] The current standard method is based on ELISA.[96]

Sensitive detection of HBsAg was achieved with electrochemical immunosensors of various architectures,[97,98] using amplification strategies based on nanomaterials that allowed improvement of the detection limit by a factor of 100 compared to the standard ELISA test.[91] An immunosensor for Hepatitis B antigen was developed by modifying a nanoporous Au electrode with HRP-labeled antibody–Au nanoparticles bioconjugates.[91] Increased sensitivity was due to the high electroactive area of the nanoporous Au electrode and Au nanoparticles, allowing for efficient immobilization of HRP-labelled secondary antibodies (one Au nanoparticle could be covered by several HRP-labelled antibody molecules). This sensor was characterized by a detection limit of 2.3 pg/mL and a linear range between 0.01 and 1.0 ng/mL (see Figure 4.6). A similar approach was reported by other authors,[93] while others preferred to use ionic liquids for enhancing the performances of immunosensors for Hepatitis B antigen.[90]

A novel, portable and integrated electrochemical biosensor for the sensitive and selective detection of HBsAg was recently described.[95] The device is based on a sandwich immunoassay put in place by combining immunochromatographic separation on a test strip with electrochemical detection by SWV using a screen-printed electrode (see Figure 4.7). After optimization, the sensor was characterized by a detection limit of 0.3 ng/mL (S/N = 3) HBsAg and a wide

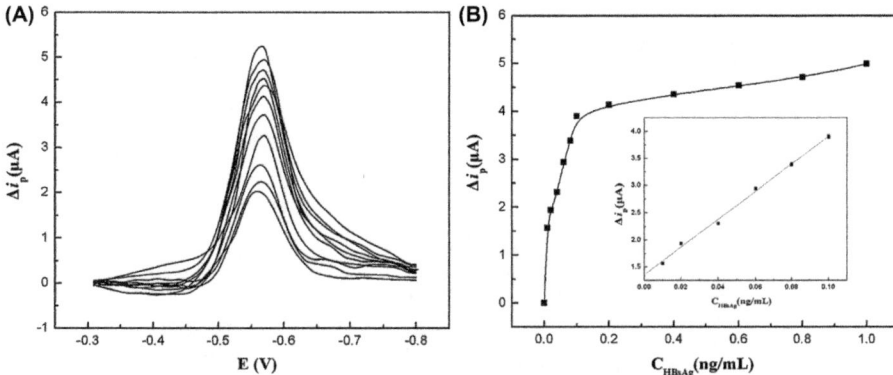

Figure 4.6   Analysis of different concentrations of HBsAg in the range 0.01–1.0 ng/mL by DPV with an electrochemical immunosensor (A). Calibration curves for sandwich immunoassay of HBsAg (B). Plot of DPV peak current *vs.* HBsAg concentration from 0 to 0.1 ng/mL (inset in B).
Reprinted from Talanta, Vol. 30, Ding, C., Li, H., Hu, K., and Lin, J.-M. Electrochemical immunoassay of hepatitis B surface antigen by the amplification of Au nanoparticles based on the nanoporous Au electrode, 1385–1391, (2010), with permission from Elsevier.

linear range (1–500 ng/mL). The sensor was applied to the detection of HBsAg in spiked human-plasma samples. The average recovery was 91.3% and the device was proven to be more sensitive than a colorimetric immunochromatographic test-strip assay.

Besides biosensors for HBsAg, significant research was focused on the electrochemical sensing of hepatitis B virus DNA (after reverse transcription PCR), by monitoring the hybridization event between the target and a probe DNA.[94,99,100]

A label-free DNA sensor was fabricated by coating a single-walled carbon-nanotube (SWCNT) array with Au nanoparticles followed by self-assembly of single-stranded probe DNA.[3] Sensitive detection of complementary hepatitis B ssDNA was achieved by EIS. The principle exploits the variation in the charge-transfer resistance of the modified electrode as a result of DNA hybridization at the sensor surface. The authors have investigated both aligned and random SWCNT arrays, which led to similar results.

Magnetic beads are increasingly used in bioassays as they bring a huge simplification in sample preparation and allow for a high degree of flexibility in the assays.[92,101] An electrochemical genomagnetic assay for the detection of specific HBV DNA included a synthetic 20-mer probe immobilized onto magnetic beads.[101] After incubation with the target DNA, the DNA hybrids were removed from the magnetic beads by washing with NaOH and were analysed by DPV using a pencil graphite electrode. The detection was focused on the guanine oxidation signal at +1.0 V by DPV. The signal only appears following hybridization, as guanine was only contained by the target DNA sequence and not by the synthetic probe.

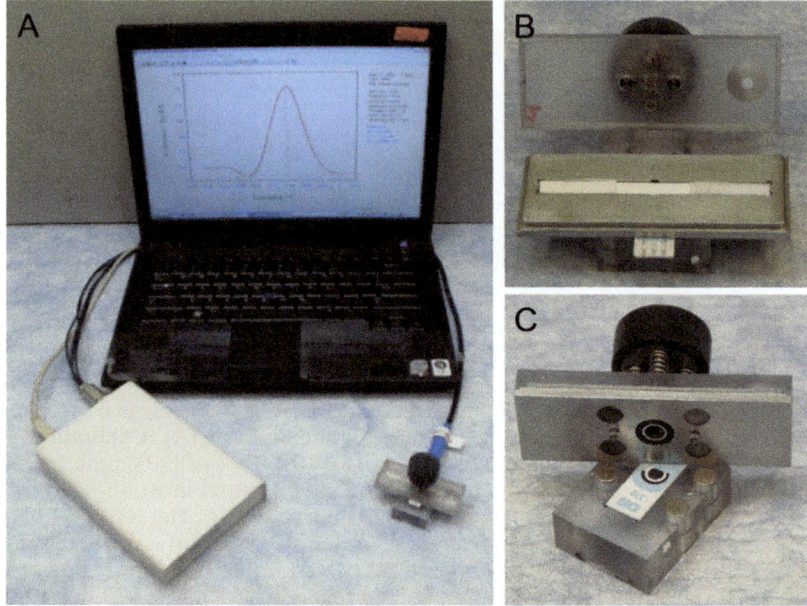

Figure 4.7   Portable electrochemical detection system for HBsAg consisting in an
integrated electrochemical biosensor, a portable electrochemical analyzer
and a laptop (A). The integrated electrochemical biosensor contains an
immunochromatographic strip for the specific recognition (B) and an
electrochemical cell with screen-printed electrodes for sensitive detection (C).
Reprinted from Talanta, Vol. 94, Zou, Z.-X., Wang, J., Wang, H., Li, Y.-Q.,
and Lin, Y. An integrated electrochemical device based on immunochromatographic test strip and enzyme labels for sensitive detection of disease-
related biomarkers, 58–64, (2012) with permission from Elsevier.

Currently, one direction of research in DNA biosensors, exemplified by
sensors for HBV DNA, focuses on the use of electroactive compounds such as
cobalt phenanthroline,[102] methylene blue[103] and $[Os(bpy)_2Cl_2]^+$,[104] as well as
novel redox probes such as bis(benzimidazole)cadmium(II) dinitrate
$(Cd(bzim)_2(NO_3)_2)$[105] for monitoring the hybridization event.

Multiplexed detection of hepatitis virus antigens was recently reported using
an integrated automatic electrochemical immunosensor array.[106] It allows for
the simultaneous detection of 5 types of hepatitis virus antigens (*i.e.* hepatitis A,
hepatitis B, hepatitis C, hepatitis D, and hepatitis E) in 5 min. The design of the
immunoassay relies on a direct format where the corresponding antibodies are
immobilized on a protein A/Au nanoparticle support and the detection of
antigens is done by potentiometry. The analytical signal is represented by the
variation in the electrical potential (ΔE) between the antibody-modified sensor
and a reference electrode as result of the antigen–antibody interaction. A cut-
off value of 2 mV was assigned to discriminate between positive and negative
results based on testing serum samples that were negative for hepatitis by
standard ELISA. The detection limit of the sensor array was around 1.0 ng/mL

for all virus antigens and the linear range extended up to 350 ng/mL. The intra- and interassay precisions of the sensor array were $\leq 7.3\%$ and 8.1%, respectively. A very good correlation was found between the results provided by the sensor and standard ELISA for the analysis of 43 serum samples.

Detection limits and linear range reported in the literature for HBV DNA vary for different lengths of target DNA sequence.[92,101,107] Thus, for a 21-bp of a Hepatitis B DNA sequence the charge-transfer resistance measured by EIS [3] was linearly correlated with the logarithm of target HBV ssDNA concentration over the range between 1.0×10^{-18} M and 1.0×10^{-12} M with a detection limit of 1 attomole/mL.

Analysis time ranges from under half an hour[90,106] to several hours,[91] depending on the testing format (direct *versus* sandwich) and the number of steps involved. In an effort to make the leap from a laboratory concept to practical applications, an increasing number of studies are focusing on the development of electrochemical biosensors aiming on improved reproducibility and precision of the method as well as applications to real samples. The intra- and interassay coefficients of variation reported for different levels of HBsAg were below 8%, which is adequate considering the manual steps involved in sensor preparation.[90,91,93] The selectivity of the immunoassay for HBsAg was studied by incubating the sensor with solutions of antigen in the presence of potentially interfering compounds in various concentrations.[91,93]

Electrochemical sensors developed in recent years have been challenged with serum samples and compared to commercial ELISA tests to establish their usefulness for the diagnosis of Hepatitis B. Careful design of the sensor architecture with consideration of nonspecific adsorption allowed observation of a high degree of correlation between the results provided by the biosensors and current standard methods.[90,91,93,95]

4.5.2   Electrochemical Sensors for Hepatitis C

Infection with HCV affects approximately 170 million people worldwide and represents the major cause of parentally transmitted non-A, non-B hepatitis infection.[108,109] Many people suffering from chronic Hepatitis C infection are asymptomatic. If left untreated, chronic infection leads to liver cirrhosis and hepatocellular carcinoma. There are 6 known genotypes of HCV and multiple subtypes, with Genotype 1 (subtypes 1a and 1b) being the most prevalent genotypes worldwide. Since the discovery of the virus in 1989, the progress in diagnosing Hepatitis C has been impressive.

Currently, diagnosis of hepatitis C is performed by serological testing of hepatitis C antibodies (first by ELISA, followed by confirmation test with recombinant immunoblot assay – RIBA) and molecular assays that detect HCV RNA. Antibody-based tests indicate past as well as present infections, while the detection of HCV RNA is performed to confirm a current infection and to follow the response to a treatment. Genotyping and viral load determination are also very important at this stage in order to assign the adequate treatment and monitor patient's response. HCV genotype and subtype can be determined by various commercially available assays based on different methods such as direct sequence analysis, reverse hybridization, and genotype-specific real-time PCR.[110,111] On the other hand, several ELISA and RIBA assays are commercialized and approved by regulatory agencies in different countries. For example, the OraQuick HCV rapid antibody test was approved by the FDA in 2010. It uses whole blood samples, it is portable, easy to use and provides an answer in 40 min.[112] Biosensors are mainly intended for the same market of point-of-care tests, therefore having to compete with these more traditional and robust tests.

Detection of HCV with biosensors relies on the use of antibodies,[106] DNA/PNAs and whole cells engineered to contain antibodies or antigens related to the virus. The advantage of using PNAs *versus* DNA probes relies in the uncharged, flexible structure of PNA and their resistance to nuclease and protease activity. PNA hybridizes with complementary nucleic acids. Intercalators as well as electroactive molecules (*e.g.* methylene blue) have been used to detect the formation of a hybrid with target DNA or of a triplex with double-stranded DNA.[113,114]

A novel amperometric genosensor for qualitative HCV RNA detection and genotyping (HCV 1, 2A/C, 2B and 3) was obtained by immobilizing a 18-mer DNA probe on functionalized graphite.[115] The sensor was incubated with target biotin-labelled complementary DNA, obtained from HCV RNA in serum by reverse transcriptase-linked polymerase chain reaction (RT-PCR). Detection of the hybridization event was achieved by further incubation of the sensor with an avidin–peroxidase conjugate and amperometry at $-0.45\,V$ *versus* Ag/AgCl using the H_2O_2-KI system. The current intensity was proportional to the amount of HCV amplicons in the samples. In this architecture, the affinity system biotin-(strept)avidin was used to immobilize the DNA probe on the modified graphite electrode and to attach the enzyme to the hybridized target DNA. By using affinity-based immobilization, the authors achieved both a controlled orientation and strong binding of DNA probes on the electrode surface. To anchor streptavidin on the electrode, the protein was entrapped within a siloxane–poly-(propylene oxide) (PPO) nanocomposites prepared by the sol–gel method. The detection limit reported for all virus genotypes was around 1 ng/mL[115] and the results were highly reproducible (RSD of 0.2–1.2% for positive and negative controls tested five times on different days).

To establish a correlation between a positive/negative result and the absolute value of the analytical response obtained with the genosensor, a cut-off value needs to be defined. For example, this value was set at the current intensity average plus twice the standard deviation obtained for samples deemed as HCV RNA negative and for negative controls analysed by the standard kit.[115]

Simultaneous detection and genotyping of HCV RNA (genotypes 1, 2A/C, 2B and 3) was demonstrated with an amperometric genosensor for eight serum samples from HCV-infected patients.[115] By immobilizing different probes (HCV 1, 2A/C, 2B and 3 oligonucleotides) on the PPO electrode it was possible to distinguish positive and negative sera samples. The amperometric sensor gave comparable results to the commercial test with spectrophotometric detection (Amplicor®, from Roche) and proved more sensitive.

Another approach reported in the literature refers to the bioelectric recognition assay (BERA) that registers the electrical response of whole cultured cells upon binding viruses such as HCV inducing a change of their physiology.[116] Vero fibroblast cells modified by electroinsertion of HBV specific antigens (HBsAg) in their membranes responded in a fast (45 s) and specific manner to HBV-positive samples from a pool of 133 clinical blood serum samples. The response was due to changes in the cell membrane potential that were measured by appropriate microelectrodes immobilized in an alginate matrix. This assay is a promising one-step, fast and label-free assay, however, it is not as sensitive as the DNA sensors yet and the response does not correlate in a linear manner with the viral load.[117]

4.6 Electrochemical Sensors for the Detection of Rheumatoid Arthritis Biomarkers

Rheumatoid arthritis (RA) is a chronic, systemic autoimmune disease that causes the inflammation of joints and surrounding tissue. It affects about 1% of world's population. The diagnosis of RA is based on clinical symptoms supported by X-rays, patient history and serological tests. Various autoantibodies have been proposed as biomarkers for RA: rheumatoid factor, antikeratin, anticitrullinated peptides, anti-RA33, anti-Sa, and anti-p68 autoantibodies.[118,119]

At present, RA is diagnosed based on the criteria developed in 2009 by the American College of Rheumatology/European League Against Rheumatism collaborative initiative.[120] According to these criteria, rheumatoid factor (RF) and the anticyclic citrullinated peptide (CCP) antibody are the most important main serological markers.

RF is an autoantibody directed against the Fc region of human IgG and it is typically detected by nephelometry using a latex agglutination assay. Auto-antibodies to citrullinated proteins or peptides (filaggrin, fibrin or vimentin) are considered the best biomarkers for the early detection of RA as they are present in patients years before the apparition of clinical symptoms.[121,122] Citrulline is formed enzymatically by deimination of arginine. The CCP antibody is an antibody directed against citrulline that appears early on in RA patients. Anti-CCP assays are performed by plate-based ELISA available from several manufacturers. A lateral flow immunoassay test for the rapid detection of RA is commercialized by Euro-Diagnostica AB. Automated, faster analysis of anti-CCP antibodies (*e.g.* by Electrochemiluminescence (ECL) on Elecsys® analysis systems) is also commercially available nowadays.[123]

It should be noted that new autoantibodies are being discovered and studied as potential markers of the disease. They hold promise for the future as currently about one-third of RA patients are seronegative for both RF and CCP antibody.[124,125]

Besides RF and anti-CCP antibodies, the macrophage migration inhibitory factor (MIF) has been suggested as a potential biomarker for RA.[126]

There are not many biosensors for the detection of RA. The presence of anticitrullinated protein antibodies specific to RA was studied with SPR microarray imaging[127] down to a detection limit of 0.5 pM. A quartz crystal microbalance (QCM) method, employing a single-walled carbon-nanotube-based immunosensor proved very sensitive (detection limit in the femtomole range).[128] Fluorescence-based and sensitive diagnostic assay of RA was successfully performed using ZnO nanorods.[118] Practical applications of such biosensors are hampered by the ease of operation and by portability issues.

An electrochemical immunosensor based on the detection of the anti-CFFCP1 antibody RA biomarker was recently reported.[129] The device is composed of a thin-film metal electrode coated with a multiwalled-carbon-nanotube–polystyrene (MWCNT–PS) composite, on top of which a citrullinated specific peptide receptor was covalently attached. After incubation of the modified electrodes with serum, the sensors were further incubated with anti-IgG secondary antibodies labelled with HRP and the detection was done by amperometry using the 3,3′,5,5′-tetramethylbenzidine (TMB)/H_2O_2 system. Compared to ELISA, the detection limits achieved with this biosensor were similar and the analysis time was shorter. Otherwise, the sensor fabrication and the incubation steps specific to a sandwich assay are limiting the practical applications of such a device. A remarkable fact about this research is the step forward made by its authors to characterize the reproducibility of sensors from different batches over a 3-month period. They emphasized that for 146 electrodes the coefficient of variation of the intensity of current *versus* a classic redox species (1.0 mM ferricyanide) was 27.5%. This was correlated with the high degree of manual intervention in sensor fabrication. The yield of sensors within 10% variation was about 70%, which is still far from adequate reproducibility for practical applications.

A novel electrochemical immunosensor for MIF was developed by functionalizing a Au electrode coated with Au nanoparticles, titanium dioxide nanoparticles and thionine (NGP-NTiP-Thi) with anti-MIF antibodies of the IgM type.[130] The IgM immunosensor was characterized by a linear range between 0.03 and 230 ng/mL and a detection limit of 0.02 ng/mL being also applied for testing MIF in sera of RA patients.

4.7   Electrochemical Sensors for the Detection of Celiac Disease Biomarkers

Celiac disease or gluten enteropathy affects many people worldwide (as much as 1% of the population of US). Gluten and similar proteins from wheat, rye and barley provoke an exacerbated immune response in genetically predisposed individuals that ultimately leads to chronic inflammation and damage to the upper part of the small intestine.

The only standard accepted worldwide for diagnosing celiac disease remains intestinal biopsy with confirmation of the diagnosis by retesting after a gluten-free diet. However, serological tests for IgA antitransglutaminase 2 (IgA anti-TG2) or antiendomysial antibody, total IgA and IgG anti-tTG2 (antitissue transglutaminase 2), antiendomysial or antigliadin antibody are performed first as they indicate an increased probability of celiac disease.[131,132] In the initial screening, IgA isotypes are preferred due to their higher sensitivity, however, some of the patients present IgA deficiency and IgG-based tests need to be performed instead. Best serological markers of celiac disease are antibodies (IgA anti-TG2 or antiendomysial) directed to the same antigen. Due to their lower sensitivity and specificity antigliadin antibodies were not used in the diagnosis of celiac disease until recently, when tests based on deamidated gliadin peptides instead of whole gliadin protein have emerged. These allow for similar results as those using IgA anti-TG2 tests. Novel antibodies against deamidated gliadin peptide (DGP) are considered especially valuable for the diagnosis of celiac disease, in those individuals with selective IgA deficiency. A recent study[133] compared 4 different strategies for the serologic diagnosis of celiac disease and concluded, by using two commercial assays, that determining IgA anti-tTG and IgG anti-DGP in all patients performed better than the other strategies.

Several electrochemical biosensors for the diagnosis of celiac disease have been described so far based on impedimetric,[134] amperometric[135] and voltammetric techniques.[4,136] The main characteristics of these sensors are summarized in Table 4.1.

A variety of sensor architectures has been considered in electrochemical devices intended for diagnosis of celiac disease. While some authors embedded a relatively high degree of complexity in the materials and technique used[136,137] others tried to rationalize the design of the sensor to achieve best analytical performances. For example, coating a Au electrode with a bipodal thiol containing ethylene glycol groups and carboxylic end groups allowed for a controlled, covalent immobilization of the bioreceptor[135,138] while insuring both low nonspecific adsorption and good diffusion of electroactive substrates to electrode surface.[139]

Multiplexing and a high degree of flexibility of the assay were reported using a dual screen-printed electrode, where one electrode was modified with gliadin and the other one with tTG. By using IgA or IgG as secondary antibody in a sandwich-type assay, this device allowed determination of the IgA and IgG-type antigliadin and anti-tTG antibodies in real patient's samples.[136]

In order to compare these new sensors with commercial ELISA the researchers used calibration solutions of antibodies from the ELISA kit to build the calibration curve with their electrochemical biosensor.[4,135,138] A cut-off value was defined for the electrochemical signal, allowing to classify samples as positive/negative.[4,140]

Correlation with ELISA was generally good[141] even excellent in some cases.[4,135,138] A more rigorous analysis including more serum dilutions and a higher number of patient samples is needed for a clear conclusion.

4.8 Electrochemical Sensors for the Detection of Urinary Tract Infection Biomarkers

Urinary tract infection (UTI) is one of the most common infections, causing severe distress to affected individuals and being associated with high healthcare

Table 4.1 Electrochemical biosensors for the diagnosis of celiac disease.

Biorecognition element	Immobilization matrix	Electrochemical method	Detection limit	Ref.
TG	On screen-printed Au electrode coated with a polyelectrolyte	EIS (3-amino-9-ethylcarbazole)	n/a[a]	136
tTG	Covalent attachment to Au coated with DT2[b] thiol	Amperometric, TMB/H_2O_2	390 ng/mL for rabbit anti-tTG antibodies	137
tTG	Physical adsorption on screen-printed carbon electrodes modified with MWCNT and Au nanoparticles	CV, anodic redissolution of silver	2.45 U/mL for tTG IgA; 2.95 U/mL for tTG IgG	138
gliadin	Physical adsorption on screen-printed carbon electrodes modified with MWCNT and Au nanoparticles	CV, anodic redissolution of silver	3.16 U/mL for AGA[c] IgA; 2.82 U/mL for AGA IgG	138
tTG	Physical adsorption on screen printed carbon electrodes modified with MWCNT and Au nanoparticles	CV, anodic redissolution of silver	n/a	4
tTG	Physical adsorption on graphite–epoxy composite electrodes	Amperometric, hydroquinone/H_2O_2	n/a	139
gliadin	Covalent attachment to Au coated with DT2 thiol	Amperometric, TMB[c]/H_2O_2	46 ng/mL	140
gliadin	Physical adsorption on screen-printed carbon electrodes modified with MWCNT and Au nanoparticles	CV, anodic redissolution of silver	9.1 U/mL for AGA IgA; 9.0 U/mL for AGA IgG	141
IgA	Covalent attachment to Au coated with DT2 thiol	Amperometric, o-dianisidine/H_2O_2	142 ng/mL	140
DGP	Physical adsorption on screen-printed carbon electrodes modified with MWCNT and Au nanoparticles	CV, anodic redissolution of silver	n/a	142

[a]Not available.
[b](22-(3,5-bis((6 mercaptohexyl)oxy)phenyl)-3,6,9,12,15,18,21-heptaoxadocosanoic acid (DT2).
[c]Antigliadin antibody.

and social costs. It is estimated that UTI affects about 40% of women at some point in their lives.

These infections represent 20–50% of hospital-acquired infections in ICU units and in the same time the most common disease for which antibiotics are prescribed in long-term care facilities. UTI is a common source of fever in children and, if left untreated, can induce renal scarring, hypertension, pre-eclampsia, and end-stage renal disease.[142] Diagnosis of UTI is made based on symptoms (*e.g.* frequency of urination, dysuria, and pyuria) and on the presence of more than 104 cfu/mL uropathogens in urine. *Escherichia coli* is by far the most common uropathogen causing more than 50% of the cases.[143] Other uropathogens, causing less than 10% of infections each are *Klebsiella pneumonia, Enterococcus faecalis, Proteus mirabilis, Staphylococcus saprophyticus* and *Pseudomonas aeruginosa.*

Conventional methods for the detection of *E. coli* include plate count, multiple-tube fermentation (MTF), membrane filter (MF), *etc.*, which require a few hours up to 2 days to be completed.

Diagnosis of UTI is one area where electrochemical biosensors for point-of-care devices could bring a significant contribution. A recent review[144] provides an optimistic prognosis on the use of electrochemical biosensors to improve UTI diagnosis. The optimism stems from the compatibility of such devices with microfluidics and the demonstrated applicability of some electrochemical sensors for point-of-care testing (*e.g.* glucose sensor or the I-STAT handheld blood analyzer developed by Abbott).

The main areas of development of biosensors for UTI diagnosis include: uropathogen identification,[145] testing the antimicrobial susceptibility[146] to select the most appropriate antibiotics and differentiating actual infections from asymptomatic bacteriuria.[147] The latter was recently achieved with integrated platforms containing both nucleic acid probes and protein biomarkers on the same sensor array.

Most recent progress regarding biosensors that have been applied to the analysis of clinical samples for UTI diagnosis will be further discussed.

In a 2005 study, a novel electrochemical 16-sensor array was combined with a set of oligonucleotide probes applied to nylon membranes in a dot-blot format.[148] The probes allowed discrimination among 16S genes from 11 different species of uropathogenic bacteria. An oligonucleotide labeled with peroxidase was used as detector probe and the detection relied on a reaction catalysed by HRP. This earlier research showed that such molecular hybridization approaches are amenable to fast room-temperature determinations.

An integrated platform has been described combining a step of bacterial lysis with sensitive electrochemical detection using a nanostructured electrode on an analysis chip. Unpurified lysates in buffer and urine were used to determine bacteria in a total time of 30 min, with a reported detection limit of 103 cfu/mL.[149]

The UTI Sensor Array[145] is an example of a biosensor for uropathogen identification based on an electrochemical 16-sensor array containing immobilized DNA probes targeting the 16S rRNA of the most common uropathogens. The eight probes (immobilized in duplicate) include universal as well as genus- and species-specific ones. Specific recognition of bacterial 16S rRNA was achieved by sandwich hybridization using capture probes (immobilized on the electrode) and pair detector oligonucleotides (free in solution). To amplify the hybridization signal and to translate it into an electrochemical signal the fluorescein-labelled detector probe is incubated with an HRP-conjugated antifluorescein antibody. Compared to urine culture, the detection limit reported for this biosensor is within the clinical cut-off (104 cfu/mL), the overall sensitivity was 92% and the specificity 97%. The biosensor provided results in 1 h.

The same group further developed the UTI sensor array for application to antimicrobial susceptibility testing.[146] An integrated sensor platform able of multiplex detection of uropathogens and lactoferrrin (as biomarker of the host immune response) illustrates further the versatility of this concept and the huge potential for UTI diagnosis.[147] Lactoferrin is a protein derived from white blood cells (WBC) and was researched as a biomarker of pyuria (the presence of white blood cells in urine). As a characteristic of UTI, pyuria is generally determined by urine microscopy, although there is also a dipstick test measuring the enzymatic activity of leukocyte esterase (LE).

A sandwich amperometric immunoassay for detection of lactoferrin achieved a detection limit of 145 pg/ml. Based on the analysis of 111 clinical samples, the concentration of lactoferrin was shown to be correlated with the results of urine microscopy for WBC,[147] supporting the choice of this protein as a biomarker of pyuria.

In the new generation of biosensors, the detection of pathogens is integrated with that of protein biomarker (lactoferrin) on the same analytical chip. The principles behind this approach rely on sandwich hybridization of capture and detector oligonucleotides to the target 16S rRNA from uropathogens and on a sandwich assay with capture and detector antibodies for the detection of protein. Electrochemical detection is achieved by using detector probes and antibodies labelled with HRP. A classic amperometric detection of the TMB/H_2O_2 reaction catalysed by HRP is employed to translate the specific recognition events into quantifiable signals.[150]

A prospective clinical study conducted in 2009 on 116 patients emphasized the possibility to identify uropathogens using an electrochemical biosensor platform, with similar results as standard urine cultures[145] within the short time frame characteristic to point-of-care testing. A study aiming to clinical validation was reported recently with an integrated sensor array for determining both nucleic acids and biomarker lactoferrin.[150] This sensor array is considered an example of the next generation of electrochemical devices for UTI diagnosis.

Compared to the commercial glucose sensor or the pregnancy test, developing a rapid test for diagnosing UTI is deemed to be more complicated, as the matrix is complex and parameters such as pH and cell count could vary in large limits. Sample preparation remains a limiting factor for the development of disposable biosensors for UTI, a problem that could find a solution by using electrokinetics or microfluidic technology. For example, a multifunctional electrode platform was used for both electrochemical sensing and to enhance the signal by electrokinetics by *in situ* stirring and heating. By this approach the sensitivity of the 16S rRNA hybridization assay for UTI diagnosis was increased by an order of magnitude while the analysis time was drastically reduced.[151]

Versatility, low cost and the potential for multiplexing have all been enumerated as advantages that drive the development of biosensors for urinary-tract infections.[144]

4.9 Challenges in the Use of Electrochemical Sensors in Diagnostics

Although many papers are published on electrochemical sensors for diagnostics, relatively few sensors have real chances to reach end users as commercially available products. The main challenges still faced by electrochemical sensors in their way towards a significant acceptance in clinical settings are detailed below.

Electrochemical sensors detecting biomarkers must reliably deal with real clinical samples. The literature on such electrochemical sensors reflects a continuous race for better detection limits and only scarce interest in assuring a selectivity that allows use with real samples. There are many sensors tested only with standard solutions or tested with too low concentrations of possible interfering species, but only few sensors tested with a reasonable number of real samples. Nonspecific binding is probably the biggest problem that must be more carefully addressed because samples commonly contain very low concentrations of biomarker in the presence of very high concentrations of other proteins.

Electrochemical sensors detecting biomarkers must display stability; Stability is another poorly investigated sensor parameter. Single-use sensors must have reasonable storage stability, while sensors for multiple uses must also display operational stability. Yet another possibility is to have sensors simple enough to be assembled and then calibrated by end users. However, most of the sensors are far from such simplicity.

Electrochemical sensors must be appropriate for mass production; The electrochemical sensors described in the scientific literature are most often manually made through processes that are not amenable for automation and thus mass production.

Electrochemical sensors must compete with existing methods in terms of sample throughput (and convenience in use, in general). Unlike sensors to be used for point-of-care testing (when a response time of a few minutes is desired next to reliability), sensors to be used in clinical laboratories must have competitive sample throughput. There are two possibilities to achieve this. The first is to have electrochemical sensors with very short response times (few minutes instead of hours). The second possibility is to use arrays of several sensors reporting on several markers or on the same marker but in several samples. Even the most advanced sensor arrays described in the literature up to now are not able to equal the sample throughput of experiments made on common ELISA microplates.

New success stories will be added to that of the portable, electrochemical glucose meters, once these challenges are eliminated.

4.10   Conclusions

Adding simplicity, speed, and cost effectiveness to the reliability of clinical laboratory tests is often very challenging. In the attempt to overcome this challenge, many electrochemical sensors were developed for the detection of biomarkers associated to some of the major human diseases. Specificity of these sensors was ensured by using biorecognition elements of very different nature ranging from small molecules (*e.g.* folic acid to recognize cancer cells through their folate receptors) to biomacromolecules (*e.g.* antibodies and more recently aptamers). Sensitivity of these sensors was warranted by using different electrochemical methods and (more recently) nanomaterials. There are several challenges still to overcome by electrochemical sensor in order to be used in diagnostics. On the positive side, results are extremely promising judged on the achieved response times, detection limits, and sensitivities (often surpassing those of classic methods). The correlation between results provided with electrochemical biosensors and standard methods (*e.g.* ELISA) was generally good (but in some cases the number of clinical samples was too low to draw a firm conclusion). Combining different concepts on the same analytical platform (*e.g.* DNA hybridization and immunosensing), and lab-on-a-chip approaches integrating sample preparation and detection steps are proof of impressive progress in the field and also of huge potential. Collaborative projects bringing together sensor developers, clinicians and end users (clinical laboratories) could be the driving force for the future development of electrochemical sensors for clinical diagnosis.

References

1. A. J. Bard and L. R. Faulkner, *Electrochemical Methods: Fundamentals and Applications*, John Wiley & Sons, Inc., Hoboken, NJ, 2nd edn, 2001.
2. E. P. Randviir and C. E. Banks, *Anal. Methods*, 2013, **5**, 1098–1115.
3. S. Wang, L. Li, H. Jin, T. Yang, W. Bao, S. Huang and J. Wang, *Biosens. Bioelectron.*, 2013, **41**, 205–210.
4. M. M. P. S. Neves, M. B. González-García, H. P. A. Nouws and A. Costa-García, *Biosens. Bioelectron.*, 2012, **31**, 95–100.
5. D. W. Kimmel, G. LeBlanc, M. E. Meschievitz and D. E. Cliffel, *Analyt. Chem.*, 2012, **84**, 685–707.
6. P. D'Orazio, *Clin. Chim. Acta*, 2011, **412**, 1749–1761.
7. F. J. Gruhl, B. E. Rapp, and K. Länge, in *Advances in Biochemical Engineering/Biotechnology 2011*, Springer, Berlin, Heidelberg, 2011, pp. 1–34.
8. A. Chen and S. Chatterjee, *Chem. Soc. Rev.*, 2013, DOI: 10.1039/C3CS35518G.
9. A. Merkoçi, *Electroanal.*, 2013, **25**, 15–27.
10. G.-J. Zhang and Y. Ning, *Anal. Chim. Acta*, 2012, **749**, 1–15.
11. J. Wang, *Biosens. Bioelectron.*, 2006, **21**, 1887–1892.

12. F. Ricci, G. Adornetto and G. Palleschi, *Electrochim. Acta*, 2012, **84**, 74–83.
13. A. Schlichtiger, P. B. Luppa, D. Neumeier and M. Thaler, *Bioanal. Rev.*, 2012, **4**, 75–86.
14. B. V. Chikkaveeraiah, A. A. Bhirde, N. Y. Morgan, H. S. Eden and X. Chen, *ACS Nano*, 2012, **6**, 6546–6561.
15. S. K. Arya and S. Bhansali, *Chem. Rev.*, 2011, **111**, 6783–6809.
16. J. Li, S. Li and C. F. Yang, *Electroanal.*, 2012, **24**, 2213–2229.
17. V. L. W. Go, *Cancer*, 1976, **37**, 562–566.
18. H. Tang, J. Chen, L. Nie, Y. Kuang and S. Yao, *Biosens. Bioelectron.*, 2007, **22**, 1061–1067.
19. T. Yasukawa, Y. Yoshimoto, T. Goto and F. Mizutani, *Biosens. Bioelectron.*, 2012, **37**, 19–23.
20. Z. Dai, J. Chen, F. Yan and H. Ju, *Cancer Detect. Prev.*, 2005, **29**, 233–240.
21. A. Bélanger, H. van Halbeek, H. C. Graves, K. Grandbois, T. A. Stamey, L. Huang, I. Poppe and F. Labrie, *Prostate*, 1995, **27**, 187–197.
22. I. M. Thompson, D. K. Pauler, P. J. Goodman, C. M. Tangen, M. S. Lucia, H. L. Parnes, L. M. Minasian, L. G. Ford, S. M. Lippman, E. D. Crawford, J. J. Crowley and C. A. Coltman Jr, *N. Engl. J. Med.*, 2004, **350**, 2239–2246.
23. R. S. Porter and J. L. Kaplan, ed., *The Merck Manual of Diagnosis and Therapy*, Merck Sharp & Dohme Corp., a subsidiary of Merck & Co., Inc., Whitehouse Station, N.J., online edition., 2010.
24. V. Mani, B. V. Chikkaveeraiah, V. Patel, J. S. Gutkind and J. F. Rusling, *ACS Nano*, 2009, **3**, 585–594.
25. H. Li, Q. Wei, G. Wang, M. Yang, F. Qu and Z. Qian, *Biosens. Bioelectron.*, 2011, **26**, 3044–3049.
26. H. Wang, Y. Zhang, H. Yu, D. Wu, H. Ma, H. Li, B. Du and Q. Wei, *Anal. Biochem.*, 2013, **434**, 123–127.
27. G. J. Kato and C. V. Dang, *FASEB J.*, 1992, **6**, 3065–3072.
28. J. K. Field, D. A. Spandidos, P. M. Stell, E. D. Vaughan, G. I. Evan and J. P. Moore, *Oncogene*, 1989, **4**, 1463–1468.
29. J.-L. He, Y.-F. Tian, Z. Cao, W. Zou and X. Sun, *Sens. Actuators, B*, **181**, 835–841.
30. R. Singh and D. Bandyopadhyay, *Cancer Biol. Ther.*, 2007, **6**, 481–486.
31. E. Szabo, in *Lung Cancer*, B. Driscoll and J. M. Walker, Humana Press, 2003, vol. 74, pp. 251–258.
32. F. Ma, C. Ho, A. K. H. Cheng, and H.-Z. Yu, *Electrochim. Acta*, http://dx.doi.org/10.1016/j.electacta.2013.02.088.
33. S. Song, L. Wang, J. Li, J. Zhao and C. Fan, *Trac-Trends Analyt. Chem.*, 2008, **27**, 108–117.
34. J. M. Catanzaro, J. L. Guerriero, J. Liu, E. Ullman, N. Sheshadri, J. J. Chen and W.-X. Zong, *PLoS ONE*, 2011, **6**, e19096.
35. Murakami, *Int. J. Oncol.*, 2010, **36**, 1395–1400.

36. D. Wu, H. Fan, Y. Li, Y. Zhang, H. Liang and Q. Wei, *Biosens. Bio-electron.*, **46**, 91–96.
37. M. Capello, S. Ferri-Borgogno, P. Cappello and F. Novelli, *FEBS J.*, 2011, **278**, 1064–1074.
38. J. A. Ho, H.-C. Chang, N.-Y. Shih, L.-C. Wu, Y.-F. Chang, C.-C. Chen and C. Chou, *Analyt. Chem.*, 2010, **82**, 5944–5950.
39. Y. Levav-Cohen, S. Haupt and Y. Haupt, *Growth Factors*, 2005, **23**, 183–192.
40. R. Elshafey, C. Tlili, A. Abulrob, A. C. Tavares and M. Zourob, *Biosens. Bioelectron.*, 2013, **39**, 220–225.
41. D. Tang, L. Hou, R. Niessner, M. Xu, Z. Gao and D. Knopp, *Biosens. Bioelectron.*, **46**, 37–43.
42. R. Malhotra, V. Patel, B. V. Chikkaveeraiah, B. S. Munge, S. C. Cheong, R. B. Zain, M. T. Abraham, D. K. Dey, J. S. Gutkind and J. F. Rusling, *Analyt. Chem.*, 2012, **84**, 6249–6255.
43. B. V. Chikkaveeraiah, A. Bhirde, R. Malhotra, V. Patel, J. S. Gutkind and J. F. Rusling, *Analyt. Chem.*, 2009, **81**, 9129–9134.
44. P. A. J. Muller and K. H. Vousden, *Nature Cell Biol.*, 2013, **15**, 2–8.
45. M. Hollstein, D. Sidransky, B. Vogelstein and C. C. Harris, *Science*, 1991, **253**, 49–53.
46. E. Farjami, L. Clima, K. Gothelf and E. E. Ferapontova, *Analyt. Chem.*, 2011, **83**, 1594–1602.
47. X. Wang, X. Wang, X. Wang, F. Chen, K. Zhu, Q. Xu and M. Tang, *Anal. Chim. Acta*, 2013, **765**, 63–69.
48. O. Y. F. Henry, J. L. A. Sanchez and C. K. O'Sullivan, *Biosens. Bioelectron.*, 2010, **26**, 1500–1506.
49. D. Shangguan, Y. Li, Z. Tang, Z. C. Cao, H. W. Chen, P. Mallikaratchy, K. Sefah, C. J. Yang and W. Tan, *PNAS*, 2006, **103**, 11838–11843.
50. Z. Tang, D. Shangguan, K. Wang, H. Shi, K. Sefah, P. Mallikratchy, H. W. Chen, Y. Li and W. Tan, *Analyt. Chem.*, 2007, **79**, 4900–4907.
51. L. Y. Feng, Y. Chen, J. S. Ren and X. G. Qu, *Biomaterials*, 2011, **32**, 2930–2937.
52. P. Chandra, H.-B. Noh and Y.-B. Shim, *Chem. Commun.*, 2013, **49**, 1900–1902.
53. X. Zhu, J. Yang, M. Liu, Y. Wu, Z. Shen and G. Li, *Anal. Chim. Acta*, 2013, **764**, 59–63.
54. J. Liu, Y. Qin, D. Li, T. Wang, Y. Liu, J. Wang and E. Wang, *Biosens. Bioelectron.*, 2013, **41**, 436–441.
55. C. Hu, D.-P. Yang, Z. Wang, P. Huang, X. Wang, D. Chen, D. Cui, M. Yang and N. Jia, *Biosens. Bioelectron.*, 2013, **41**, 656–662.
56. Y. Wu, P. Xue, Y. Kang and K. M. Hui, *Analyt. Chem.*, 2013, **85**, 3166–3173.
57. A. Venkatanarayanan, T. E. Keyes and R. J. Forster, *Analyt. Chem.*, 2013, **85**, 2216–2222.
58. Y. Zhu, P. Chandra and Y.-B. Shim, *Analyt. Chem.*, 2013, **85**, 1058–1064.
59. I. Ivanov, J. Stojcic, A. Stanimirovic, E. Sargent, R. K. Nam and S. O. Kelley, *Analyt. Chem.*, 2013, **85**, 398–403.

60. G. T. Wilson and F. A. Schaller, *J. Am. Osteopath. Assoc.*, 2008, **108**, 344–349.
61. U. Friess and M. Stark, *Anal. Bioanal. Chem.*, 2009, **393**, 1453–1462.
62. A. Qureshi, Y. Gurbuz and J. H. Niazi, *Sens. Actuators, B*, 2012, **171–172**, 62–76.
63. L. Babuin and A. S. Jaffe, *Can. Med. Assoc. J.*, 2005, **173**, 1191–1202.
64. S. Sharma, P. G. Jackson and J. Makan, *J. Clin. Pathol.*, 2004, **57**, 1025–1026.
65. A. Periyakaruppan, R. P. Gandhiraman, M. Meyyappan, and J. E. Koehne, *Analyt. Chem.*, 2013.
66. S. Ko, B. Kim, S.-S. Jo, S. Y. Oh and J.-K. Park, *Biosens. Bioelectron.*, 2007, **23**, 51–59.
67. L.-N. Feng, Z.-P. Bian, J. Peng, F. Jiang, G.-H. Yang, Y.-D. Zhu, D. Yang, L.-P. Jiang and J.-J. Zhu, *Analyt. Chem.*, 2012, **84**, 7810–7815.
68. H. Qi, X. Qiu, D. Xie, C. Ling, Q. Gao, and C. Zhang, *Analyt. Chem.*, 2013, DOI: 10.1021/ac4005259.
69. F. Li, Y. Yu, H. Cui, D. Yang and Z. Bian, *Analyst*, 2013, **138**, 1844–1850.
70. B. V. M Silva, I. T. Cavalcanti, A. B. Mattos, P. Moura, M. D. P. T. Sotomayor and R. F. Dutra, *Biosens. Bioelectron.*, 2010, **26**, 1062–1067.
71. B. E.-F. de Ávila, V. Escamilla-Gómez, S. Campuzano, M. Pedrero and J. M. Pingarrón, *Electroanal.*, 2013, **25**, 51–58.
72. K. Lewandrowski, A. Chen and J. Januzzi, *Am. J. Clin. Pathol.*, 2002, **118**, S93.
73. K. M. Eggers, J. Oldgren, A. Nordenskjöld and B. Lindahl, *Am. Heart J.*, 2004, **148**, 574–581.
74. E. Suprun, T. Bulko, A. Lisitsa, O. Gnedenko, A. Ivanov, V. Shumyantseva and A. Archakov, *Biosens. Bioelectron.*, 2010, **25**, 1694–1698.
75. I. Lee, X. Luo, X. T. Cui and M. Yun, *Biosens. Bioelectron.*, 2011, **26**, 3297–3302.
76. M. J. Stone, M. R. Waterman, D. Harimoto, G. Murray, N. Willson, M. R. Platt, G. Blomqvist and J. T. Willerson, *Br. Heart J.*, 1977, **39**, 375–380.
77. P. M. Ridker, *Circulation*, 2003, **107**, 363–369.
78. X. Chen, Y. Wang, J. Zhou, W. Yan, X. Li and J.-J. Zhu, *Analyt. Chem.*, 2008, **80**, 2133–2140.
79. K. A. Sepkowitz, *N. Engl. J. Med.*, 2001, **344**, 1764–1772.
80. L. Zheng, L. Jia, B. Li, B. Situ, Q. Liu, Q. Wang and N. Gan, *Molecules*, 2012, **17**, 5988–6000.
81. K. Teeparuksapun, M. Hedström, E. Y. Wong, S. Tang, I. K. Hewlett and B. Mattiasson, *Analyt. Chem.*, 2010, **82**, 8406–8411.
82. K. Kerman and H.-B. Kraatz, *Analyst*, 2009, **134**, 2400–2404.
83. K. A. Mahmoud and J. H. T. Luong, *Analyt. Chem.*, 2008, **80**, 7056–7062.

84. K. A. Mahmoud, S. Hrapovic and J. H. T. Luong, *ACS Nano*, 2008, **2**, 1051–1057.
85. C. Esseghaier, A. Ng and M. Zourob, *Biosens. Bioelectron.*, 2013, **41**, 335–341.
86. S.-Y. Niu, S.-S. Zhang, L. Wang and X.-M. Li, *J. Electroanal. Chem.*, 2006, **597**, 111–118.
87. X. Chen, C.-Y. Hong, Y.-H. Lin, J.-H. Chen, G.-N. Chen and H.-H. Yang, *Analyt. Chem.*, 2012, **84**, 8277–8283.
88. D. Zhang, Y. Peng, H. Qi, Q. Gao and C. Zhang, *Biosens. Bioelectron.*, 2010, **25**, 1088–1094.
89. A. Patterson, F. Caprio, A. Vallée-Bélisle, D. Moscone, K. W. Plaxco, G. Palleschi and F. Ricci, *Analyt. Chem.*, 2010, **82**, 9109–9115.
90. D. Tang, H. Li and J. Liao, *Microfluid. Nanofluid.*, 2009, **6**, 403–409.
91. C. Ding, H. Li, K. Hu and J.-M. Lin, *Talanta*, 2010, **80**, 1385–1391.
92. H. Hanaee, H. Ghourchian and A.-A. Ziaee, *Anal. Biochem.*, 2007, **370**, 195–200.
93. H. Liu, Y. Yang, P. Chen and Z. Zhong, *Biochem. Eng. J.*, 2009, **45**, 107–112.
94. W. M. Hassen, C. Chaix, A. Abdelghani, F. Bessueille, D. Leonard and N. Jaffrezic-Renault, *Sens. Actuators, B*, 2008, **134**, 755–760.
95. Z.-X. Zou, J. Wang, H. Wang, Y.-Q. Li and Y. Lin, *Talanta*, 2012, **94**, 58–64.
96. J. Pál, L. Pálinkás, Z. Nyárády, T. Czömpöly, I. Marczinovits, G. Lustyik, Y. Saleh Ali, G. Berencsi, R. Chen, R. Varró, A. Pár and P. Németh, *J. Immunol. Methods*, 2005, **306**, 183–192.
97. D. Tang, R. Yuan, Y. Chai, Y. Fu, J. Dai, Y. Liu and X. Zhong, *Biosens. Bioelectron.*, 2005, **21**, 539–548.
98. T. Konry, B. Hadad, Y. Shemer-Avni, S. Cosnier and R. S. Marks, *Talanta*, 2008, **75**, 564–571.
99. C. Ding, F. Zhao, M. Zhang and S. Zhang, *Bioelectrochem.*, 2008, **72**, 28–33.
100. N. Liu, G.-J. Li, S.-F. Liu and S.-S. Zhang, *Sens. Actuators, B*, 2008, **133**, 582–587.
101. A. Erdem, D. Ozkan Ariksoysal, H. Karadeniz, P. Kara, A. Sengonul, A. A. Sayiner and M. Ozsoz, *Electrochem. Commun.*, 2005, **7**, 815–820.
102. A. Erdem, K. Kerman, B. Meric, U. S. Akarca and M. Ozsoz, *Electroanal.*, 1999, **11**, 586–587.
103. A. Erdem, K. Kerman, B. Meric, U. S. Akarca and M. Ozsoz, *Anal. Chim. Acta*, 2000, **422**, 139–149.
104. H.-X. Ju, Y.-K. Ye, J.-H. Zhao and Y.-L. Zhu, *Anal. Biochem.*, 2003, **313**, 255–261.
105. X. Li, Y. Chen and X. Huang, *J. Inorg. Biochem.*, 2007, **101**, 918–924.
106. J. Tang, B. Su, D. Tang and G. Chen, *Biosens. Bioelectron.*, 2010, **25**, 2657–2662.
107. A. Erdem, M. Muti, H. Karadeniz, G. Congur and E. Canavar, *Colloids Surf., B*, 2012, **95**, 222–228.

108. European Association for the Study of the Liver, *J. Hepatol.*, 2011, **55**, 245–264.

109. D. Lavanchy, *Clin. Microbiol. Infect.*, 2011, **17**, 107–115.

110. M. G. Ghany, D. B. Strader, D. L. Thomas and L. B. Seeff, *Hepatology*, 2009, **49**, 1335–1374.

111. S. Chevaliez and J.-M. Pawlotsky, *Best Pract. Res., Clin. Gastroenterol.*, 2008, **22**, 1031–1048.

112. S. R. Lee, G. D. Yearwood, G. B. Guillon, L. A. Kurtz, M. Fischl, T. Friel, C. A. Berne and K. W. Kardos, *J. Clin. Virol.*, 2010, **48**, 15–17.

113. M. S. Hejazi, M. H. Pournaghi-Azar and F. Ahour, *Anal. Biochem.*, 2010, **399**, 118–124.

114. M. H. Pournaghi-Azar, F. Ahour and M. S. Hejazi, *Anal. Bioanal. Chem.*, 2010, **397**, 3581–3587.

115. C. dos, S. Riccardi, K. Dahmouche, C. V. Santilli, P. I. da Costa and H. Yamanaka, *Talanta*, 2006, **70**, 637–643.

116. S. Kintzios, E. Pistola, J. Konstas, F. Bem, T. Matakiadis, N. Alexandropoulos, I. Biselis and R. Levin, *Biosens. Bioelectron.*, 2001, **16**, 467–480.

117. A. Perdikaris, N. Alexandropoulos and S. Kintzios, *Sensors*, 2009, **9**, 2176–2186.

118. K.-Y. Ahn, K. Kwon, J. Huh, G. T. Kim, E. B. Lee, D. Park and J. Lee, *Biosens. Bioelectron.*, 2011, **28**, 378–385.

119. K. Conrad, D. Roggenbuck, D. Reinhold and T. Dörner, *Autoimmun. Rev.*, 2010, **9**, 431–435.

120. D. Aletaha, T. Neogi, A. J. Silman, J. Funovits, D. T. Felson, C. O. Bingham, N. S. Birnbaum, G. R. Burmester, V. P. Bykerk and M. D. Cohen, *Arthritis Rheum.*, 2010, **62**, 2569–2581.

121. G. Pruijn, A. Wiik and W. van Venrooij, *Arthritis Res. Ther.*, 2010, **12**, 203.

122. A. S. Wiik, W. J. van Venrooij and G. J. M. Pruijn, *Autoimmun Rev.*, 2010, **10**, 90–93.

123. B. Cai, L. Wang, J. Liu and W. Feng, *Clin. Biochem.*, 2011, **44**, 989–993.

124. K. Somers, P. Geusens, D. Elewaut, F. De Keyser, J.-L. Rummens, M. Coenen, M. Blom, P. Stinissen and V. Somers, *J. Autoimmun.*, 2011, **36**, 33–46.

125. L. A. Trouw and M. Mahler, *Autoimmun. Rev.*, 2012, **12**, 318–322.

126. E. F. Morand, R. Bucala and M. Leech, *Arthritis Rheum.*, 2003, **48**, 291–299.

127. A. M. C. Lokate, J. B. Beusink, G. A. J. Besselink, G. J. M. Pruijn and R. B. M. Schasfoort, *J. Am. Chem. Soc.*, 2007, **129**, 14013–14018.

128. K. A. Drouvalakis, S. Bangsaruntip, W. Hueber, L. G. Kozar, P. J. Utz and H. Dai, *Biosens. Bioelectron.*, 2008, **23**, 1413–1421.

129. M. de Gracia Villa, C. Jiménez-Jorquera, I. Haro, M. J. Gomara, R. Sanmartí, C. Fernández-Sánchez and E. Mendoza, *Biosens. Bioelectron.*, 2011, **27**, 113–118.

130. S. Li, R. Zhang, P. Li, W. Yi, Z. Zhang, S. Chen, S. Su, L. Zhao and C. Hu, *Int. Immunopharmacol.*, 2008, **8**, 859–865.

131. C. Briani, D. Samaroo and A. Alaedini, *Autoimmun. Rev.*, 2008, **7**, 644–650.

132. N. K. Gatselis, K. Zachou, G. L. Norman, G. Tzellas, M. Speletas, S. Gabeta, A. Germenis, G. K. Koukoulis and G. N. Dalekos, *Clin. Chim. Acta*, 2012, **413**, 1683–1688.

133. P. Vermeersch, K. Geboes, G. Mariën, I. Hoffman, M. Hiele and X. Bossuyt, *Clin. Chim. Acta*, 2012, **413**, 1761–1767.

134. T. Balkenhohl and F. Lisdat, *Anal. Chim. Acta*, 2007, **597**, 50–57.

135. S. Dulay, P. Lozano-Sánchez, E. Iwuoha, I. Katakis and C. K. O'Sullivan, *Biosens. Bioelectron.*, 2011, **26**, 3852–3856.

136. M. M. P. S. Neves, M. B. González-García, C. Delerue-Matos, and A. Costa-García, *Sens. Actuators, B*, 2013, DOI: 10.1016/j.snb.2012.09.019.

137. M. Ortiz, A. Fragoso and C. K. O'Sullivan, *Analyt. Chem.*, 2011, **83**, 2931–2938.

138. L. C. Rosales-Rivera, J. L. Acero-Sánchez, P. Lozano-Sánchez, I. Katakis and C. K. O'Sullivan, *Biosens. Bioelectron.*, 2011, **26**, 4471–4476.

139. L. C. Rosales-Rivera, J. L. Acero-Sánchez, P. Lozano-Sánchez, I. Katakis, and C. K. O'Sullivan, *Bios. Bioelectron.*, 2012, **33** , 134–138.

140. M. Pividori, A. Lermo, A. Bonanni, S. Alegret and M. del Valle, *Anal. Biochem.*, 2009, **388**, 229–234.

141. M. M. P. S. Neves, M. B. González-García, A. Santos-Silva and A. Costa-García, *Sens. Actuators, B*, 2012, **163**, 253–259.

142. V. A. Heffner and M. H. Gorelick, *Clin. Ped. Emerg. Med.*, 2008, **9**, 233–237.

143. N. S. Sheerin, *Medicine*, 2011, **39**, 384–389.

144. K. E. Mach, P. K. Wong and J. C. Liao, *Trends Pharmacol. Sci.*, 2011, **32**, 330–336.

145. K. E. Mach, C. B. Du, H. Phull, D. A. Haake, M.-C. Shih, E. J. Baron and J. C. Liao, *J. Urol.*, 2009, **182**, 2735–2741.

146. K. E. Mach, R. Mohan, E. J. Baron, M.-C. Shih, V. Gau, P. K. Wong and J. C. Liao, *J. Urol.*, 2011, **185**, 148–153.

147. Y. Pan, G. A. Sonn, M. L. Y. Sin, K. E. Mach, M.-C. Shih, V. Gau, P. K. Wong and J. C. Liao, *Biosens. Bioelectron.*, 2010, **26**, 649–654.

148. C.-P. Sun, J. C. Liao, Y.-H. Zhang, V. Gau, M. Mastali, J. T. Babbitt, W. S. Grundfest, B. M. Churchill, E. R. B. McCabe and D. A. Haake, *Mol. Genet. Metab.*, 2005, **84**, 90–99.

149. B. Lam, Z. Fang, E. H. Sargent and S. O. Kelley, *Analyt. Chem.*, 2012, **84**, 21–25.

150. R. Mohan, K. E. Mach, M. Bercovici, Y. Pan, L. Dhulipala, P. K. Wong and J. C. Liao, *PLoS ONE*, 2011, **6**, e26846.

151. M. L. Y. Sin, T. Liu, J. D. Pyne, V. Gau, J. C. Liao and P. K. Wong, *Analyt. Chem.*, 2012, **84**, 2702–2707.

CHAPTER 5

In Vivo Sensors for Continuous Monitoring of Blood Gases, Glucose, and Lactate: Biocompatibility Challenges and Potential Solutions

MEGAN C. FROST,*[a] ALEXANDER K. WOLF[b] AND MARK E. MEYERHOFF[b]

[a] Department of Biomedical Engineering, Michigan Technological University, Houghton, MI 49931-1295, USA; [b] Department of Chemistry, The University of Michigan, Ann Arbor, MI 48109-1055, USA
*Email: mcfrost@mtu.edu

5.1 Introduction

Advances in electrochemical and optical sensors for specific ions (H^+, Na^+, K^+, Ca^{+2}, Cl^-, gases (PO_2, PCO_2), and low molecular weight biomolecules (*e.g.*, glucose, lactate, creatinine, urea, *etc.*) over the past 40 years have revolutionized the measurement of such species in the clinical setting, enabling rapid, near patient/point-of-care testing of such critical care analytes in undiluted blood samples.[1,2] However, nearly all existing sensor-based measurement systems require discrete samples of blood. Only a few commercial systems are designed for continuous monitoring capability either *in vivo* (for glucose monitoring in subcutaneous fluid[3,4]) or for short periods in blood

RSC Detection Science Series No. 2
Detection Challenges in Clinical Diagnostics
Edited by Pankaj Vadgama and Serban Peteu
© The Royal Society of Chemistry 2013
Published by the Royal Society of Chemistry, www.rsc.org

within the bypass blood loops used during open-heart surgery (where blood clotting is prevented with high levels of anticoagulant).[5,6] Even the existing implantable sensors for glucose that are commercially available have significant limitations in their accuracy, requiring frequent calibration and confirmation of results *via* separate *in vitro* measurements of blood glucose using conventional glucometer test strips.[7–10]

Although there have been enormous efforts over the past 25 years by many companies to develop catheter-type sensors that could monitor blood gases (oxygen, carbon dioxide and pH) within arteries of critically ill patients using miniaturized electrochemical or optical sensors,[11–14] all of these efforts failed to produce a successful product. The lack of success in developing implantable chemical sensors that can provide reliable analytical results, especially for the blood gas test panel (PO_2, PCO_2, and pH), as well as for glucose and lactate (note: lactate levels are a major indicator of status for intensive-care patients[15–17]) can be attributed primarily to the biological response to the indwelling sensors. In this chapter, we summarize the status of implantable chemical sensor technology for these analytes, describe the biocompatibility problems inherent to both intravascular (placed in the blood stream) and subcutaneous (implanted under the skin) chemical sensors, and discuss new approaches aimed at solving these fundamental problems that might enable greatly enhanced analytical performance for indwelling chemical sensing devices.

5.2 Design of *In Vivo* Sensors

The basic requirements for sensors potentially useful for continuous real-time, *in vivo* measurements are: a) be of very small size (<1 mm o.d.) for easy placement within a blood vessel (*i.e.*, artery for blood gases/pH; vein for glucose/lacate) or under the skin in subcutaneous space (primarily for glucose); b) exhibit long-term analytical signal stability (limiting frequency of recalibration); c) cannot exude toxic chemicals into the patient; and d) provide reliable analyte values that correlate with conventional *in vitro* tests that use discrete blood samples, ideally such that analyte values reported by the implanted sensor can be used in place of discrete blood-test measurements to initiate appropriate medical therapy.

5.2.1 Sensing PO_2/PCO_2/pH in Blood

Miniaturized versions of both electrochemical (potentiometric and amperometric) and optical sensors have been employed to prepare intravascular devices. Figure 5.1 illustrates the overall concept of intravascular blood-gas/pH sensors to monitor patients continuously. Optical sensors have certain advantages over electrochemical devices, including (1) ease of miniaturization using modern optical fibers; (2) potentially less electronic noise (no transduction wires); (3) potential long-term stability using ratiometric measurements at multiple wavelengths; and (4) no need for a separate reference electrode. These advantages promoted the development of optical sensor technology initially for the design of intravascular blood gas sensors using multiple optical fibers.[11] In

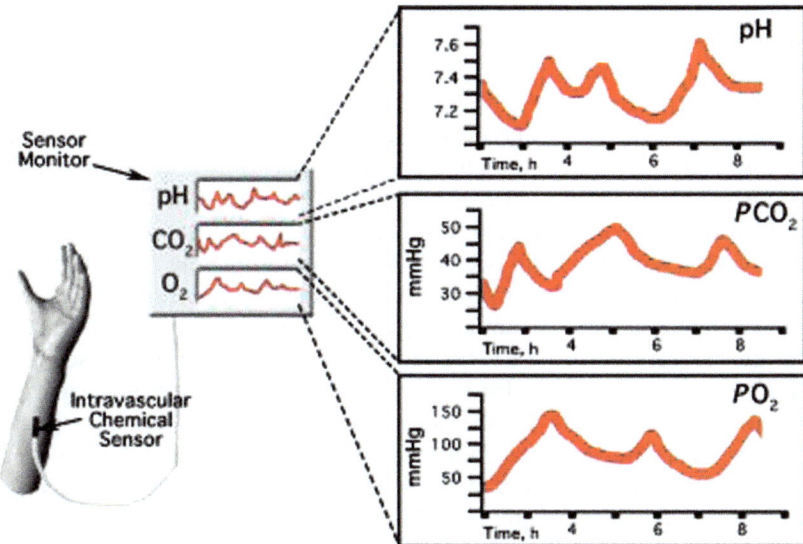

Figure 5.1   Overall concept of *in vivo* blood gas/pH patient monitoring system based on use of intravascular chemical sensors. Sensor inserted in to artery would be able to continuously monitor oxygen, carbon dioxide and pH levels in the atrial blood.

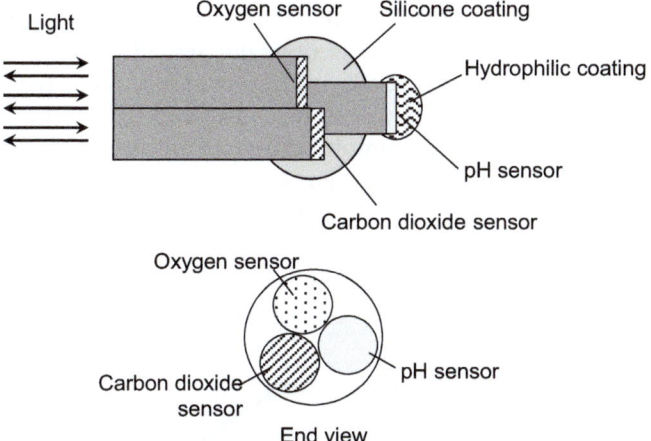

Figure 5.2   Schematic of intravascular optical blood-gas/pH sensing catheter configuration that uses three optical fibers to carry light to and from fluorescent dyes in sensing layers that enables optical detection of pH, PCO_2 and PO_2. This configuration is based on US Patent # 5,047,627 (1991).

such systems, light is passed to and from the sensing site that contains immobilized dyes sensitive to the analyte species by the optical fibers and measures (by reflectance) the fluorescence or phosphorescence originating from the immobilized indicator at the distal end of the fibers (see Figure 5.2 for multifiber design for intravascular blood gas/pH sensing originally patented by Abbott

Laboratories). Optical fiber sensors for PO_2 measurements are typically based on the immobilization of certain organic dyes, such as pyrene, diphenylphenanthrene, phenanthrene, fluoranthene, or metal ligand complexes, such as ruthenium[II] tris[dipyridine], or Pt and Pd metalloporphyrins within hydrophobic polymer films (*e.g.*, silicone rubber) in which oxygen is very soluble.[18,19] The fluorescence or phosphorescence of such species at a given wavelength is often quenched in the presence of paramagnetic species, including molecular oxygen. In the case of embedded fluorescent dyes, the intensity of the emitted fluorescence of such films will decrease in proportion to the PO_2 of the sample in contact with the polymer film. Optical-fiber oxygen sensors have also been developed using a luminophore platinum (II) mesotetrakis(pentafluorophenyl)porphyrin (PtTFPP) complex embedded in different sol-gels.[20,21] The fluorescence of the PtTFPP is quenched in the presence of oxygen.

Optical sensors suitable for the determination of PCO_2 employ optical pH transducers (with immobilized indicators) as inner transducers in an arrangement quite similar to the classic Stow–Severinghaus style electrochemical PCO_2 sensor design. In this design, the pH-sensing layer is behind an outer gas permeable coating (*e.g.*, silicone rubber) that prevents sample proton access to the pH sensing layer (see Figure 5.3).[22] The addition of bicarbonate salt within the pH-sensing hydrogel layer creates the required electrolyte film layer, which varies in pH depending on the partial pressure of PCO_2 in equilibrium with the film. As the partial pressure of PCO_2 in the sample increases, the pH of the bicarbonate layer decreases, and the corresponding decrease in the concentration of the deprotonated form of the indicator (or increase in the concentration of protonated form) is optically sensed. An alternate optical intravascular PCO_2 sensor was developed by Jin, *et al.*[23] using the fluorescence indicator 1-hydroxypyrene-3,6,8-trisulfonate that directly interacts with CO_2.

Optical pH sensors require immobilization of appropriate pH indicators within thin layers of hydrophilic polymers (*e.g.*, hydrogels) because equilibrium access of protons to the indicator is essential. Fluorescein or 8-hydroxy-1,2,6-pyrene trisulfonate (HPTS) have often been used as the indicators and the fluorescence of the protonated or deprotonated form of the dyes can be used for

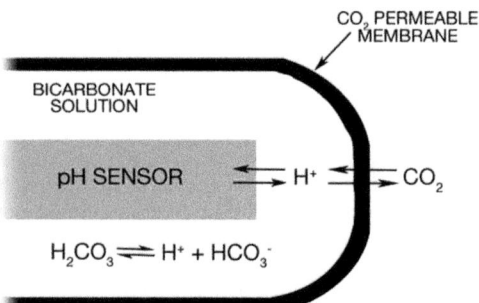

Figure 5.3   Basic *in vivo* PCO_2 sensing configuration that can be employed to detect carbon dioxide levels in blood using inner electrochemical or optical pH sensor.

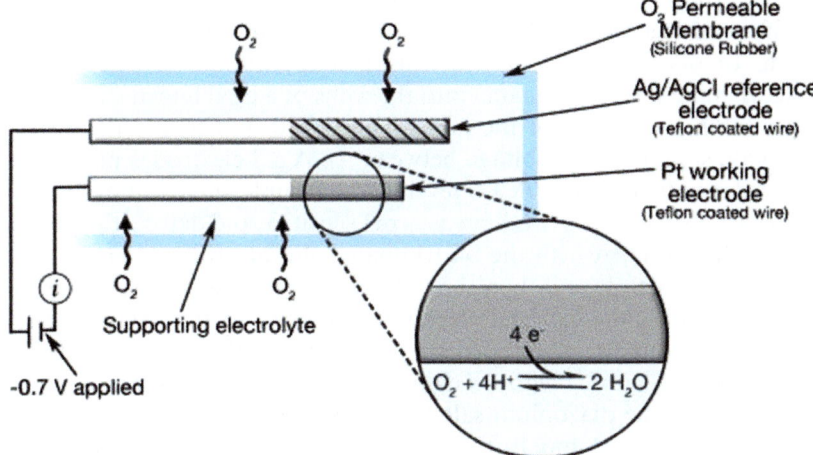

Figure 5.4   General design of implantable electrochemical PO_2 sensing catheter, where oxygen in sample permeates outer tubing wall of the catheter and is reduced electrochemically at working Pt wire electrode. The current flow is directly proportional to the PO_2 level in the blood stream.

sensing purposes.[24] Jin and coworkers report the fabrication of an intra-arterial pH sensor based on fluorescence using N-allyl-4-piperazinyl-1,8-napthalimide and 2-hydroxyethyl methacrylate.[25] The sensors were tested in rabbits and showed appropriate response times ($t_{90} = 13.7 \pm 0.8$ s to 109.61 $\pm$ 4.06 s, depending on the conditions of the pH change) that are quite useful for real-time monitoring. One problem with respect to using immobilized indicators for accurate physiological pH measurements is the effect of ionic strength on the pKa of the indicator.[26] Because optical sensors measure the concentration of protonated or deprotonated dye as an indirect measure of hydrogen ion activity, variations in the ionic strength of the physiological sample have been known to influence the accuracy of the pH measurement.

    While many of the prototypes for implantable blood gas monitoring developed in the 1980s/1990s were based on optical sensors that could be easily inserted in the radial artery of patients, miniaturized electrochemical sensors have also been described for *in vivo* sensing of blood gases.[11–13] Electrochemical oxygen sensing catheters are based on the Clark-style oxygen sensing configuration (Figure 5.4), in which a platinum wire along with a Ag/AgCl or other reference electrode are placed within a narrow gas permeable tubing filled with an electrolyte solution.[11] Oxygen within the blood stream diffuses through the walls of the tubing and is reduced electrochemically at the Pt working electrode *via* application of a -0.6 to -0.7 V *vs.* the Ag/AgCl reference electrode. The resulting current in the circuit is proportional to the PO_2 levels in the blood.

    Electrochemical PCO_2 sensors suitable for intravascular sensing are typically prepared with miniature pH sensors, ideally nonglass potentiometric devices, made with metal oxides[27] or ionophore-based polymer membrane devices[28,29] that contain modified filling solutions. Carbon dioxide diffusion through the walls of a catheter tubing changes the pH within an inner bicarbonate filling solution. In one proposed *in vivo* sensor configuration, the pH ionophore tridodecylamine was impregnated into the walls of a dual lumen silicone rubber catheter tubing, with one of the lumens filled with buffer and the other with bicarbonate solution. The voltage between Ag/AgCl electrodes placed in each lumen is dependent on the PCO_2 level in the sample surrounding the tubing, since the wall between the two lumens is pH sensitive and detects the pH change within the lumen filled with the bicarbonate solution. If an external reference electrode is also employed, then the voltage between the Ag/AgCl wire in the buffered lumen *vs.* the external reference is related to the pH of the sample phase. Other recent work focused on electrochemical pH sensing includes a report by Makos, *et al.*[30] who developed a modified carbon-fiber electrode by the reduction of the diazonium salt Fast Blue RR on the fiber. The resultant sensor was capable of monitoring pH changes as small as 0.005 pH units. Huang and coworkers[31] report the fabrication of a flexible pH sensor utilizing a polymer substrate with an iridium oxide sensing layer, a sol-gel matrix, and an electroplated AgCl reference electrode. The ability to conform to a variety of shapes offers a wide range of potential applications.

5.2.2 Sensing Glucose in Subcutaneous Tissue

The vast majority of sensors that can be implanted subcutaneously for continuous glucose monitoring (CGM) and lactate monitoring are electro-chemical devices.[7,32–40] For glucose sensors, the enzyme glucose oxidase is most often employed as an immobilized catalyst on the surface of a suitable electro-chemical transducer. Some configurations use amperometric oxygen sensors with glucose oxidase immobilized as layer on the outer surface. Glucose present in the bathing solution at the sensor reacts with oxygen *via* the enzyme, lowering the partial pressure of oxygen immediately adjacent to the outer gas-permeable membrane of the oxygen sensor. The decrease in PO_2 and concomitant current decrease is proportional to the level of glucose in the surrounding blood or tissue. However, typically, a second nonenzyme covered PO_2 sensor is required to monitor the level of PO_2 present in the sample phase. Similar configurations based on optical PO_2 sensors have also been reported.[41]

More often, implantable glucose sensors are fabricated using electrochemical oxidation of hydrogen peroxide (H_2O_2), produced by the immobilized glucose oxidase catalyzed reaction of glucose and oxygen. This is normally accomplished using a small surface area of platinum wire as the anode and an applied voltage of +0.6 to +0.7 V *vs.* an external Ag/AgCl electrode. Other physiological species (*e.g.*, ascorbic acid, uric acid, *etc.*) can be oxidized at this potential, so additional layers of membranes/coatings that decrease access of these species to the platinum surface are required. Further, the outermost layer of such sensors requires a polymeric coating (illustrated in Figure 5.5) that restricts glucose diffusion[42] so that the glucose concentrations within the underlying immobilized enzyme layer are always less than the lowest levels of

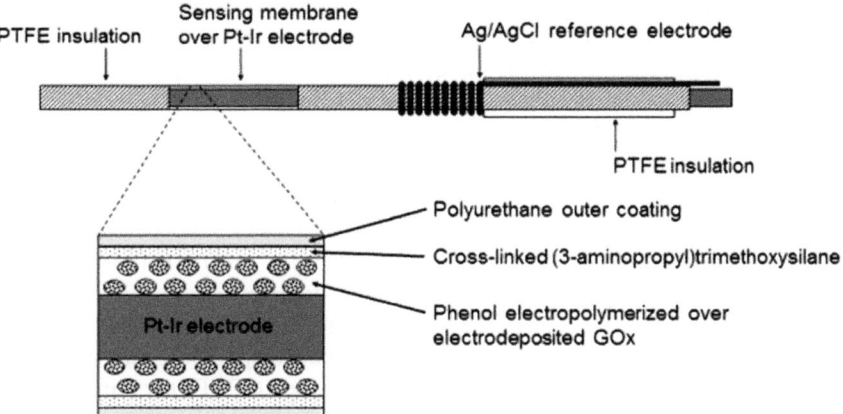

Figure 5.5   One design of subcutaneous or intravascular glucose sensors (diameter < 0.6 mm), with glucose oxidase (GOx) immobilized over surface of Pt/Ir wire and a coiled Ag/AgCl reference electrode. The outer polyurethane layer decreases the rate of glucose access to enzyme layer, while allowing rapid diffusion of oxygen, that enables the sensor to exhibit a linear response to high levels of glucose (to 20 mM). The Pt/Ir electrode is poised at +0.65 V *vs.* Ag/AgCl to oxidize hydrogen peroxide generated from the enzymatic reaction.
With permission from reference 11.

oxygen that might be present in the *in vivo* sample phase. This requirement greatly decreases the sensitivity of the devices to variations in localized oxygen levels and also enables the sensors to respond linearly to abnormally high levels of glucose (up to 20 mM) that might be present in blood or tissue of patents.

Another approach taken to prepare *in vivo* glucose sensors using immobilized redox hydrogels in which tethered mediator sites (*e.g.*, ferrocene or Os(III) redox species) are able to accept electrons from the reduced form of the immobilized glucose oxidase enzyme after reaction with glucose.[43] These reduced mediators are then oxidized and hence the steady-state current is directly proportional to glucose concentration. The advantage of the mediator-based devices is that they partially reduce the need for outer membranes that carefully control the flux of glucose relative to oxygen. Further, the applied potential to the inner conductive electrode material can be less positive to reoxidize the mediator (compared to hydrogen peroxide), reducing the need for exclusion layers to block potential oxidizable species present in the *in vivo* milieu.

There have been a few optical-based glucose sensors reported with small size requirements required for implantation.[44] These are usually fluorescent sensors that utilize glucose-binding proteins and fluorophore-labeled saccharides in a competitive binding mode behind membranes that are permeable to glucose.

Recent developments in electrochemical glucose sensors have branched in several directions from the second-generation sensors described above. Some continue to utilize the glucose oxidase/hydrogen peroxide enzyme electrodes, but they feature novel outer layers to limit glucose diffusion to extend the linear range and to extend the stability and lifetime of the sensors. Others have worked to eliminate electron-transfer mediators through the use of carbon nanotubes. Amperometric glucose sensors developed by Tipnis *et al.*[45] follow the standard design of immobilized glucose oxidase to generate hydrogen peroxide for oxidation. Glucose concentration is still determined from the measured steady-state current, but these sensors incorporate a layer-by-layer (LBL) assembled film as an outer coat to both limit glucose diffusion and minimize calcification, which can clog sensor pores or crack the outer membrane. LBL coatings of humic acids (HAs) and ferric chloride ($FeCl_3$) of various thicknesses were compared to optimize the number of layers to tune glucose diffusion from the bulk to the sensor surface.[45] Lin *et al.* constructed sensors that utilized low density carbon nanotubes (CNT) for direct electron transfer between the enzyme and the electrode surface so that sensors neither required an electron mediator nor an outer glucose limiting layer.[46,47] These sensors had a large linear range (2–30 mM). Even more promising are sensors fabricated with carbon nanotube fibers, spun into nanoyarns.[48] The fibers are unwound at the tip to create a "brush" onto which glucose oxidase is immobilized and an outer polyurethane outer membrane is applied. These sensors exhibit limits of detection as low as 25 µM, although such low detection levels are not required for blood glucose measurements. Using nickel or copper nanoparticles to make a CNT array also lowered the limit of detection.[48]

Recent glucose sensors have also once again utilized optical detection to take advantage of technological developments in optical materials and the high degree of reversibility characteristic of optical sensors. Tierney *et al.* fabricated glucose sensors that function by covalent binding of an acrylamide-based hydrogel to the tip of an optical fiber.[49] The local glucose concentration swelled and contracted the phenylboronic acid within the hydrogel, changing the optical length, which was measured by the optical fiber. What especially sets these optical sensors apart is their elimination of the need for a fluorescent molecule or tag. The sensors had high reversibility but could have interferences from glycoproteins, polysaccharides, elevated blood levels of lactate, and the common anticoagulants heparin and citrate. Sensor function was also significantly temperature dependent, as the difference between room and body temperature significantly affected the equilibrium swelling degree. Glucose sensors developed by Paek *et al.* utilize a reversible glucose binder, rather than a catalytic enzyme reaction that consumes local glucose and optical detection, that would not suffer signal noise electromagnetic fields, and could use a detector that was spatially separated from the sensor.[50] The sensor functioned by competitive binding between glucose and a ligand conjugate for the immobilized glucose-selective protein (concanavalin A), which caused a detectable wavelength shift in the optical sensor. However, in order or preserve sensor lifetime for CGM and recycle the ligand conjugate, the sensing area must be separated from the sample by a semipermeable membrane selective for glucose diffusion. All of these sensors show promise in helping to extend the linear working range of glucose sensors and replace the sensitive glucose oxidase enzyme; however, the

Table 5.1 The current commercially available *in vivo* electrochemical glucose sensors, sensor lifetime, and the suggested frequency of recalibration.

Company	Sensor	Implantation time	Calibration frequency
Abbott	FreeStyle Navigator	5 days	3 times on first day of use (1, 3, 24 h) not re-calibrated until new sensor implanted
Medtronic	Guardian	3 days	3–4 finger pricks/day (recommended), 1 per 12 hours required
	Enlite	6 days	
Dexcom	Seven Plus CGM	7 days	1 per 12 hours

biocompatibility of these sensors to thrombosis and immune response from long-term *in vivo* implantation has yet to be assessed.

Despite tremendous advances in sensor design, however, none of these designs have yet achieved widespread use, and all existing commercial *in vivo* glucose sensors are either based on electrochemical detection of hydrogen peroxide or redox-mediator species. Indeed, Table 5.1 lists the current commercially available electrochemical glucose sensors, the lifetime of the sensor *in vivo*, and the suggested frequency of recalibration. The only FDA approved devices (by Medtronic-Minimed, Abbott Inc., Dexcom Inc.) require recalibration *in vivo* several hours after implantation (*via* repeated *in vitro* blood tests by the diabetic patient).[7–10] Even with such measures, the outputs of the devices are still not reliable enough for the patient to exclusively use values for monitoring glucose levels to decide on insulin doses.

5.2.3 Sensing Lactate in Blood

As mentioned above, continuous lactate measurements would be most useful in intensive-care units within hospitals since elevated lactate levels in blood are a negative predictor of a good outcome for such patients. Indeed, many critical-care doctors believe that trends in blood-lactate levels provide the best indicator of tissue oxygenation in patients, and when lactate levels continue to rise, therapy is failing, and patient survival rate is greatly decreased.[15–17] Hence, an intravenous catheter that could provide reliable real-time lactate concentrations would be highly desirable. Several indwelling lactate sensors have been reported,[51,52] and most are based on a design quite similar to the aforementioned glucose sensors except that lactate oxidase is used as the immobilized enzyme. The enzyme produces hydrogen peroxide, and thus oxidation of this product species at a platinum, or other electrode material, provides the transduction method. Since levels of lactate are generally lower than glucose, controlling fluxes of lactate into the immobilized enzyme, relative to oxygen, is not quite as stringent, although still important.

Recent developments in lactate sensors have also utilized both electrochemical and optical detection methods. While glucose sensor development focused on applications for *in vivo* continuous monitoring for diabetic patients and patients under tight glycemic control, several recent lactate sensors have been designed for *in vitro* measurements of cell cultures, as increased lactate production has been identified as a cancer marker. A potentiometric sensor based on zinc-oxide nanorods was developed recently by Ibupoto *et al.* The ZnO nanorods were grown on gold-coated glass and lactate oxidase was subsequently immobilized on the surface.[53] The sensors showed excellent linear response range, stability for more than three weeks, and high selectivity over potential interfering species such as ascorbic acid, urea, glucose, galactose and Mg^{2+} and Ca^{2+} ions. However, the sensor's response decreased at pH > 8 and with temperatures above room temperature, necessitating a stable operating pH and temperature for the accurate and consistent measurements needed for a continuous monitoring system.

Zheng, *et al.* developed an optical-fiber nanosensor based on lactate dehydrogenase capable of measuring extracellular lactate concentration at the single-cell level. Lactate dehydrogenase immobilized at the tip of an aluminum-coated, heated/pulled optical fiber generates NADH as a fluorescent byproduct, directly proportional to the local lactate concentration. When excited by 360-nm UV light, the NADH fluoresces at 460 nm, allowing extremely sensitive and selective detection. While this promises to be an excellent technique for identifying cancerous cells *in vitro*, no plans have been made to use the sensors for implanted *in vivo* measurements.[54] Another very recent lactate sensor designed by Martín *et al.* utilizes genetically encoded Förster resonance energy transfer (FRET) and fluorescence-based detection to measure local lactate extracellular concentration.[55] The sensor design proved extremely selective and very sensitive (single cell), but was pH sensitive and difficult to fabricate. Future *in vivo* lactate sensors, designed for continuous monitoring of patients to indicate health of critically ill patients may still utilize electrochemical detection, but could also benefit from the methods designed to improve the biocompatibility of implanted *in vivo* glucose measurements.

5.3   Biocompatibility Issues that Influence *In Vivo* Sensor Performance and Reliability

Several research teams actually fabricated prototype blood-gas/pH-sensing catheter products that were evaluated first in animals and then in humans.[13,56–59] Despite claims of success by some investigators, significant problems remain, both for intravascular and subcutaneous devices. A common observation for some sensors during *in vivo* studies is the performance pattern illustrated in Figure 5.6. Specifically, after a given period of time, sensor output signals often begin to deviate significantly from the values obtained for discrete blood samples analyzed *in vitro* on conventional laboratory instruments. The pattern is always the same, with intravascular sensor values for pH and PO_2 always lower than corresponding *in vitro* measurements and PCO_2 levels always higher than the corresponding *in vitro* values. This behavior was initially

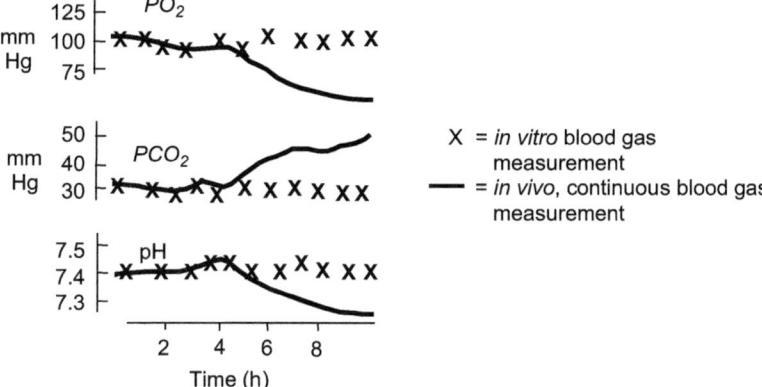

Figure 5.6   Overall concept of *in vivo* blood gas/pH patient monitoring system based on use of intravascular chemical sensors. Sensor inserted in to artery would be able to continuously monitor oxygen, carbon dioxide and pH levels in the atrial blood.

thought to arise exclusively from thrombus formation on the surface of implanted sensors, in which platelets and other entrapped metabolically active cells create a local environment that differs in PO_2, PCO_2 and pH levels compared to the bulk blood (due to cellular respiration). Although this effect will certainly be accentuated by the formation of a thick thrombus layer, even the adhesion of a single layer of metabolically active cells, *e.g.* platelets, could cause a local change in analyte concentrations immediately adjacent to the surface of the device. Similar errors can be observed for glucose measurements, given the ability of cells to metabolize glucose, creating lower values near the surface of cell coated sensors.

5.3.1   Details of Biological Response in Blood

Platelet adhesion and activation on foreign surfaces, such as the outer coatings of intravascular sensors, is primarily governed by the ability of platelets to recognize certain proteins that adsorb on the device surfaces once they are inserted into the bloodstream. Two proteins, von Willebrand's Factor (vWF) and fibrinogen, play key roles in this process.[60] vWF is a circulating protein that readily adsorbs to collagen and foreign materials that do not an have intact endothelial layer. Upon adsorption, vWF will bind with glycoprotein Ib (GPIb) on the surface of platelets to form the GPIb-V-IX complex, which allows platelet adhesion to the device surface.[61] Adherent and activated platelets release a host of factors that promote more platelet adhesion and accelerate the coagulation cascade including prostaglandins (PGH_2 and PGG_2), Ca^{2+}, and thromboxane A_2.[62] Thomboxane A_2 promotes the presentation of GP IIb/IIIa on the surface of platelets that actively binds circulating fibrinogen, linking platelets together and locally converting fibrinogen to insoluble fibrin.[62] Fibrin

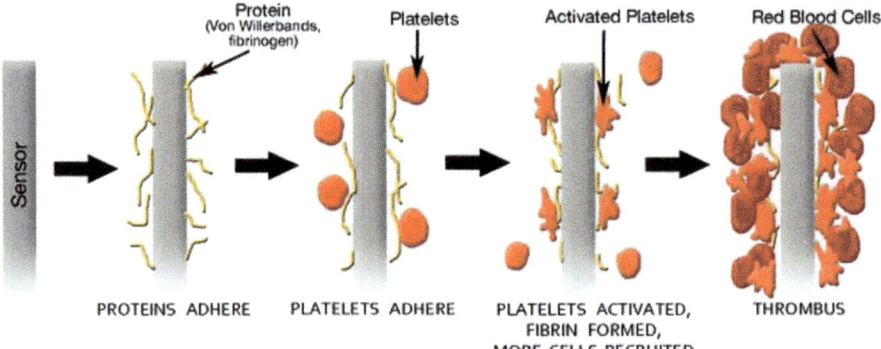

Figure 5.7   Physiological response to the outer surface of a sensor placed into the blood stream, ultimately causing activation of platelets and potentially thrombus formation on all or part of the sensor's surface. The coatings of cells on sensor surface can cause significant errors in the measurement of the target analytes.

entangles blood cells and forms the physical barrier that covers devices placed in blood. In addition to the role of platelets in coagulation, coagulation factor XII and platelet factor 3 (which is released by adherent platelets) will bind directly to the foreign surface and also convert prothombin to thrombin, which will further promote the conversion of fibrinogen to fibrin (see Figure 5.7).[62] As mentioned above, even if significant blood clots are not formed, the presence of just a single layer of active platelets and other potential cells can alter the analyte concentrations near the surface of the implanted sensor, yielding inaccurate values compared to the concentration measured in the bulk blood phase. Within the realm of designing intravascular sensors, it is critical to be aware that the dynamic nature of this surface interaction is a significant challenge to obtaining accurate measurements of physiologically important analytes.

5.3.2   Details of Biological Response in Subcutaneous Tissue

Sensors developed for real-time subcutaneous measurements, most notably electrochemical glucose sensors,[12–21] do not suffer from accuracy issues related to platelet activation and thrombus formation. However, such sensors are subject to their own unique biocompatibility problems that can influence the analytical performance of the implanted devices.[63,64] Specifically, a significant tissue inflammatory response to a large "foreign body" (*i.e.*, the sensor) can lead to encapsulation of the devices within a sheath of leukocytes, macrophages, and fibroblasts (see Figure 5.8). This sheath can alter local analyte levels (*e.g.*, glucose) *via* metabolic reactions (*e.g.*, inflamed tissues consume glucose at an accelerated rate[12,32,65,66]) and also greatly perturb mass transfer of the analyte to the sensor surface as well as between the blood and the interstitial

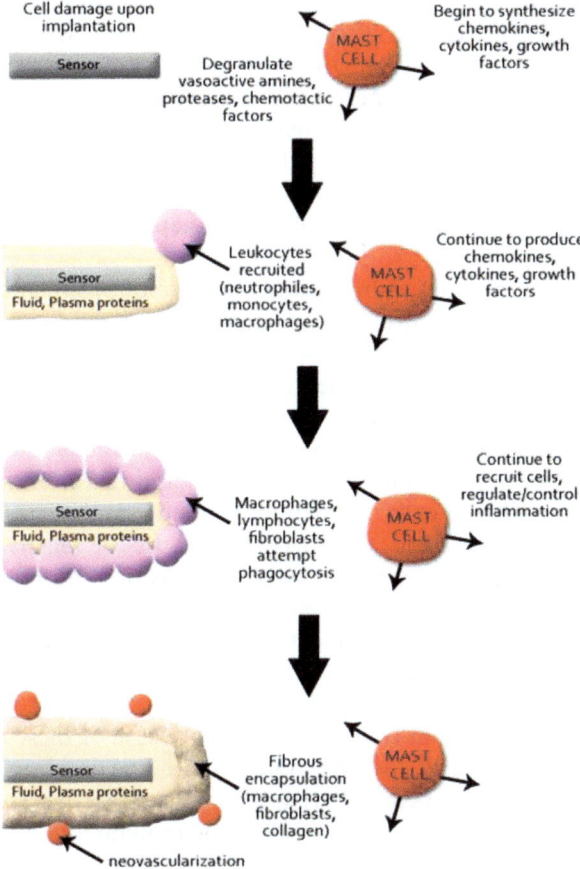

Figure 5.8 Schematic of inflammatory response process that takes place when a sensor, such as a glucose sensor, is placed subcutaneously. The increase in inflammatory cells in the area and the ultimate encapsulation of the device in a fibrous sheath creates a situation where levels of the analyte near the sensor surface do not reflect the concentrations in the blood.

fluid region. Such changes can effectively alter the calibration curve (*e.g.*, change sensitivity) for the device and can also greatly affect the lag time of the sensors response to varying glucose levels in the bulk blood.[12,32,67]

When foreign materials are placed in subcutaneous tissue, the inflammatory response poses significant challenges to stable sensor performance. Upon implantation of a sensor, an acute response occurs that lasts from minutes up to days.[68] As is the case with intravascular sensors, nearly instantly, proteins adsorb to the surface of the device. Due to the need to pass through vascularized tissue to place the sensors in subcutaneous tissue, both plasma and tissue proteins such as albumin, fibrinogen, complement factors, globulins, fibronectin, and vitronectin adsorb to the device.[68] Mast cells (MCs) present in subcutaneous tissue play a central role in regulating inflammation.[69,70] Upon injury, MCs will release a series of stored proinflammatory agents including vasoactive amine, chemotactic factors and proteases. As a result of sustained stimulation, they will then synthesize and continue to release pro-inflammatory mediators that will recruit macrophages and lymphocytes. Additionally, they release vascular endothelial cell growth factor (VEGF) that is important for angiogenesis at the site of injury. MCs will also release cytokines that recruit and activate fibroblasts that will eventually participate in fibrous encapsulation.[69]

Following initial protein adsorption and release of proinflammatory agents, neutrophils are recruited to the site of implantation and begin attempting to phagocytose and degrade the foreign device (in this case, the sensor).[71] This involves the release lysosomal proteases, reactive oxygen species, and chemokines that serve as potent signals to recruit cells to the wound site. Neutrophils will persist at the wound for 1–2 days before they are replaced by monocytes/macrophages (monocytes excite the vasculature and mature into macrophages in tissue near the site of sensor implantation) and lymphocytes.[71] The macrophages become activated and release IL-12 and IL-23, reactive nitrogen and oxygen species (such as nitric oxide (NO), superoxides and H_2O_2), and proinflammatory cytokines.[72] Activated macrophages also continue to release chemoattractants to recruit more monocytes/macrophages. It has been shown that there is heterogeneous activation of macrophages, dependent on many factors that include the nature of the protein layer present and environmental signals that recruit monocytes from circulation.[73] The release of these active agents is intended to kill bacteria at the implantation site and biodegrade the foreign material that is in the subcutaneous tissue. These reactive species will additionally damage healthy tissue surrounding the implant and can lead to a state of chronic inflammation that ultimately leads to fibrous encapsulation of the sensor.

As the macrophages adhere to the protein-coated sensor, they undergo a cytoskeletal rearrangement in order to spread over the surface of the device. Adherent macrophages then fuse to form foreign-body giant cells (FBGCs) that further attempt to biodegrade the implanted device. After approximately 1–2 weeks, chronic inflammation leads to the deposition of a collagen matrix by fibroblasts to encapsulate the implant and isolate it from native tissue.[74] The collagen layer thickness is partially dependent on factors such as the implant texture, the protein layer adhered to the implant, and cells adhered to the protein layer. Additionally, movement of the sensor over time can change the thickness and level of inflammation around the implant. It is important to recognize that the material properties, topography, and placement *in vivo* affect the biological response to the implant, but the biological response itself will further influence long-term tissue response.[74] Again, this is a complicated and dynamic response that changes over time. This continuous change creates a highly challenging analytical environment in which reliable measurements must be made in order to make the use of implanted subcutaneous sensors for *in vivo* monitoring a clinical reality.

5.4   Strategies for Mediating Biological Response to Implanted Sensors

The commercial development of implantable chemical sensors for routine blood monitoring in humans has, at this point in time, reached an impasse of sorts. With only three commercial subcutaneous glucose sensors approved for limited use and no commercially available intravascular sensors, it appears that the combination of problems related to cell adhesion, thrombus formation, inflammatory response, *etc.* yielding poor *in vivo* sensor performance is a formidable challenge to overcome. Key to ultimate success in this field will be developing coatings that can prevent/resist the *in vivo* biocompatibility issues that play a dramatic role in the performance of implanted sensors. Fortunately, there has been significant fundamental research recently to address these issues. In the following section, we discuss how various new coatings may provide a solution to the dynamic biocompatibility problems that have plagued all *in vivo* sensors developed to date.

5.4.1   Materials Development for Improved Biological Response to Implanted Sensors

A great deal of research over the last two decades has focused on designing materials that can be used to coat or fabricate sensors that are able to reduce the *in vivo* biological response that is responsible for loss of sensor function over time. An excellent review by Wisniewski and Reichert[75] details approaches that have been investigated to modify the materials surface properties to achieve this goal. Briefly, this review discusses use of hydrogels such as crosslinked poly(hydroxyethyl methacrylate) (PHEMA) or poly(ethylene glycol) (PEG) that allow water-soluble analytes to diffuse to the sensing element, and the hydrophilic nature of these materials is thought to improve biocompatibility. Phospholipid surface modification has been used in an attempt to mimic phospholipids present on cells in the body, thereby decreasing physiological response toward the sensor. Another approach is to apply thin Nafion (a chemically inert anionic compound that contains both hydrophobic and hydrophilic properties) coatings that show little adsorption of molecules from solution or use surfactants applied to sensor surfaces that have shown reduced protein adsorption. The use of natural materials such as modified cellulose has been shown to reduce compliment activation, covalent attachment of anti-fouling agents such as phenylenediamine, diols, and glycols seem to enhance biocompatibility. Additionally, using diamond-like carbons and controlling the topography of sensor coats also enhances biocompatibility and stability over time.

The development of biomimetic hydrogels for sensor coatings involves PHEMA-based membranes with PEG incorporated and crosslinked with tetraethyleneglycol diacrylate. Such a material has been reported by Justin and coworkers.[76] This UV-cured material was applied to microdisc electrode arrays, and it was shown that the diffusion coefficient of analyte was decreased, and pH response of the material from 6.1 to 8.8 caused a change in the impedance signal obtained. This material is expected to improve biocompatibility of the coated sensor, but tuning of the porosity and crosslinking density is needed to preserve the sensor response. Silica based sol-gels (SG) have also been developed[77] that include the incorporation of either PEG (SG-PEG), heparin (SG-HEP), dextran sulfate (SG-DS), Nafion (SG-NAF), or polystyrene sulfonate (SG-PSS). The composite materials have been tested for toxicity *in vitro* with fibroblast cultures. The effect on cell proliferation was dependent on which additive was included. SG-DS showed the most promising results as glucose sensor coating when tested in BSA and serum, slowing the growth of fibroblasts on the surface of the hybrid material, while also allowing the best glucose response when coated on an amperometric sensor.

Work focused on controlling topography and stability of sensor coatings by Ju and coworkers[78] includes the development of a 3-dimensional porous type-I collagen scaffold crosslinked by nordihydroguaiaretic acid (NDGA) or glutaraldehyde (GA) that was applied to subcutaneous glucose sensors that were implanted for up to 28 days. The NDGA crosslinked collagen coatings showed a decrease in inflammation compared to bare implanted sensors. Wang and coworkers[79] developed electrospun fibroporous polyurethane coatings for implantable glucose sensors. The authors were able to control the dimensions of the electrospun fibers and their density on the sensors to secure certain advantages including a stronger mechanical stability, a greater ability to elongate to accommodate the swelling enzyme layer on the sensor, a reduction in mechanical weak spots along the length of the coating, fibers that have excellent interconnected porosity, and a fiber structure that mimics natural extracellular matrix structure. Although this material has not been tested *in vivo* yet, the authors point out that *in vitro* sensor performance was not adversely affected by the electrospun coating, and there is huge potential to apply biomimetic electrospun fibers to sensors once functioning of the fiber coating is validated. Ai, *et al.*[80] developed a perfluorocarboxylic acid ionomer (PFCI) that showed reduced cracking compared to the commonly used perfluorosulfonate ionomer membrane (PFSI). The advantage was realized because PFCI possess higher crystallinity and smaller ion clusters compared to PFSI, thereby reducing mineralization *in vivo* and reducing cracking, while maintaining the inert benefits of perfluoro materials.

5.4.2 Active Releasing Materials for Controlling Biological Response

Although the development of static materials that are more cognizant of potential biological response and biocompatibility problems is useful, it is unlikely that changing the chemistry of the material itself will alleviate biofouling and subsequent sensor failure when placed in the harsh, dynamic physiological environment. Work has also been undertaken that relies on more biomimetic approaches such as inclusion of specific surface active agents such as CD47 and drug eluting materials, both targeted at controlling explicit aspects of the inflammatory response. Stachelek, *et al.*[81] developed a material that appended a recombinant CD47 to poly(vinyl chloride) (PVC) or Tecothane, a polyether polyurethane (PU). CD 47 is a transmembrane protein that binds to specific receptors on leukocytes and macrophages, which causes a downregulation of inflammatory cell attachment. The CD47-PU material was tested in a rat subdermal implant model for 70 days and showed significantly reduced number of attached cells associated with both acute and chronic inflammation.

Several researchers have also proposed development of systems that will release the anti-inflammatory agent, dexamethasone (DX), and/or the angiogenic agent, vascular endothelial growth factor (VEGF). Norton *et al.*[82] developed a hydrogel system based on 2-hydroxyethyl methacrylate that released both DX and VEGF. The idea was to decrease inflammation and encourage angiogenesis around an implanted glucose sensor. Decreasing inflammation should decrease the thickness of the fibrous capsule and allow a stable background current to be reached more quickly, and increasing vasculature should help increase glucose flux to the sensor. It was found that DX alone did decrease inflammation after 2 weeks of implantation in rats. VEGF alone did increase vascularity of the fibrous capsule, but when DX and VEGF are released at the same time from the hydrogel, while there is still a decrease in inflammation, there is no increase in vascularization. After six weeks, there is no observable difference between the drug-releasing hydrogels and control implants, indicating that the positive effect is not maintained once the drug-release reservoir is depleted. Sung, *et al.*[83] then investigated a similar approach by sequentially delivering DX and VEGF from a crosslinked PEG. The release of DX is faster ($k_{DX} = 0.5/\text{day}$) and VEGF is slower ($k_{VEGF} = 0.4/\text{day}$). The difference in delivery rate is enough to see a marked decrease in inflammation and increase in vascular density in an *ex ova* chick embryo after 8 days. This is significant because it points to the temporal importance of delivering agents in the appropriate sequence, thus more closely mimicking normal physiological conditions.

Bhardwaj and coworkers[84] tested the temporal effect of DX over much longer term implants by fabricating different poly(lactic-*co*-glycolic acid) (PLGA) microspheres that release DX for 30 days or PLGA/poly vinyl alcohol (PVA) composite microspheres that release DX over a 3-month period. The microspheres were implanted in rats to test for both acute and chronic inflammation. Sustained release of DX was achieved by combining microspheres, and the continued release did show suppression of inflammation for 3 months, while spheres that exhausted their drug after 30 days lost the ability to decrease inflammation. The same group then modified this approach and developed a system that dispersed PLGA microspheres in a crosslinked PVA hydrogel.[85] The motivation for this was that many different drugs in addition to the DX that was tested could be encapsulated in PLGA microspheres to impart a wide array of drug-release capability, and the PVA allows both analyte to diffuse to the sensor surface and drug to diffuse into the tissue immediately surrounding the sensor. The "smart" PLGA microsphere/PVA composite system showed good proof of concept tissue response when tested in both type-1 and type-2 diabetic rats for 1 month. Importantly, this system also recognizes the importance of tailoring the implanted material to the site of implantation and the specific medical condition under which the sensors are being used and holds the potential for incorporating additional biological mediators into the sensor coating.

Another very promising active agent for release from implanted sensors is nitric oxide (NO). Nitric oxide is a free radical-gas that is released endogenously in many physiological processes.[86] Namely, within the context of indwelling sensors, NO is a potent antithrombotic agent[87] and plays a role in mediating the inflammatory response in subcutaneous tissue.[88] NO is an important signaling molecule in shifting inflammation from acute stages to potentially helping in enhancing wound healing. If NO can be released from intravascular or subcutaneous sensors at the appropriate levels and with good temporal control, evidence suggests that NO may radically reduce the biological response to these indwelling sensors.[89] A class of NO donor that has been extensively used for improving the biological response toward implanted sensors is diazeniumdiolates.[87–90] Figure 5.9 shows the structure of some examples of these molecules that have been used in polymeric materials to impart NO release to materials for sensor fabrication. Each diazeniumdiolate molecule will release 2 molecules of NO *via* thermal decomposition at 37 °C or a proton-mediated decomposition mechanism. The rate of decomposition can be controlled by changing the molecular structure of the parent compound used to form the diazeniumdiolate and/or by tuning the water uptake and pH of the polymer matrix in which the molecule is embedded. Initial work dispersed (Z)-1-[N-methyl-N-[6-(N-methylammoniohexyl)amino]]-diazen-1-ium-1,2-diolate (DMHD/N$_2$O$_2$), evenly through a PDMS matrix coated on an amperometric oxygen sensor.[90] This material showed greatly reduced thromobogenicity,

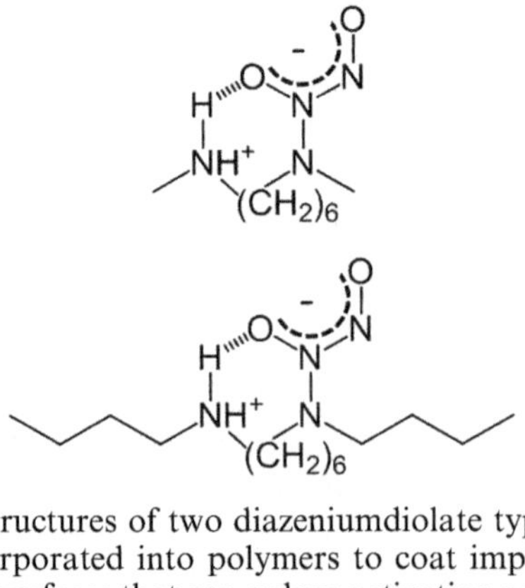

Figure 5.9   Chemical structures of two diazeniumdiolate type NO donor species that can be incorporated into polymers to coat implantable sensors, creating NO release surfaces that can reduce activation and adhesion of platelets in the blood stream and also decrease inflammatory response for subcutaneous sensors. (A) diazeniumdiolated dimethylhexanediamine (B) diazeniumdiolated dibutyhexanediamine.

but it was found that the hydrophilic $DMHD/N_2O_2$ leached out of the polymer matrix, presenting potential problems by allowing the parent diamine compound used to form the diazeniumdiolate (and present after NO release) to circulate in the body. Some diamine compounds are known to be carcinogenic.[91]

Because of the extremely promising results for inhibiting thrombus formation on the surface of the blood contacting device, work began to either covalently link the NO donor to the backbone of the polymer matrix used or to make the donor lipophilic enough that it would not exit the hydrophobic polymer matrix to enter the aqueous environment of blood. To this end, modified PDMS materials were developed that contained pendant secondary amines that could be converted to diazeniumdiolates and result in materials with different NO release properties. Zhang and coworkers[92] developed $DACA/N_2O_2$ and $TACA/N_2O_2$ PDMS that used diamine and triamino-containing crosslinking agents to react with hydroxyl-terminated PDMS chains. Once the crosslinked polymer was cured, it could be converted to the diazeniumdiolate by reacting in 5 atm of NO. Polyurethanes capable of releasing NO were reported by Reynolds, *et al.*[93] They developed two different methods to synthesize polyurethanes, containing pendent amines in different positions along the backbone of the PU that could be diazeniumdiolated, and these polyurethans released NO for up to 6 days in aqueous solution at 37 °C. These PDMSs and PUs are routinely used as membranes for sensors (O_2, CO_2, glucose, lactate, *etc.*) and should be useful for improving the biological response to indwelling sensors.

Another approach to impart NO-release properties to polymers used for sensor fabrication is to develop a more lipophilic analog of $DMHD/N_2O_2$ that replaces the methyl side groups with butyl side groups $(DBHD/N_2O_2)$[94] in an effort to alter the partition coefficient enough to keep the NO donor within the PDMS matrix in which it is dispersed. There are two distinct advantages to blending a discrete NO donor into the polymer matrix being used to fabricate a sensor over modifying materials to covalently attach NO donors. First, it provides the ability to alter the amounts of the NO donor blended into the material, depending on the level and reservoir of NO donor desired for a given application. Secondly, the base polymer matrix can be varied without changing the fundamental synthetic chemistry of the polymer itself. A limitation to this approach, however, is that the NO donor can still diffuse out of the polymer, albeit a greatly reduced risk compared to $DMHD/N_2O_2$. An alternative, which allows similar advantages but reduces leaching, is the development of derivatized particle fillers with NO-releasing surface chemistry, such that these particles could be incorporated into different polymers, thereby imparting NO release capability to the material without changing the fundamental chemistry used to synthesize the matrix. These particles also have a much higher potential loading capacity and release of NO owing to the fact that they have more potential NO donating groups than just single molecules. Zhang and coworkers[95] developed a series of silica particles modified by covalent attachment of alkylamines to their surface that could be converted into diazeniumdiolates and subsequently blended into different polymers. This yielded mechanically entrapped NO donors that were unable to leach from the polymer surface. Frost and Meyerhoff[96] also developed fumed silica particles derivatized with different S-nitrosothiols, another class of NO donor (*e.g.*, S-nitroso-cysteine, S-nitroso-N-acetylcysteine, and S-nitroso-N-acetylpenicillamine) that could impart NO-release ability with different release profiles compared to the diazeniumdiolate-based particle/polymer composites in response to Cu^+, or visible light when blended into plasticized PVC films.

There has also been extensive work to develop NO-releasing sol-gels (*i.e.* xerogels) for potential implantable sensor applications.[97–100] Shin and Schoenfisch[97] note that these are attractive materials because they are synthesized under mild conditions and are chemically flexible and generally porous, facilitating diffusion of analytes through the matrix to the sensor. Marxer and coworkers[98,99] report the synthesis of sol-gels prepared by the hydrolysis and condensation of amine-functionalized alkoxysilanes and alkyltrimethoxysilanes. The crosslinked materials were exposed to 5 atm of NO to form diazeniumdiolate moieties covalently linked to the material. Four different aminosilanes (isobutyltrimethoxysilane, (aminoethylaminomethyl)-phenethyltrimethoxysilane, (aminohexyl)aminopropyltrimethoxy-silane and N-[3-(trimethoxysilyl)propyl]diethylenetriamine) all had different storage and release properties, demonstrating the ability to achieve a wide range of NO-release profiles from this class of material. Shin *et al.*[100] embedded NO-releasing sol-gel particles into a polyurethane matrix to use as an NO-release coating for glucose sensors. Roughly 10–200 µm NO-releasing particles were obtained by grinding cured sol-gel monoliths synthesized by using (amino-hexyl)aminopropyltrimethoxysilane reacted with methyltrimethoxysilane or isobutyltrimethoxysilane. This further demonstrates the utility of these materials and the approach of incorporating NO release particles into a wide variety of polymer matrices to obtain composite materials that are able to release physiologically relevant levels of NO.

These materials that release bioactive agents show immense promise in controlling and mediating biological response around implanted sensors. One limitation all these materials possess, however, is a finite reservoir of active agent that is able to be released from materials preloaded with molecules such as VEGF, DX or NO. Since these materials also release their agents continuously once release is initiated, this limits the lifetime of a device. NO is, consequently, released even when it may not be needed to limit thrombosis, or even when a lower level of release may be adequate, and thus the total reservoir is exhausted sooner than need be. This unnecessary NO release limits the effective lifetime of the sensor.

In an effort to impart dynamic control over the level of NO released, RSNO-derivatized fumed silica particles imbedded in a hydrophobic polymer matrix showed dynamic NO release when illuminated with different intensities of light.[101] The hydrophobic polymer matrix excludes transition metal ions and ascorbate ions such that NO release is only initiated by illumination, thereby providing a dynamic on-off trigger for controlling NO release. Gierke *et al.*[102]

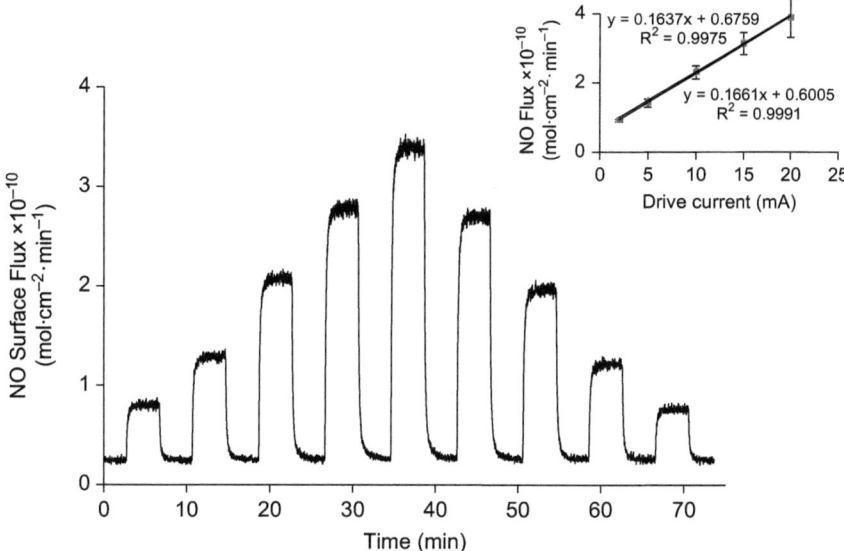

Figure 5.10   NO generated from S-nitroso-N-acetylpenicillamine-polydimethylsilox-ane (SNAP-PDMS) coated on declad region of a 500 μm poly(methyl-methacrylate) (PMMA) optical fiber with drive current turned on and off and increasing with each step. The pulses used in this example were in 4 min intervals, light off and then light on with increasing light applied in each subsequent step from 2, 5, 10, 15 and 20 mA of I_{drive} applied and then 15, 10, 5, 2 and 0 mA applied to the LED *via* the ED. The corresponding surface fluxes of NO generated was 0.95 ±0.04, 1.50 ±0.13, 2.37 ±0.33, 3.19 ±0.45, 3.88 ±0.57, 3.13 ±0.13, 2.30 ±0.19, 1.42 ±0.12, and 0.91±0.04 (all×10^{-10} mol·cm^{-2}·min^{-1}), respectively. The inset shows that the surface flux of NO generated is linearly related to the drive current applied to the LED illuminating the coated optical fiber and is identical whether I_{drive} is applied form 0–20 mA or 20–0 mA. Measurements were made at 22 °C with chemiluminescence detection. (From ref. 103 with permission.)

then developed a modified polydimethylsiloxane with S-nitroso-N-acetyl-penicillamine (SNAP-PDMS) covalently attached to the crosslinking agent used to form the PDMS. This material, when coated onto declad optical fibers, enables the precise control over the level and duration of NO released in re-sponse to a programmed sequence of light generated by a wirelessly controlled LED light source (see Figure 5.10).[103] This system will permit control over NO release and allow detailed information to be obtained regarding the level and duration of NO needed to ensure controlled biological response in both blood contacting and tissue contacting sensor applications. Riccio and cow-orkers[104,105] developed RSNO-containing xerogels that also releases NO in response to illumination using incandescent bulbs of different wattages. These types of materials that are able to release NO in a dynamically controllable manner may offer a means to more closely mimic the physiological release of NO and have the potential to allow implanted sensors to reach a stable, interfacial blood/tissue response that may ultimately allow for widespread clinical utility.

5.4.3   Materials to Mimic Biological Form and Function

To more fully mimic biological tissue such as intact endothelium, Zhou and Meyerhoff[106] created a trilayer system that combined NO release with surface-immobilized heparin. The idea is to more closely mimic the dual action of antiplatelet NO and anticoagulant heparin. Wu and coworkers[107] later combined a continuously releasing NO PurSil® layer top coated with a CarboSil® with surface-bound thrombomodulin and heparin that were shown to be biologically active. As material development produces more matrices that exhibit better biocompatibility, it is likely that combining these improved materials properties with controlled active agent release and immobilized bioactive molecules will lead to greatly enhanced performance of indwelling sensors.

An additional step toward creating longer-term implants would be to eliminate the dependence on the reservoir of active agent present in the implanted device. Klueh and coworkers[108] developed a system in which they genetically engineering cells to produce excess levels of VEGF. These cells were entrapped in the tip of a glucose sensor. The goal was to increase vascularization around the implanted sensor to allow ample diffusion of glucose to the sensor. When implanted in *ex ova* chick embryos, they were able to demonstrate a dramatic increase in the density of blood vessels present around the implant. This approach relies on the production of VEGF by cells *in situ* rather than a preloading of the sensor with the active agent. Another approach to avoid preloading donor molecules was demonstrated by Oh and Meyerhoff.[109] They developed a lipophilic catalytic copper-containing species that was incorporated into poly(vinyl chloride) (PVC) and demonstrated that RSNOs present in the bathing solution surrounding the material were decomposed to release NO catalytically. Hwang and Meyerhoff[110] then took a similar copper cyclen structure and tethered it to the surface of biomedical-grade polyurethane that was capable of spontaneously generating NO when in contact with blood. This approach offers the advantage of using already existing medical grade polymers and imparting them with NO-release properties. Cha and Meyerhoff[111] then extended this by developing an immobilized organoselenium moiety that was able to catalytically generate NO from RSNOs present in plasma. These catalytic systems should be capable of nearly indefinite production of NO from their polymer surfaces because RSNOs are endogenously produced in the body and circulate in blood and intestinal fluid. As long as adequate levels of the endogenous RSNOs and reducing equivalents are present, the material will continue to generate physiological NO levels at the polymer/biological interface.

5.5   Summary

The goal of developing implantable blood gas, glucose, and lactate sensors that can provide reliable, real-time clinically relevant results remains elusive.

While great advances have been made in the development of miniaturized electrochemical and optical sensors that have adequate sensitivity, selectivity and stability when operated *in vitro*, the placement of such devices in the *in vivo* environment creates a physiological response that greatly influences the quality of analytical results obtained. As summarized above, progress in this area mandates the development of appropriate sensor coating materials that can mitigate the normal activation of platelets and other cells when devices are placed intravascularly and the inhibition of the inflammatory response when sensors are placed subcutaneously. These coatings must be capable of diminishing the normal physiological responses to the implanted sensors, yet they cannot adversely affect analytical response properties, including response times, sensitivity to the target analyte, and selectivity of the sensor. A wide range of new, passive and active biomaterials coatings have been devised in recent years specifically to target the *in vivo* sensing challenge. In the end, it is likely that a combination of immobilized agents and slow chemical release strategies will be needed to resolve the physiological response issues that have limited the analytical performance of implanted sensors to date. Continued research in this area during the coming years will hopefully achieve success and provide an array of new *in vivo* sensor tools that will enhance the quality of health care and disease management for millions of people worldwide.

References

1. V. Gubala, L. F. Harris, A. J. Ricco, M. X. Tan and D. E. Williams, *Anal. Chem.*, 2012, **84**, 487.
2. P. St. Louis, *Clin. Biochem.*, 2000, **33**, 427.
3. F. B. Myers and L. P. Lee, *Lab Chip*, 2008, **8**, 2015.
4. G. McGarraugh and R. Bergenstal, *Diab. Technol. Ther.*, 2009, **11**, 145.
5. S. Gelsomino, R. Lorusso, U. Livi, S. Romagnoli, S. M. Romano, R. Carella, F. Lucà, G. Billè, F. Matteucci, A. Renzulli, G. Bolotin, G. De Cicco, P. Stefàno, J. Maessen and G. F. Gensini, *BMC Anesthesiol.*, 2011, **11**, 1.
6. D. H. Bailey, E. J. da Silva and T. H. Cluton-Brock, *Anesthesia*, 2011, **66**, 889.
7. V. R. Kondepati and H. M. Heise, *Anal. Bioanal. Chem.*, 2007, **388**, 545.
8. C. M. Girardin, C. Huot, M. Gonthier and E. Delvin, *Clin. Biochem.*, 2009, **42**, 136.
9. E. Cheyne and D. Kerr, *Diabet. Metab. Res. Rev.*, 2002, **18**(Suppl 1), S43.
10. T. Kubiak, N. Hermanns, H. J. Schreckling, B Kulzer and T. Haak, *Diabetic Med.*, 2004, **21**, 487.
11. M. C. Frost and M. E. Meyerhoff, *Current Opin. Chem. Biol.*, 2002, **6**, 633.
12. M. C. Frost and M. E. Meyerhoff, *Anal. Chem.*, 2006, **78**, 7370.
13. M. Ganter and A. Zollinger, *Brit. J. Anaesth.*, 2003, **91**, 397.
14. M. E. Meyerhoff, *Trends Analyt. Chem.*, 1993, **12**, 257.
15. P. Holloway, S. Benham and A. St. John, *Clin. Chim. Acta.*, 2001, **307**, 9.

16. Shapiro, N. Howell, M. Talmor, D. Nathanson, L. Lisbon, A. Wolfe and R. Weiss, *J. Ann. Emerg. Med.*, 2005, **45**, 524.
17. F. Valenza, G. Aletti, T. Fossali, G. Chevallard, F. Sacconi, M. Irace and L. Gattinoni, *Crit. Care*, 2005, **9**, 588.
18. C. K. Mahutte, C. S. H. Sasson, J. R. Muro, D. R. Hansmann, T. P. Maxwell, W. W. Miller and M. Yafuso, *J. Clin. Monitor.*, 1990, **6**, 147.
19. Y. Amao, *Microchim. Acta*, 2003, **143**, 1.
20. T.-S. Yeh, C.-S. Chu and Y.-L. Lo, Sens, *Actuators B*, 2006, **119**, 701.
21. C.-S. Chu and Y.-L. Lo, Sens, *Actuators B*, 2010, **151**, 83.
22. J. W. Severinghaus, *Ann. NY Acad. Sci.*, 1968, **148**, 115.
23. W. Jin, J. Jiang, Y. Song and C. Bai, *Res. Physiol. Neurobiol.*, 2012, **180**, 141.
24. D. A. Nivens, M. V. Schiza and M. Angel, *Talanta*, 2002, **58**, 543.
25. W. Jin, J. Jiang, X. Wang, X. Zhu, G. Wang, Y. Song and C. Bai, *Res. Physiol. Neurobiol.*, 2011, **177**, 183.
26. H. Offenbacher, O. S. Wolfbeis and E. Furlinger, *Sens. Actuators*, 1986, **9**, 73.
27. P. J. Kinlen, J. E. Heider and D. E. Hubbard, *Sens. Actuators B*, 1994, **22**, 13.
28. M. Telting-Diaz, M. E. Collison and M. E. Meyerhoff, *Anal. Chem.*, 1994, **66**, 576.
29. R. K. Meruva and M. E. Meyerhoff, *Biosens. Bioelectron.*, 1997, **13**, 201.
30. M. A. Makos, D. M. Omiatek, A. G. Eweing and M. L. Heien, *Langmuir*, 2010, **26**, 10386.
31. W.-D. Huang, H. Cao, S. Deb, M. Chiao and J. C. Chiao, Sens, *Actuators A*, 2011, **169**, 1.
32. G. S. Wilson and R. Gifford, *Biosens. Bioelectron.*, 2005, **20**, 2388.
33. J. W. Mo and W. Smart, *Front. Biosci.*, 2004, **9**, 3384.
34. W. K. Ward, L. B. Jansen, E. Anderson, G. Reach, J. C. Klein and G. S. Wilson, *Biosens. Bioelectron.*, 2002, **17**, 181.
35. J. Wang, *Electroanalysis*, 2001, **13**, 983–988.
36. T. Koschchinsky and L. Heinemann, *Diabet. Metab. Res. Rev.*, 2001, **17**, 113.
37. C. Choleau, J. C. Klein, G. Reach, B. Aussedat, V. Demaria-Pesce, G. S. Wilson, R. Gifford and W. K. Ward, *Biosens. Bioelectron.*, 2002, **17**, 647.
38. P. Waeger and M. Hummel, *Diabet. Stoffwechsel und Herz*, 2008, **17**, 385.
39. N. Tubiana-Rufi, J. P. Riveline and D. Dardari, *Diabet. Metab*, 2007, **33**, 415.
40. I. B. Hirsch, D. Armstrong, R. M. Bergenstal, B. Buckingham, B. P. Childs, W.L. Clarke, A. Peters and H. Wolpert, *Diabet. Technol. Ther.*, 2008, **10**, 232.
41. R. J. McNichols and G. L. Cote, *J. Biomed. Opt.*, 2000, **5**, 5.
42. J. Wang, *Talanta*, 2008, **75**, 636.
43. A. Heller, *Current Opin. Chem. Biol.*, 2006, **6**, 664.

44. J. S. Marvin and H. W. Hellinga, *J. Am. Chem. Soc.*, 1998, **120**, 7.
45. R. Tipnis, M. Vaddiraju, F. Jain, D. Burgess and F. Papadimitrakopoulos, *J. Diabet. Sci. Technol.*, 2007, **2**, 193.
46. Y. Lin, S. Taylor, H. Li, K. A. S. Fernando, L. Qu, W. Wang, L. Gu, B. Zhou and Y. P. Sun, *J. Mater. Chem.*, 2004, **14**, 527.
47. Y. H. Lin, F. Lu, Y. Tu and Z. F. Ren, *Nano Lett.*, 2004, **4**, 191.
48. Z. Zhu, L. Garcia-Gancedo, A. Flewett, H. Xie, F. Moussy and W. Milne, *Sensors*, 2012, **12**, 5996.
49. S. Tierney, B. Falch, D. Hjelme and B. Stokke, *Anal. Chem.*, 2009, **81**, 3630.
50. S. Paek, I. Cho, D. Kim, J. Jeon, G. Lim and S. Paek, *Biosens. Biolectron.*, 2013, **40**, 38.
51. Y. Hu, Y. Zhang and G. S. Wilson, *Anal. Chim. Acta*, 1993, **281**, 503.
52. D. A. Baker and D. A. Gough, *Anal. Chem.*, 1995, **67**, 1536.
53. Z. Ibupoto, S. Shah, K. Kuhn and M. Willander, *Sensors*, 2012, **12**, 2456.
54. X. Zheng, H. Yang and C. Li, *Anal. Chem.*, 2010, **82**, 5082.
55. A. Martín, S. Ceballo, I. Ruminot, R. Lerchundi, W. Frommer and L. Barros, *Plos One*, 2013, **8**, 1.
56. B. Venkatesh, T. H. Clutton-Brock and S. P. Hendry, *J. Med. Eng. Technol.*, 1994, **18**, 165.
57. B. Venkatesh, T. H. Clutton-Brock and S. P. Hendry, *J. Cardiothorac. Vasc. Anesth.*, 1995, **9**, 412.
58. B. Venkatesh, T. H. Clutton-Brock and S. P. Hendry, *Crit. Care Med.*, 1994, **22**, 588.
59. S. Divers, W. Marshall, and R. Foster-Smith, in *Proceedings of the Conference on Electrolytes, Blood Gases and Other Critical Analytes* Burritt, M. S. Sena, S. F. D'Orazio, P., Eds., AACC: Washington, D.C., 1992 pp. 1–9.
60. M. B. Gorbet and M. V. Sefton, *Biomaterials*, 2004, **25**, 5681.
61. D. Dörmann, K. J. Clemetson and B. E. Kehrel, *Blood*, 2000, **96**, 2469.
62. K. C. Dee D. A. Puleo R. Bizios, *An Introduction to Tissue-Biomaterial Interactions*, John Wiley & Sons, Inc., Hoboken, New Jersey, USA, 2002, **4**, pp 68–81.
63. S. P. Nichols, A. Koh, W. L. Storm, J. H. Shin and M. H. Schoenfisch, *Chem. Rev.*, 2013, **113**, 2528.
64. Z. Zhu, L. Garcia-Gancedo, A. J. Flewitt, H. Xie, F. Moussy and W. I. Milne, *Sensors*, 2012, **12**, 5996.
65. J. M. Anderson, *Cardiovasc. Path.*, 1993, **2**, 33S.
66. J. M. Daley, J. D. Shearer, B. Mastrofrancesco and M. D. Caldwell, *Surgery*, 1990, **107**, 187.
67. D. A. Baker and D. A. Gough, *Anal. Chem.*, 1996, **68**, 1292.
68. J. M. Anderson, A. Rodriguez and D. T. Chang, *Semin. Immunol.*, 2008, **20**, 86.
69. U. Klueh, M. Kaur, Y. Qiao and D. L. Kreutzer, *Biomaterials*, 2010, **31**, 4540.
70. P. T. Thevenot, D. W. Baker, H. Weng, M.-W. Sun and L. Tang, *Biomaterials*, 2011, **32**, 8394.

71. J. D. Bryers, C. M. Giachelli and B. D. Ratner, *Biotechnol. Bioeng.*, 2012, **109**, 1898.
72. D. T. Chang, J. A. Jones, H. Meyerson, E. Colton, I. K. Kwon, T. Matsuda and J. M. Anderson, *J. Biomed. Mater. Res.*, 2008, **87A**, 676.
73. J. M. Anderson, *Ann. Rev. Mater. Res.*, 2001, **31**, 81.
74. K. L. Helton, B. D. Ratner and N. A. Wisniewski, *J. Diabet. Sci. Technol.*, 2011, **5**, 632.
75. N. Wisniewski and M. Reichert, *Colloid Surf. B: Biointerfaces*, 2000, **18**, 197.
76. Justin, G. Finely, S. Rahman and A. R. A. Guiseppi-Elie, *A. Biomed. Microdevices*, 2009, **11**, 103.
77. A. Kros, M. Gerritsen, V. S. I. Sprakel, N. A. J. M. Sommerdijk, J. A. Jansen and R. J. M. Nolte, *Sens. Actuators B*, 2001, **81**, 68.
78. Y. M. Ju, B. Yu, L. West, Y. Moussy and F. Moussy, *J. Biomed. Mater. Res.*, 2009, **92A**, 650.
79. N. Wang, K. Burugapalli, W. Song, J. Halls, F. Moussy, A. Ray and Y. Zheng, *Biomaterials*, 2013, **34**, 888.
80. F. Ai, Q. Wang, W. Z. Yuan, H. Li, X. Chen, L. Yang, Y. Zhang and S. Pei, *J. Mater. Sci.*, 2012, **47**, 5181.
81. Stachelek, S. J. Finely, M. J. Alferiev, I. S. Wang, F. Tsai, R. K. Eckells, E. C. Tomczyk, N. Connolly, J. M. Discher, D. E. Eckmann and D. M. Levy, *R.J. Biomaterials*, 2011, **32**, 4317.
82. L. W. Norton, H. E. Koschwanez, N. A. Wisniewski, B. Klitzman and W. M. Reichert, *J. Mater. Res.*, 2007, **81A**, 858.
83. J. Sung, P. W. Barone, H. Kong and M. S. Strano, *Biomaterials*, 2009, **30**, 622.
84. U. Bhardwaj, R. Sura, F. Papadimitrakopoulos and D. J. Burgess, *Int. J. Pharmaceut.*, 2010, **384**, 78.
85. Wang, Y. Papadimitrakopoulos, F. Burgess *J. Control. Release* 2013, in press. (http://dx.doi.org/10.1016/j.jconrel.2012.12.028).
86. S. Moncada, *Ann. NY Acad. Sci.*, 1997, **811**, 60.
87. M. W. Radomski, R. M. J. Palmer and S. Moncada, *Biochem. Biophys. Res. Commun.*, 1987, **148**, 1482–1489.
88. S. P. Nichols, A. Koh, N. L. Brown, M. B. Rose, B. Sun, D. L. Slomnerg, D. A. Riccio, B. Klitzman and M. H. Schoenfisch, *Biomaterials*, 2012, **33**, 6305.
89. Y. Wu and M. E. Meyerhoff, *Talanta*, 2008, **75**, 642.
90. K. A. Mowery, M. H. Schoenfisch, N. Baliga, J. A. Wahr and M. E. Meyerhoff, *Electroanalysis*, 1999, **11**, 681.
91. K. A. Mowery, M. H. Schoenfisch, J. E. Saavedra, L. K. Keefer and M. E. Meyerhoff, *Biomaterials*, 2000, **21**, 9.
92. H. Zhang, G. M. Annich, J. Miskulin, K. Osterholzer, S. I. Merz, R. H. Bartlett and M. E. Meyerhoff, *Biomaterials*, 2002, **23**, 1485.
93. M. M. Reynolds, J. A. Hrabie, B. K. Oh, J. K. Politis, M. L. Citro, L. K. Keefer and M. E. Meyerhoff, *Biomacromolecules*, 2006, **7**, 987.

94. T. C. Major, D. O. Brant, M. M. Reynolds, R. H. Bartlett, M. E. Meyerhoff, H. Handa and G. M. Annich, *Biomaterials*, 2010, **31**, 2736.
95. H. Zhang, G. M. Annich, J. Miskulin, K. Staniewwicz, K. Osterholzer, S. I. Merz, R. H. Bartlett and M. E. Meyerhoff, *J. Am. Chem. Soc.*, 2003, **125**, 5015.
96. M. C. Frost and M. E. Meyerhoff, *J. Biomed. Mater., Res.*, 2005, **72A**, 409.
97. J. H. Shin and M. H. Schoenfisch, *Analyst*, 2006, **131**, 609.
98. S. M. Marxer, A. R. Rothrock, B. J. Nablo, M. E. Robbins and M. H. Schoenfisch, *Chem. Mater.*, 2003, **15**, 4193.
99. S. M. Marxer, M. E. Robbins and M. Schoenfisch, *H Analyst*, 2005, **130**, 206.
100. J. H. Shin, S. M. Marxer and M. H. Schoenfisch, *Anal. Chem.*, 2004, **76**, 4543.
101. M. C. Frost and M. E. Meyerhoff, *J. Am. Chem. Soc.*, 2004, **126**, 1348.
102. G. E. Gierke, M. Nielsen and M. C. Frost, *Sci. Technol. Adv. Mater.*, 2011, **12**, 055007.
103. M. A. Starrett, M. Nielsen, D. M. Smeenge, G. E. Romanowicz and M. C. Frost, *Nitric Oxide*, 2012, **27**, 228.
104. D. A. Riccio, K. P. Dobmeier, E. M. Hetrick, B. J. Privett, H. S. Paul and M. H. Schoenfisch, *Biomaterials*, 2009, **30**, 4494.
105. D. A. Riccio, P. N. Coneski, S. P. Nichols, A. D. Broadnax and M. H. Schoenfisch, *Appl. Mater. Interfaces*, 2012, **4**, 796.
106. Z. Zhou and M. E. Meyerhoff, *Biomaterials*, 2005, **26**, 6506.
107. B. Wu, B. Gerlitz, B. W. Grinnell and M. E. Meyerhoff, *Biomaterials*, 2007, **28**, 4047.
108. U. Klueh, D. I. Dorsky and D. L. Kreutzer, *Biomaterials*, 2005, **26**, 1155.
109. B. K. Oh and M. E. Meyerhoff, *J. Am. Chem. Soc.*, 2003, **125**, 9552.
110. S. Hwang and M. E. Meyerhoff, *Biomaterials*, 2008, **29**, 2443.
111. W. Cha and M. E. Meyerhoff, *Biomaterials*, 2007, **28**, 19.

CHAPTER 6

Peroxynitrite Electrochemical Quantification: Recent Advances and Challenges

SERBAN F. PETEU*[a] AND SABINE SZUNERITS[b]

[a] National Institute for Research and Development in Chemistry and Petrochemistry, 202 Splaiul Independentei, Sector 6, 060021 Bucuresti, Romania; [b] Institut de Recherche Interdisciplinaire (IRI, USR 3078), Université Lille 1, Parc de la Haute Borne, 50 Avenue de Halley, BP 70478, 59658 Villeneuve d'Ascq, France
*Email: sfpeteu@umich.edu

6.1 Introduction

The ageing population worldwide is affected by a high incidence of devastating diseases, metabolic, cardiovascular and neurodegenerative to name a few, caused by age-dependent formation of *free radicals*.[1–3] Free radicals include atoms, ions, molecules with unpaired valence electrons or open electron shell, typically highly reactive and short lived.[4–6]

Clinical evidence shows the so-called reactive nitrogen and oxygen species (RNOS), which include free radicals and peroxynitrite ($ONOO^-$) to play a fundamental role in cell signaling, for example being produced to maintain their integrity when challenged by environmentally unsafe exposures such as mechanical stress, UV radiation, toxins in air or water, bacteria, or viruses.[1] When these species are, however, in excess, the steady state maintained by physiological processes (homeostasis) is disturbed. At this point, the *nitro*

RSC Detection Science Series No. 2
Detection Challenges in Clinical Diagnostics
Edited by Pankaj Vadgama and Serban Peteu
© The Royal Society of Chemistry 2013
Published by the Royal Society of Chemistry, www.rsc.org

oxidative metabolic stress develops and its actions over time leads to modifications of bio-macromolecules and also interfere with the redox signaling processes.[4–8]

Already established in the last decade as a powerful *nitrating, nitrosating and oxidative "triple agent"* for cellular constituents, peroxynitrite (ONOO⁻ or PON for short) is also being clinically established to exert a variety of detrimental, cytotoxic effects in cells and tissues, both *in vitro* and in living organisms.[4,10,11] Here, ONOO⁻ is typically formed *via* the diffusion-controlled reaction between superoxide radical ($O_2^{\bullet-}$) and nitric oxide (NO).[4–8] The PON anion is a short-lived species, a potent inducer of cell death, thus perhaps rightfully hailed as the "ugly side" of nitric oxide and with a very short life time, that is only about one second at physiological pH[11]. By contrast, NO is acting as a "good" molecule, whose low-level production is important, including in protecting organs, such as the liver, from ischemic damage. Consequently, the trio ($^{\bullet}NO$; $O_2^{\bullet-}$; $ONOO^-$) was notably dubbed[5] as "the Good, the Bad and the Ugly". Its fate[50] is illustrated in Scheme 6.1.

Starting from the 1990s with the seminal publications by J. S. Beckman, H. Ischiropoulos, R. Radi, and their groups and others[4–6] suggested the biological formation of peroxynitrite by the near diffusion-controlled reaction *in vivo* between nitric oxide and superoxide radicals with their implications for oxidative injury.[4,5,7,12–14,50]

Ever since, the literature constantly provides solid clinical evidence showing that endogenously generated peroxynitrite is correlated with acute cytotoxicity and is thus incriminated in various pathologies and diseases.[4–6] In fact, the discovery of the peroxynitrite biological formation, role and fate, arguably continues and accomplishes the nitric oxide crucial discovery as a signaling molecule in the cardiovascular system[15] pioneered by R.F. Furchgott, L.J. Ignarro and F. Murad, the 1998 Physiology-Medicine Nobel laureates.

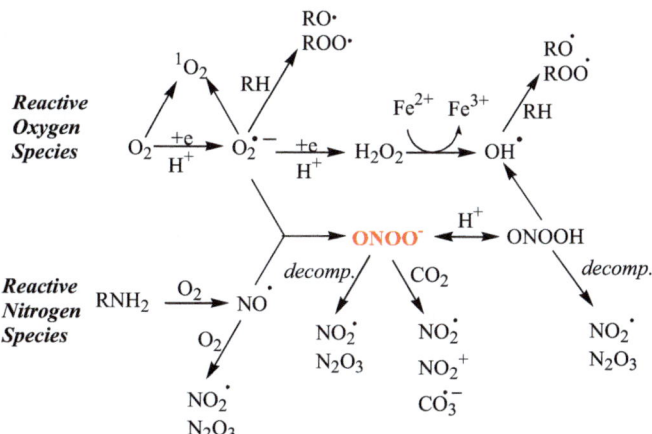

Scheme 6.1   The principal RNOS derived from the biological conversion of oxygen into superoxide ion and nitric oxide. Redrawn with the kind permission of the American Chemical Society.

Before dealing with peroxynitrite detection methods and tough challenges, it seems highly appropriate to briefly review its role in biological systems.[8,47,48] Herein, we table very succinctly the published clinical evidence for peroxynitrite toxicity, especially to emphasize the vital need for the *in vivo* detection of this analyte. The pathogenic effects of peroxynitrite have been aptly reviewed and readers are referred to these articles.[4-6]

Furthermore, it is equally important to discuss efforts and strategies employed by others to attenuate the toxic effects of ONOO⁻.[7,10,35-37,43] Some of these potentially therapeutic methods are based on peroxynitrite isomerization

(A) FeTMPyP

(B) MnTDE-2-IMP⁵⁺

(C) FeTPPS

(D)

(E)

R=

TCPP

TPPS

TM-4-PyP

TE-2-PyP

4-alkylpyridium

2,2'-dialkylimidazolinium [R₁= (CH₂)ₙH]
2,2'-dimethoxyethylimidazolinium [R₁= (CH₂)₂OCH₃]

FP15

to nitrate using metallo-macrocycle molecules and the catalytic reduction of PON to nitrite and are examined extensively on models of several devastating diseases models. An interesting fact is that same classes of metallo-macrocycles *decomposing or destroying* peroxynitrite are arguably the ones that are also *electrochemically sensitive* to PON, since in both cases catalytic processes are involved. *So, one can imagine perhaps designing PON sensitive films, with "detect-and-destroy"* double capabilities. Some of such PON decomposition catalysts are shown in Scheme 6.2.

6.2 Challenges Confronted in the Accurate Quantification of Peroxynitrite

Peroxynitrite is an extremely reactive molecule involved in numerous and fast reaction pathways, thus the exogenous addition (from outside sources) of $ONOO^-$ into (multi)cellular model systems is less than ideal to study its biological effects.[6] There are a number of reviews that outline the advances and challenges when confronted to PON detection.[4-7,47-50] Clinical and pharmacological studies have discovered that PON is generated *in vivo* at *steady, low rates* often quite the opposite of any analytical approach,[4,5] where chemists employ mostly *short bolus exposures* to characterize the PON-sensitive chemical sensors.[50] Thus, a gradual permeation of PON using an automatic variable-rate delivery system (*i.e.* motor-driven syringe) seems better to model the *in vivo* reality.[6]

The chemical instability of PON is an added difficulty to be considered, with special weight for *in vivo* studies – where some investigators prefer the so-called

---

Scheme 6.2   Peroxynitrite decomposition catalysts: **(A)** FeTMPyP = Fe (III)meso-tetra(N-methyl-4-pyridyl) porphine chloride; **(B)** MnTDE-2-ImP^{5+} = Mn(III) meso-tetrakis(N,N'-diethylimidazolium-2-yl) porphyrin; and **(C)** FeTPPS = 5,10,15,20-tetrakis(4-sulfonatophenyl)porphyrinato Fe (III), chloride. **(D)** The general structure of several PON decomposition catalysts including several Mn and Fe porphyrins.[6] The metallo-porphyrins group that have alkylpyridyl substituents are represented by TE-2-PyP (tetrakis((N-ethyl) pyridynium-2-yl) 1 porphyrin) and TMPyP (tetrakis((N-methyl) pyridynium-4-yl) 1 porphyrin). These readily decompose PON. By changing substituents position and alkyl chain length, one can influence both hydrophobicity and redox potential. **(E)** Also, several specific R groups are specified.[6] The metalloporphyrins such as Mn(III) tetrakis [N-N'-diethylimida-zolium-2-yl]porphyrin MnTDE-2-ImP^5 contain dialkylimidazoline substituents as seen in Scheme 2B where R$_1$ is en ethyl group. Other acronyms are TCPP (tetrakis 4-carboxylato-phenyl porphyrin); TPPS (5,10,15,20-tetrakis-4'-sulfonatophenyl porphyrin). Finally, the FeCl tetrakis-2-(triethylene glycol monomethyl ether) pyridyl porphyrin FP15 contains a triethylene glycol monomethyl ether pyridynium moiety as the R-group linked to the porphyrin ring. The acronyms in bold are referred to in Table 6.1.

Table 6.1   Peroxynitrite   pathological   effects   and   potential   therapeutic
strategies.

Diseases	*Pathogenic Effects of Peroxynitrite Established Clinically in Disease Models*	*PON Decomposition Catalysts Potentially Therapeutic*
Cardiac diseases • Cardiovascular effects *in vitro*/ *in vivo* • Myocardial ischemia/ reperfusion injury • Myocarditis • Cardiac allograft acute rejection • Chronic heart failure	PON has been shown to trigger apoptosis in cardiomyocytes[16,17] endothelial[18] and vascular smooth muscle cells.[19,20] Reperfusion (treatment to reduce myocardial damage resulted from infarction, organ transplantation) leads to additional tissue injury facilitated by RNOS[21,22] upon reperfusion, including by an apparent PON formation increase.[23,24] Myocarditis has been linked to increased nitro-tyrosine immune-reactivity, indicating a pathogenic role of PON formation.[25,26,64]	**FP15** reduced myocardial infarct size in infarction porcine model.[38] **FeTPPS** prevented myocardial dysfunction in an inflammatory heart-failure model.[7]
Vascular diseases • Atherosclerosis • Aging • Hypertension	The pathogenic role of peroxynitrite in atherosclerosis[27] and in arterial hypertension is supported by recent findings.[28] Increased oxidative stress in aging also leads to functional inactivation of NO by elevated concentrations of O_2^- favoring an enhanced PON formation.[29]	Using **FeTPPS**, recent studies have provided additional data for role of peroxynitrite.[39] **FP15** improved vascular function, cardiac function, peripheral motor and sensory nerve conductance.[38]
Chronic Inflammation • Chronic arthritis • Inflammation of toxic origin • Chronic inflammation with cancer potential	In arthritis, the PON could be the final finisher of previously considered NO-dependent inflammation, as ascertained by a series of findings.[7] PON is suspected to play an important role in the development of organ damage and inflammation prompted by various toxic chemicals.[30] A correlation between chronic inflammation and tumorigenesis has been thought. Long-term exposures to PON nitro-oxidative stress could lead to cancer, under inflammatory conditions.[7]	**FeTMPS** exerted anti-inflammatory effects in paw-swelling models.[40,41] **FP15** improved the effect of autoimmune arthritis and colitis in murine models.[42] PON decomposition catalysts might develop into potential tools for chronic joint inflammation.[6,7]
Neurodegenerative diseases • Multiple sclerosis • Parkinson's • Alzheimer's • Huntington's	The concurrently increased nitric oxide and superoxide production associated with neurodegenerative disorders favors the *in vivo* generation of peroxynitrite.[31] Numerous findings seem to strongly support the involvement of PON in all neurodegenerative disorders.[11]	NOS inhibitors were helpful in neurodegeneration models, potentially by averting peroxynitrite formation.[7] Porphyrinic antioxidants improved the outcome in sclerosis murine models.[44]

Table 6.1 (*Continued*)

Diseases	Pathogenic Effects of Peroxynitrite Established Clinically in Disease Models	PON Decomposition Catalysts Potentially Therapeutic
Diabetes and complications • Primary diabetes • Cardio-vascular dysfunction • Neuropathy, nephropathy, retinopathy	Type 1 (insulin-dependent) diabetes is an autoimmune destruction of insulin producing beta cells, thus resulting in long-term hyperglycemia. While the real trigger is still unknown, the current findings converge for a crucial function of nitric oxide, superoxide and peroxynitrite in the pathogenesis of diabetes and associated complications[32] with PON potentially implicated in beta-cell destruction.[33,34]	Mercaptoalkylguanidines and **FP15** reduced the type 1 diabetes onset in murine models.[7,45] **FP15** improves retinal microcirculation in diabetes rodent models.[46]

The acronyms utilized are (see also Scheme 6.2):
FeTMPS = Fe (III)meso-tetra(2,4,6-trimethyl-3,5-disulfonato)porphine chloride
FeTPPS = 5,10,15,20-tetrakis(4-sulfonatophenyl)porphyrinato Fe(III), chloride
FP15 = Fe(III)tetrakis-2-(*N*-triethylene glycol monomethyl ether) pyridyl porphyrin.

"genuine" $ONOO^-$. Several synthetic methods for the preparation of per-oxynitrite have been reported and are being successfully used, including:

(a) the ozone reaction with azide ions in presence of low concentration alkali;[51]
(b) the auto-oxidation of hydroxylamine in moderately alkaline NaOH solution;[52]
(c) the nitrite reaction with acidified H_2O_2 followed by a base quench;[53]
(d) the ethoxyethyl nitrite reaction with H_2O_2 in a basic medium;[54]
(e) the UV photolysis of solid KNO_3;[55]
(f) the two-phase displacement reaction by the hydroperoxide anion, in the aqueous phase, on the isoamyl nitrite, the organic phase.[56,57]

Typically, the as-synthesized PON solution results with a double-digit milliMolar final concentration and could be stored at $-50\,^{\circ}C$ being useable for several months, with a typical loss, nonetheless, of about 2% PON per day. Nevertheless, its real concentration can be assessed by absorbance in UV-Vis at 302 nm, ($\varepsilon = 1700\ mol^{-1}\ L\ cm^{-1}$). All these experiments are mostly carried out at around $2\text{--}4\,^{\circ}C$ over an ice bath to minimize the spontaneous peroxynitrite decay.[57]

Additionally, it has to be kept in mind that synthetic PON stock solutions need a higher pH (such as pH 9.5) where the analyte is more stable and characterization can be made adequately. By contrast, at physiological pH peroxynitrite is prone to fast decomposition.[4,5,7] This is one reason that other routes for the PON formation have been looked at.

The use of donor solutions of SIN-1 (3-Morpholino-sydnonimine, also called Linsidomine) has become widely accepted as one way to overcome this limitation. SIN-1 is being used as a PON producer at physiological pH, nowadays, under ambient molecular oxygen conditions, it produces both nitric oxide and superoxide as illustrated in Scheme 6.3 and therefore is a PON donor or generator.[58–63]

SIN-1 liberates superoxide anions and nitric oxide spontaneously in solution with a 1:1 stoichiometry, thereby generating peroxynitrite continuously for a certain period of time. Indeed, in an aerobic aqueous solution SIN-1 decomposes readily to SIN-1A, which in the presence of an oxidant like oxygen forms the unstable SIN-1A radical cation. The latter liberates NO and eventually forms the stable end product 3-morpholino-iminoacetonitrile SIN-1C, as shown in Scheme 6.3.

The quantitative determination of ONOO⁻ continues to be complicated, due to a variety of intrinsic obstacles,[47–49] including

(i) inherent difficulties to accurately reproduce the true *in vivo* kinetics of PON in the model experiments *in vivo*;[7,9,10]
(ii) the potential of misinterpreting PON concentration "as determined", if the experimental conditions are not carefully optimized;[8,47]
(iii) the vast complexities of the *in vivo* real environment, as PON typically interacts with more than one target per unit time, due to its high reactivity.[10]

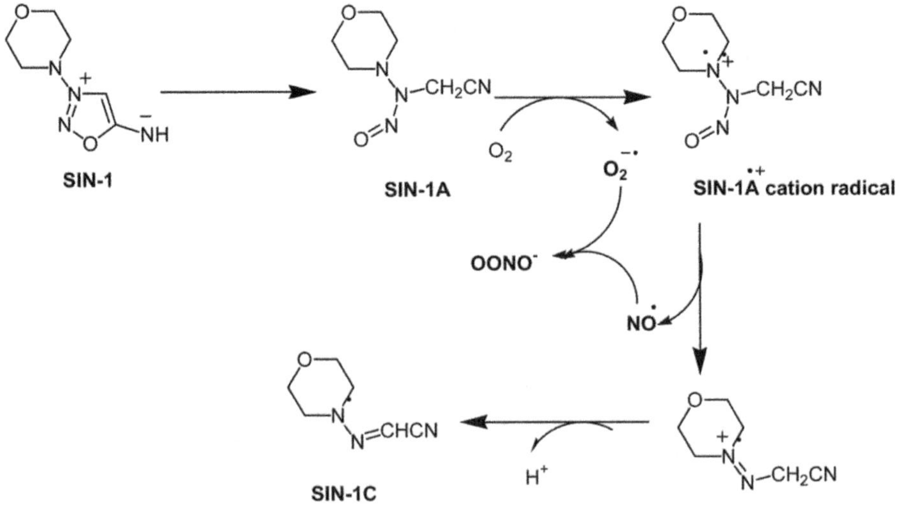

Scheme 6.3   The oxidative mechanism for ONOO⁻ release from SIN-1.

The detection arsenal for peroxynitrite as well as its precursors nitric oxide and superoxide, includes indirect biochemical assays, optical detection methods with suitable dyes, electron (paramagnetic) spin resonance spectroscopy and electrochemical sensing strategies.[8,50,71]

Biochemical assays measure PON concentrations indirectly, typically based on up- or down- stream interactions, furthermore suffering from additional unwanted sample processing, such as freezing.[65–67] Other, more sophisticated and extremely expensive techniques, such as *electron spin resonance spectroscopy* (ESR) used with spin-trapping agents, are afforded by only few select laboratories, are considered too far outside of the scope of this chapter, and plus these have been reviewed in detail elsewhere.[8,50,71]

By contrast, electrochemical detection using chemically modified electrodes[89] is considered a great analytical technique when it comes to real-time, label-free and direct measurements of these reactive species.[11,47,50] One main drawback, however, is that the selectivity sometimes implies the use of an additional membrane or film, which brings a trade-off between specificity or a slower sensor response.[68,69]

6.3 Electrochemical Interfaces and Methods for Peroxynitrite Detection

Several reviews on the PON electrochemical detection have outlined the different materials and methods used.[4,47] While the best perspectives for label-free, *in situ* detection to date seem to be offered by electroanalytical methods, surprisingly there are just a handful of papers dealing with this issue.[68–73] Furthermore, most of the reports have not comprehensively characterized the peroxynitrite-sensitive electrocatalytic matrices such as the conductive polymer-manganese ion complex[73] or the manganese (II) tetraamino-phthalocyanine coated with poly(4-vinylpyridine).[72]

Thus, motivated by the challenge to find better-performing interfaces for the electrochemical detection of peroxynitrite, the state-of-the-art in PON sensors will be discussed, including their sensitivity, response time, specificity against other electroactive molecules with selectivity to peroxyitrite. Also, the advantages and drawbacks will emphasize their potential to be used in real biological matrices. The discussion will be focused on two main aspects, the reaction mechanism and the selectivity of the respective PON-sensitive film, when available. The other important performance factors such as limit of detection, response time, sensitivity, are shown in Table 6.2.

Amatore and colleagues[68,69] have pioneered a number of interesting methods and dedicated micro- and nanosensors supported by theoretical and experimental models. From this body of work, herein will be outlined only some of the efforts leading to peroxynitrite quantification. Simultaneous detection of NO, $O_2^{•-}$ and $ONOO^-$ at different potentials was possible with a new experimental method illustrated in Figure 6.1. They reported that a single-cell oxidative burst could be in fact decomposed into each of the corresponding

Table 6.2 Electrochemical methods for peroxynitrite-sensitive interfaces.

Reference	Detection Limit (LoD), Response Time (RT), Sensitivity (S) and Selectivity
A Amatore *et al.*, *Chem. Eur. J.*, 2001 Amatore *et al.*, *Chem. Bio. Chem.*, 2006 Amatore *et al.*, *Anal. Chem.*, 2010	Special methods. The different species including PON were reported as being selectively quantified by amperometry at their respective potentials (see text)
B Xue *et al. Anal. Chem.*, 2000	LoD = 18 nM $S = 2.4 \times 10^{-3}$ nA/nM Selectivity was examined (see text)
C Wang and Chen *Talanta*, 2010	LoD = 100 nM $S = 0.22$ nA/nM Selectivity was examined (see text)
D Bedioui *et al.*, *Phys. Chem. Chem. Phys.*, 2010 Bedioui *et al.*, *NO Biol. Chem.*, 2012 Cortes *et al.*, *Electroanalysis*, 2007 Quinton, Griveau, Bedioui *Electrochem. Commun.*, 2010 Quinton *et al.*, *Lab on Chip*, 2011	LoD = 5000 nM $S = 14 \times 10^{-6}$ nA/nM Qualitative indicator of ONOO- bolus pH 7.1 Selectivity not examined (see text)
E Koh *et al.*, *Anal. Chem.*, 2010	LoD = 2 nM Selectivity determined (see text)
F Kubant *et al.*, *Electroanalysis*, 2006	LoD = 1 nM Selectivity not apparent (see text)
G Peteu *et al.*, *Biosens. Bioelectron.*, 2010 Peteu *et al.*, *An. Chim. Acta*, 2013	LoD = 200 nM TR = 5 s $S = 13 \times 10^{-3}$ nA/nM LoD = 2500 nM $S = 75 \times 10^{-3}$ nA/nM Selectivity examined (see text)
H Szunerits *et al.*, *Analyst*, 2013	LoD = 5 nM $S = 7.5$ nA/nM Selectivity not yet examined (see text)

species, specifically reduced oxygen ($O_2^{\bullet-}$, H_2O_2) and reduced nitrogen (NO, NO_2^-, $ONOO^-$). The distance between the platinized carbon microelectrode (CFE) tip and the cell surface was minimized, thus facilitating a fast detection of species despite their short life spans. Then, amperometric tests performed on fibroblast cells at different oxidative potentials, have shown that the cell – response level increased with the potential and, accordingly these responses appeared to integrate multiple electroactive species. The *in vivo* experimental data consisted of several oxidative waves, subsequently confirmed by *in vitro* experiments with same electrodes for stable solutions. Figure 6.1 shows the species H_2O_2, $ONOO^-$, NO, NO_2^- to be directly oxidized at the platinized CFEs for their respective potentials *vs.* SSCE. Each species, including the $ONOO^-$, could selectively be quantified by amperometry at their respective potentials. While this experimental algorithm allowed new investigations in carefully controlled bioenvironments, its definitive merits for unknown complex configurations *in vivo*, still remain to be seen.

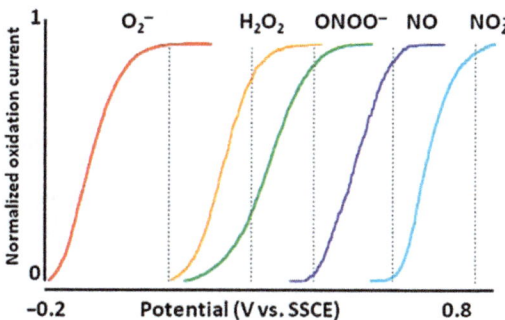

Figure 6.1 Individual steady-state voltammograms obtained at the same electrodes for the respective species O_2^-, H_2O_2, $ONOO^-$, NO, NO_2^-· including peroxynitrite.
Reproduced with kind permission from the PCCP Owner Societies.

Metallo-phthalocyanine and metallo-porphyrin complexes have been tested as peroxynitrite-sensitive electrochemical interfaces. From the family of the tetrapyrrolic complexes, the metallo-phthalocyanines (MPC) are chemically and thermally stable, have good electrocatalytic activity, with diverse co-ordination and substitutional possibilities conferring versatile physicochemical properties to MPCs.[75,76] MPCs are known to behave as shuttles/relays in a series of oxidative and reductive electron-transfer processes, even when im-mobilized on surfaces.[77–80] Furthermore, the MPCs can be readily tailored by substituent variations onto the phenyl ring, with their electroactive metal center typically conferring good electrocatalytic performance.[81] Catalytic applications of MPCs include sensing analytes such as oxygen, peroxide, thiols, dopamine, epinephrine, nitrite.[81–83] In aqueous phase, the MPC electrochemistry is illus-trated by the multiple and often reversible redox processes localized on their PC ring and metal center.[84] On the other hand, the metallo-porphyrin complexes have shown similar characteristics.[99]

The applications targeting PON detection follow. The manganese tetra-aminophthalocyanine (MnTAPC) was employed by both the groups of Jin[71] and of Bedioui[70] for the PON detection on MnTAPC-modified CFEs and Pt disc microelectrode, respectively.

Jin and colleagues[71] electropolymerized the MnTAPC onto 7-μm diameter CFEs. Based on quantum chemistry, they showed the O* atom from peroxynitrite $O-N-O-O^{*-}$ providing the lone-pair electron to the center of the Mn atom of the MnTAPC when these form an axial coordination complex. The reaction in Scheme 6.4 shows that PON was oxidized to nitric dioxygen and nitrite. The detection method employed was differential pulse amperometry (DPA), the current measured being the difference between the values at the potentials 0.20 V and 0.0 V *vs.* Ag/AgCl. The majority of the electroactive compounds did not interfere with the signal, or were signifi-cantly attenuated, since apparently these were easily oxidized at potentials poly(TAPc)Mn$^{(III)}$ (H$_2$O)+ O=N-O-O$^-$ $\longrightarrow$ [poly(TAPc)Mn$^{(III)}$-OONO]$^-$ +H$_2$O

2 [poly(TAPc)Mn$^{(III)}$-OONO]$^-$ $\longrightarrow$ 2 poly(TAPc)Mn$^{(IV)}$=O + NO$_2^-$ + NO$_2$

poly(TAPc)Mn$^{(IV)}$=O + 2e + 2H+ $\longrightarrow$ poly(TAPc)Mn$^{(III)}$ (H$_2$O)

Scheme 6.4 The detailed catalytic reaction between MnTAPC and peroxynitrite. Reproduced with kind permission from the American Chemical Society.

over 0.2 V *vs.* Ag/AgCl. Additionally, a poly(4-vynilpyridine) film (PVP) positively charged was added and it performed as a cation-repulsive barrier and as a diffusional screen for larger biomolecules. The selectivity was tested against a list of electroactive, interfering agents from biological media including: nitric oxide, nitrite, nitrate, hydrogen peroxide, ascorbic acid, uric acid, dopamine, epinephrine, glutathione, arginine. However, this improvement in selectivity typically comes at the expense of the response time, which is a drawback when fast detection is needed.

Bedioui and colleagues[70] reported a similar MnTAPC film polymerized in DMF at a Pt disc microelectrode, showing the determination of stable PON in aerobic alkaline aqueous solution of 1.2 M NaOH at pH 10.2 based on the electrocatalytic reduction current at –0.45 V *vs.* Ag/AgCl according to an overall mechanism involving a two-electron reduction of a manganese oxo intermediate [Mn^{IV} = O], as previously suggested, although the exact mechanism has not been identified. The main advantage of both of these MnTAPC-based PON electrodes is the relatively low potential applied between 0.0 V and –0.5 V *vs.* Ag/AgCl in the peroxynitrite detection. This should solve the problem with most interfering oxidizable compounds.

For another PON-sensitive method developed in the same group,[48] the detection was reported at a bare gold RDE (rotating disc electrode) at 7.1 pH in PBS. Since PON is extremely unstable in neutral solution, the current–potential curve was reconstructed by RDE amperometry in the interval (–0.5 to 0.7) V *vs.* SCE. The detection was reported to occur *via* electroreducing the conjugated peroxynitrous acid ONOOH. While the detection is claimed to be selective due to the –0.1 V operating potential, the PON bolus detection seems qualitative at best, plus the sensitivity cannot easily be calculated and that is a clear drawback of this method.

The same group published a proof of concept for an on-chip electrochemical sensor array (ESA) platform.[50] The ESA has individually addressable gold ultramicroelectrodes (UMEs) of 50 μm in diameter and counter with Ag/AgCl reference) for simultaneous screening of nitric oxide and peroxynitrite. The ESA was interfaced with cell-culture wells.

Other electrochemical approaches to detect peroxynitrite include Wang and Chen[72] who employed a film of cyanocobalamin (vitamin B$_{12}$) electropolymerized on a modified glassy carbon electrode, showing good electrocatalytic activity for PON oxidation.

$$poly(Cbl)^{(III)} + O{=}N{-}O{-}O^- \longrightarrow poly(Cbl)^{(II)} - OONO^*$$
$$poly(Cbl)^{(II)} - OONO^* -e \longrightarrow poly(Cbl)^{(II)} + ONO_2^*$$
$$2ONO_2^* \longrightarrow O_2 + 2NO_2$$

Scheme 6.5 Reaction mechanism showing the redox-active cobalt center in poly-(cyanocobalamin) film to catalyze peroxynitrite decomposition.

Earlier studies have shown that cobalt (II) tetraphenylporphyrins (Co(II)TPP) can bind the O* atom in peroxynitrite (O–N–O–O*$^-$) to form an axial coordination compound (Co(II) –OONO–TPP). The reversible binding reaction of peroxynitrite with Co(II)TPP has been confirmed by kinetic and mechanistic experiments.[72]

Because cyanocob(II)alamin – or Cbl(II) – also contains cobalt(II) porphyrin, the thought was that Cbl(II) binds peroxynitrite to form a stable Cbl(II)–OONO$^-$ coordination compound. The equations showing the postulated electrocatalysis mechanism of poly(cyanocobalamin) film toward peroxynitrite oxidation are illustrated in Scheme 6.5.

The postulated mechanism demonstrated the redox-active cobalt center in poly(cyano-cobalamin) film can catalyze peroxynitrite decomposition through the formation of intermediates, which is in agreement with the mechanism of peroxynitrite decay in basic aqueous media proposed previously.[72]

The selectivity was claimed by studying the interference from: glucose, dopamine, ascorbic acid, uric acid, xanthine, l-arginine, H_2O_2, NO and NO_3^-, NO_2^-. As a drawback, however, there were no data shown, or theoretical reasoning, in support for these specificity claims.

Koh with colleagues[73] synthesized a manganese–polymer complex film (polydithienyl-pyrrole-benzoic acid, or PDB) and electropolymerized it onto a Pt microelectrode. This (Mn-pDPB) film also had gold nanoparticles electrodeposited on the modified surface and it reportedly enhanced the PON reduction according to the reaction mechanism from Scheme 6.6.

In this work SIN-1 was employed as a peroxynitrite donor solution with a half-life of 14–26 min, during which time PON was produced continuously. This was timely confirmed with its 302-nm absorbance peak and was, furthermore, ascertained in inhibitory tests with a peroxynitrite scavenger, such as uric acid. During such tests, first the response current rose sharply after several additions of PON, then the current arrived at a plateau, only to decline abruptly following the addition of the PON scavenger uric acid.

An increase in selectivity was reported by using an outer layer of polyethyleneimine (PEI), the interferents tested included serotonin, dopamine, ascorbic acid, bilirubin, and the PON decomposition molecules. The selectivity was reported as satisfactory, allowing these UMEs to be employed for the PON determination *in vitro* on glioma cells. To fully validate this interesting work, it would be perhaps interesting to repeat the same experiments with the "genuine" peroxynitrite freshly synthesized.

Malinski and colleagues[74] utilized a manganese(III) [2–2]paracyclophenylporphyrin (MnPCP) that was electrodeposited on a 5-μm long, 250-μm

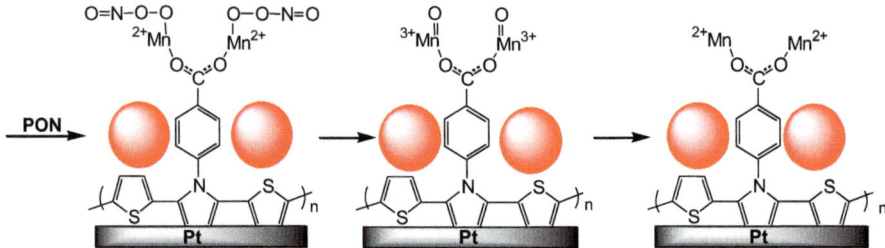

Scheme 6.6   Reaction mechanism involving the manganese-polymer complex polymerized and the gold nanoparticles electrodeposited from solution.
Reproduced with kind permission from the American Chemical Society.

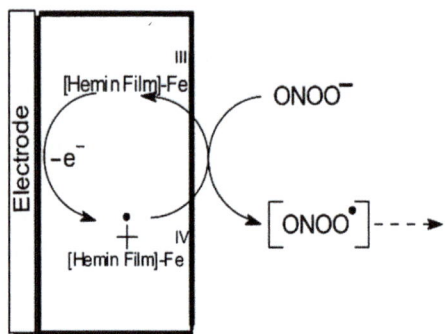

Scheme 6.7   Electrocatalytic PON oxidation scheme by the hemin-modified GC
electrode.

diameter CFE for the PON sensor. They reported the detection of PON in the
presence of NO and O_2^-.

The chrono-amperometry was conducted with a module integrating 3
working electrodes poised at different potentials, namely 0.67 V (for NO),
0.35 V (for O_2^-) and –0.35 V (ONOO$^-$), although the nature of the reference
electrode was not clear from the paper. Here, an interesting fact is that the ratio
between the concentration of NO and ONOO$^-$, that is [NO]/[ONOO-] was
reported as a potential diagnostic marker, *e.g.* in cardiovascular diseases. While
this idea is surely interesting, a disadvantage here is that some details seem
unclear and so should be clarified in the future.

Elsewhere,[57,127] the electropolymerized iron(III) chloroprotoporphyrin IX
(hemin) thin films were developed to examine the electrocatalytic oxidation of
peroxynitrite on glassy carbon and carbon-fiber microelectrodes, in both
standing solutions and in flow-through systems. The reaction scheme is illustrated in Scheme 6.7.[57]

The PON tests revealed a peak-shaped catalytic wave, indicating a fast
catalytic oxidation of peroxynitrite that was not observed with protoporphyrin-
only (without the iron) films. In reference to Scheme 6.7, this suggests the

essential role of the bound iron center/atom to allow the oxidative catalytic turnover of peroxynitrite, also consistent with other reported direct interaction with metallo-macrocycles. In one study, this interaction led to PON decomposition, with a set of iron porphyrins in solution is based at the FeIII center, and yields to the isomerization product, nitrate, through a sequence of binding, "in-cage" dissociation mechanism, and rebound, utilizing an $Fe(IV) = O$ intermediate.[57]

Additionally, a comparison was made between the properties of hemin alone with hemin-PEDOT films. Hemin film exhibited a moderate response mediated by a direct electrocatalytic oxidation of peroxynitrite. By contrast, after adding PEDOT to the matrix, the sensitivity of the hemin-PEDOT modified CFEs is strongly enhanced. This was due to the tortuous porous hybrid surface shaped as a "nanocauliflower" that is characteristic of the PEDOT grown film superimposed with the apparent synergistic effects between PEDOT and the hemin macrocycle molecules.

The specificity was verified for the main decomposition products of peroxynitrite, in this case (pH 10.5) nitrite and nitrate compared with PON for same concentration 400 μM. The peak current ratio of PON compared with NO_2^- and NO_3^- was 6.7 times higher. Certainly, more work is needed to address the selectivity against other typical interferents from biological media including: ascorbic acid, uric acid, dopamine, glutathione, arginine.[71]

The next section will examine more on the use of PEDOT and graphene to increase the response sensitivity. In addition, some just-published results on using a (reduced graphene–hemin) PON-sensitive film will be discussed in detail.

6.4 Strategies to Increase Response Sensitivity: Electroactive Polymers or Graphene

Electroactive Polymers (EAPs) are one way to increase the response/sensitivity of the electrochemical systems, also known as *Inherently Conductive Polymers* (ICPs) these two terms being used interchangeably in the literature.[85–87]

The common sense that plastics, unlike metals, do not conduct electricity was shattered in the 1970s. The discovery and development of conductive polymers was later rewarded with the 2000 Chemistry Nobel prize to A.J. Heeger, A.G. MacDiarmid and H. Shirakawa.[88]

The use of EAPs deposition will be examined next, with a focus on the electropolymerization method to apply a film or (multi)layer onto a working electrode (WE). Herein, the main goal is to exploit the EAP's potential and thus to impart its qualities to the WE and improving the electroanalytical performance and function adequately as "electrode modifiers", enhancing the WE electrocatalytic activity, while also often offering antifouling properties.[85–87]

With their intrinsic advantages, nowadays the polythiophenes may be considered the preferred choice for amperometric sensing, although the polyaniline and polypyrrole should not be undervalued for electroanalysis in general.[90,91]

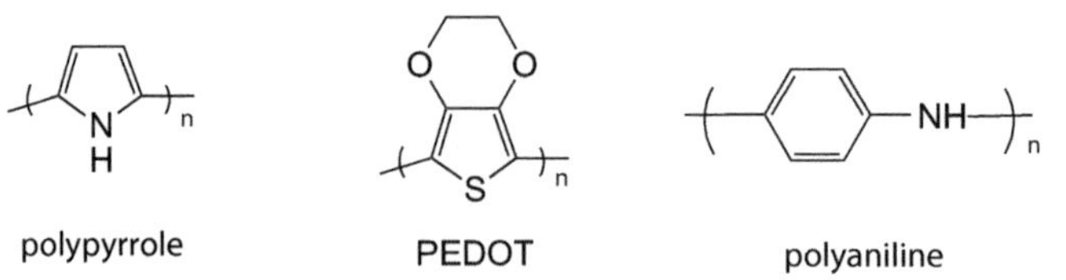

polypyrrole   PEDOT   polyaniline

Scheme 6.8   Typical conductive polymers are polypyrrole, PEDOT and polyaniline.

One reason for the attractiveness of thiophene is the arsenal of functionalization tools afforded by its monomer ring, in order to confer good conductivity and high stability to the electrode and/or to create properties such as higher/lower specific affinity or binding towards a specific analytes, or by contrast interferents.

From the thiophene family, the poly-3,4-ethylenedioxythiophene or PEDOT (Scheme 6.8), has been especially preferred for several reasons. The two oxygen atoms coupled to the thiophene rings permits this monomer to be oxidized at especially low potentials. This aspect, added to the very low steric hindrance of the condensed rings and the molecule symmetry, leads the PEDOT to experience a high conductivity and a narrow bandgap, being easily oxidized with a wide anodic potential window.[92–94] Furthermore, it was determined that its condensed ring polarity allows polymerization in aqueous solution.[87]

The advancement of nanostructured hybrid materials came recently to offer new perspectives to increase the sensor's response (sensitivity) including the use of conductive nanohybrids with an EAP complementing the electrocatalytic activity of a partner such as metal (oxide) nano or microparticles.[95–98]

Metal complexes or ions can also act as (co-)electrocatalysts, while the overall (PEDOT-metal ions/complex) hybrid film was shown to improve its stability and, furthermore, decrease its resistance to charge transfer.[87] In one such case, the iron protoporphyrin (hemin) has been employed for the detection of peroxynitrite ($ONOO^-$) electroassembled on a glassy carbon electrode.[8] When the hemin was included in a PEDOT coating, the sensitivity and linear range of the calibration plots was strongly enhanced with respect to the two single components, that is the PEDOT or the hemin, separately. Notably, the hemin–PEDOT hybrid film enjoys a fractal three-dimensional "cauliflower" porous matrix with a very large specific area/volume, similar with the one shown in Figure 6.2 as ascertained by SEM imaging.[100]

Additionally, a comparison was made between the properties of hemin alone with hemin–PEDOT films. Hemin film exhibited a moderate response (mediated a direct electrocatalytic oxidation of peroxynitrite and. By contrast, after adding PEDOT to the matrix, the sensitivity of the hemin–PEDOT-modified CFEs is being strongly enhanced. This was due to the tortuous porous hybrid surface shaped as a "nanocauliflower" that is characteristic of the PEDOT grown film superimposed with the apparent synergistic effects between PEDOT and the hemin macrocycle molecules.

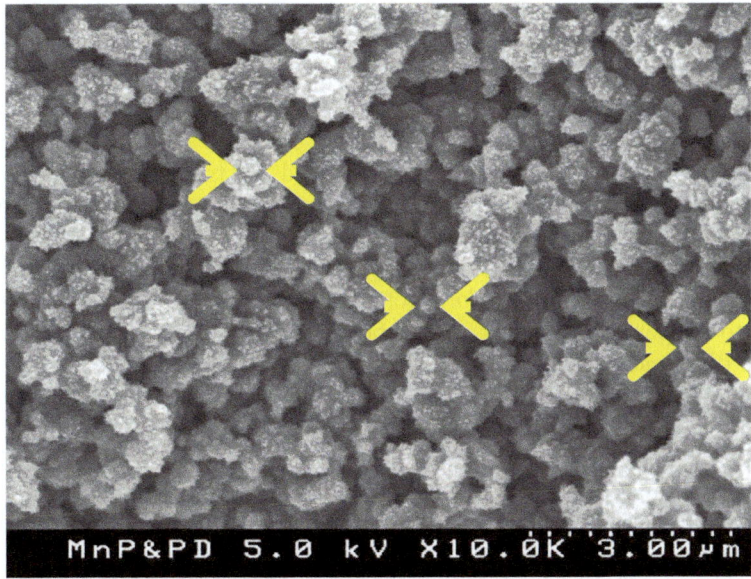

Figure 6.2   Nanohybrid of a metallo-porphyrin and PEDOT. The dimension between arrows is less than 300 nm.

Further testing has arguably ascertained synergy effects developed between the PEDOT and the hemin macrocycle molecules: this conclusion was supported by a significant lowering of the oxidation potential for the nanohybrid material, as compared with the responses from either one of the two components – when employed separately.[8,87]

Whereas such hybrid nanomaterials always seem desirable due to the ability to tailor their properties to the specifically desired application, however, the fine tuning of the properties is often the result of a "trial and error" approach.[87] Additionally, there are few publications outlining the use of PEDOT for sensors, in real biological media.

To optimize a particular electropolymerization method, the experimental matrix has tens of variables or combinations to be addressed, even when the same monomer is used, including: the nature of solvent, supporting electrolyte, monomer concentration, film thickness, the temperature and each of those or their combination can affect the usability (structural, morphological, chemical features) of the electrodeposited film.

To this aim, *the interaction at the film/electrode interface is important*, as it is considered that the very first layers of adsorbed molecules on the WE surface very strongly affect the properties of the whole WE system.[101] And finally, *in situ measurements performed by electrochemical, microgravimetric, microscopic and spectroscopic* techniques performed on the electropolymerized film can accurately characterize the thin layer formed on this conductive substrate.[102–105]

Graphene, or other carbon-based materials such as glassy carbon, diamond, diamond-like carbon, carbon nanotubes, fullerenes are another way to increase the *WE response and sensitivity.* They have attracted much interest, because of their marked structural differences and their related variety of electronic and electrochemical properties.[106–109] Their low cost, chemical stability, wide potential window, rich surface chemistry and enhanced electrocatalytic activity for a variety of redox reactions have made these materials of high importance for analytical and industrial electrochemistry. Different forms of carbon nanomaterials, including carbon nanotubes, carbon nanofibers and fullerenes have been more recently employed for electroanalytical applications and have been shown to outperform the classical carbon materials based on graphite, glassy carbon, diamond and carbon black.[110]

In contrast to other carbon-based electrodes, carbon nanotubes are typically grown from carbon-containing gases with the use of catalytic metal nanoparticles, which remain in the nanostructures even after extensive purification procedures. As a consequence, the electrochemical behavior of carbon nanotubes is dominated by these metal nanoparticles.[111] These metallic impurities are not only causing problems for the construction of reliable sensors and energy devices[112,113] but are also responsible of toxicological hazards within biological samples.[113]

The demonstration by Novoselov and Geim, the 2010 Nobel Laureates in Physics, that single layers of graphene can be isolated from graphite and identified by microscopy has pushed graphene to the forefront of research in the design of electrochemical sensors.[114,115] Most recently, heminfunctionalized graphene nanosheets have revealed PON activity.[116,117] The potential of graphene to support organic molecules such as hemin, and other porphyrin species, through $\pi - \pi$ stacking interactions[116–122] makes this material of high interest to study its resulting peroxynitrite activity.[117] The hemin-functionalized reduced graphene oxide (rGO/hemin) matrix can be formed by reduction of graphene oxide to reduced graphene oxide using hydrazine[116] or dithiothreitol[117] followed by immersion into hemin solution, or by a simultaneous reduction and chemical modification process as illustrated in Figure 6.3A.[116]

The incorporation of hemin into the RGO matrix can be revealed by the presence of the $Fe^{3+/2+}$ center of hemin[119,122,123] at $E = -0.4$ V (Figure 6.1B)., where the amount of incorporated hemin was estimated as Γ was $(3.6 \pm 0.5) \cdot 10^{-8}$ mol cm^{-2}. This value is much larger than that reported for hemin–pyrolytic graphite $(0.74 \cdot 10^{-8}$ mol cm$^{-2})$,[124] hemin–multiwalled carbon nanotubes $(0.27 \cdot 10^{-8}$ mol cm$^{-2})$[125] and is comparable to hemin immobilized on highly ordered mesoporous carbon $(1.75 \cdot 10^{-8}$ mol cm$^{-2})$.[126]

Addition of ONOO$^-$, generated from SIN-1 to a GCE/rGO/hemin electrode shows an oxidative wave at $E_{ox,1} = 1.17$ V (Figure 6.4A). This wave is close to the oxidation potential of hemin and is thus assigned to the electrochemical oxidation of hemin on the rGO platform. In the presence of ONOO$^-$, an electrocatalytic oxidations ONOO$^-$ mediated by oxidized hemin centers seems to occur (Figure 6.4B).[127] The Fe^{3+} center in the rGO/hemin film is oxidized to

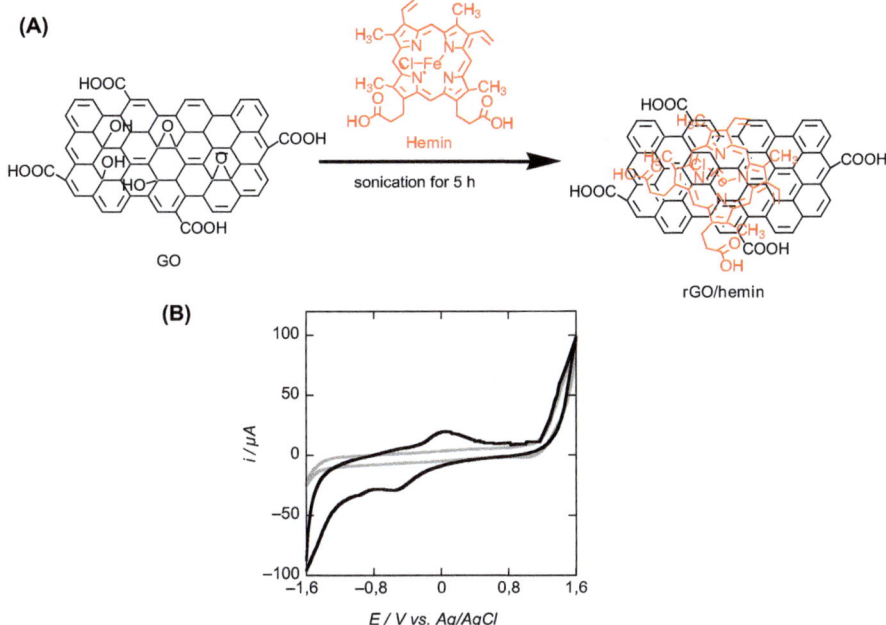

Figure 6.3   (A) Schematic representation of the synthesis of rGO/hemin; (B) Cyclic voltammograms of bare glassy carbon (black) and a rGO/hemin modified GCE electrode prepared by drop casting 5×25 μL of rGO/hemin (0.5 mg/mL) onto GCE (black): solution: N_2-saturated PBS buffer (0.1 M, pH 7.4), scan rate: 50 mV s^{-1}.

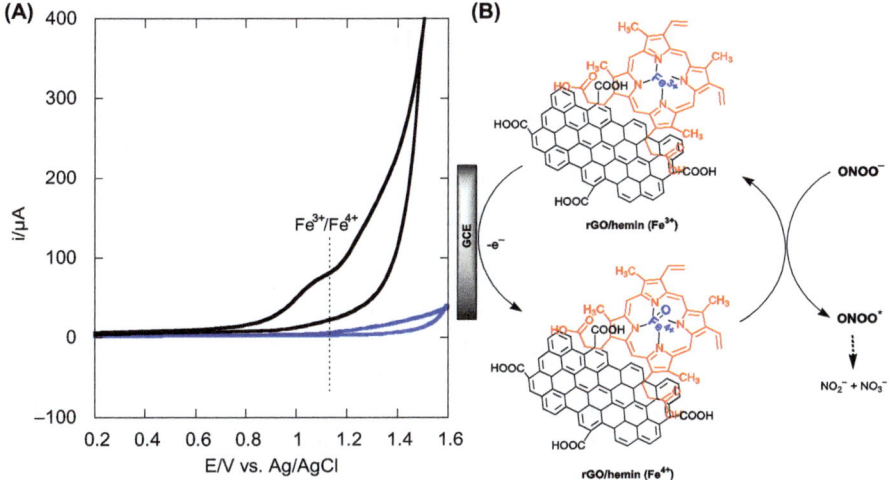

Figure 6.4   (A) Cyclic voltammograms in the absence (black) and presence of 200 μM SIN-1 (pH 7.4) on GCE electrode modified with rGO/hemin (blue); (B) Proposed mechanism for the electrocatalytic oxidation of ONOO$^-$ by rGO/hemin interface.

(A)   **(B)**

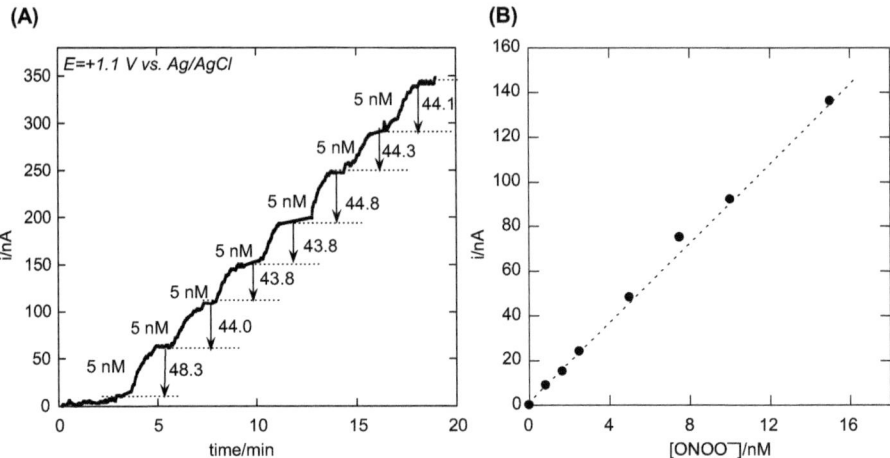

Figure 6.5   (A) Typical amperometric response curve obtained using a GCE/rGO/
hemin electrode polarized at +1.1 V *vs.* Ag/AgCl with subsequent addition of SIN-1 and, (B) Calibration curve.

a high-valent iron form (*e.g.* iron oxo intermediate, $[Fe^{4+} = O]$) electrochemically at the electrode interface, which, in the presence $ONOO^-$, is re-
reduced back to Fe^{3+} for further turnovers. The performance of the rGO/
hemin modified GCE towards the detection of peroxynitrite was evaluated by
chronoamperometry and showed a PON sensitivity of $\approx 7.5 \pm 1.5$ nA nM^{-1} with
a low nanomolar detection limit of $\approx 5 \pm 1$ nM, as illustrated in Figure 6.5.

6.5 Conclusions and Perspectives

This chapter offered an overview of the state-of-the-art in electrochemical
quantification of peroxynitrite, a powerful *nitrating, nitrosating and oxidative
agent*[4,10,11] and product of the nitric oxide reacting with superoxide radical.[4–8]
In addition, this chapter examined the real challenges to detect this analyte,
together with the most recent advances.

   In order to show the vital necessity for *in vivo* $ONOO^-$ detection, the medical
facts are purposely outlined, as a solid body of published findings has clinically
ascertained peroxynitrite to be a potent cell death inducer, in several devastating diseases.[4–6] Equally important are the strategies to develop PON decomposition catalysts as potential therapies to attenuate its adverse effects.
These therapeutic methods are based on peroxynitrite isomerization to nitrate
using metallo-macrocycle molecules and the catalytic reduction of PON to
nitrite.

   Currently, a number of these PON decomposition compounds are examined
extensively on diseases models,[7–10] including by few dedicated startups or
spinoffs. Not surprisingly, the interest from the "angel investors" to finance
such potentially successful endeavors – PON decomposition catalysts – could

be massive in the United States for example, where clearly the medical benefits of any candidates successfully vetted by the Food and Drug Administration (FDA) would bring vast scientific and financial paybacks.

Interestingly, some of the classes of metallo-macrocycles *decomposing or destroying* peroxynitrite[7–10] are arguably also *electrochemically sensitive* to PON,[57,70–72] since in both cases catalytic redox processes are involved. *So, one can imagine perhaps the synthesis of peroxynitrite-sensitive films with "detect-and-destroy" double capabilities?*

There is little doubt that PON detection poses multiple challenges, especially *in vivo*, stemming from its very short lifetime, coupled with an extremely high reactivity within multiple pathways concurrently, in real environments.[7–10,47–49] The optical detection has been available for some time now (*e.g.* oxidation of fluorescent probes, chemiluminescence, probe nitration), however, it was not a focus in this chapter. As these optical techniques tend to be indirect, rather elaborate and relatively difficult to apply real time, it is not surprising that one turns to electrochemical quantification that is conceptually simple, more convenient for real-time, label-free and direct *in situ* measurements especially in the microsensor format[89,127–130] and including for peroxynitrite.[8,57,70–74]

The state-of-the-art in the electrochemical quantification of peroxynitrite was examined in some detail. There are just a handful of groups involved, and some PON *detection limits* were reported in the *low-single digits*.[70–74,116] This would seem encouraging when associated with a *fast response time*. However, a major difficulty for implementation continues to be the *specificity* and not all authors investigated it for their PON-sensitive films. This selectivity is perhaps the most essential element to consider, particularly for *in vivo* usability. Equally important are other performance factors such as sensor *calibration*, the *geometry* of the *sensitive surface* – be it macro-, micro-, or nano*scale* – and the *noise* in the response that often has multiple roots.

The efforts to develop specificity for peroxynitrite were reviewed herein and these included: (i) using an outer membrane to reduce competing interference by means of *charge repulsion* or *size exclusion*[71] (ii) employing interrogation techniques with specific pulse profiles, such as differential pulse amperometry (DPA) or differential pulse voltammetry (DPV) that can help discriminate between competing species,[71] (iii) considering the specific redox potential of each species including PON and designing methods to separate their contribution within the sensor response signal.[68,69] However, one needs to be mindful of possible *trade-offs between specificity at a cost of slower response time*, especially since PON is a short-lived species with about 1 s lifetime.

In particular, two interesting hybrid films were examined: the polythiophene–hemin[57] and the reduced graphene oxide–hemin complex[116] and their apparent significant increase in sensor response was further scrutinized.

In the case of the PEDOT-hemin hybrid film utilized for the first time in PON detection,[8] the Fe ions acted as (co-)electrocatalysts, while the overall hybrid film seemed to improve the overall stability and furthermore enhance charge transfer.[87] Notably, the hemin-PEDOT hybrid film was imaged by SEM, as a porous matrix typically shaped as a "nanocauliflower". When the hemin was included in a PEDOT coating, the sensitivity and linear range of the calibration plots was strongly enhanced with respect to any of the two single components (PEDOT or hemin) tested separately, seemingly due to the extremely high (specific area *per* volume) ratio[100] superimposed with apparent synergy effects between PEDOT and the hemin macrocycle molecules. This conclusion was supported by a significant lowering of the oxidation potential for the PEDOT–hemin nanostructured material, compared with responses from either one of the two, separately used components.[8,87]

And finally, the rGO/hemin nanocomposite was successfully tested for the first time,[116] as a sensitive platform for the detection of peroxynitrite generated by SIN-1 in PBS. The sensitivity of the rGO/hemin-modified electrodes for peroxynitrite was $7.5 \pm 1.5\,\mathrm{nA\,nM^{-1}}$ with a low nanomolar detection limit of $5 \pm 1\,\mathrm{nM}$. Compared with the rGO formed by hydrazine reduction and postmodified with hemin, the sensitivity was more than an order of magnitude lower at $0.6\,\mathrm{nA\,nM^{-1}}$ and a detection limit more than twice larger.

The fundamental causes for such an enhancement are diverse and need to be further investigated, as several graphene features may contribute collectively to the enhanced performance, as discussed elsewhere.[116]

While more efforts are ongoing to better understand the catalytic mechanism at play, this study clearly highlights for the first time, the importance of the use of graphene-supported hemin as a general strategy for the fabrication of highly sensitive peroxynitrite sensors.[116]

Acknowledgements

Sabine Szunerits gratefully acknowledges financial support from the Centre National de Recherche Scientifique (CNRS), the University Lille 1, Nord Pas de Calais region and the Institut Universitaire de France (IUF). Serban F. Peteu is grateful for financial support from the Agentia Nationala de Cercetare Stiintifica (ANCS - UEFISCDI) through PN-II Project 184/2011. The bilateral Partenariat Hubert Curien – Program Brancusi is also acknowledged.

References

1. J. S. Dawane and V. A. Pandit, *J. Clin. Diagn. Res.*, 2012, **6**, 1796.
2. E. E. Lomonosova, M. Kirsch, U. Rauen and H. de Groot, *Free Rad. Biol. Med.*, 1998, **24**, 511.
3. M. Tecder-Unal and H. Tufan, *Free Rad. Biol. Med.*, 2002, **33**, S43.
4. W. H. Koppenol, J. J. Moreno, W. A. Pryor, H. Ischiropoulos and J. S. Beckman, *Chem. Res. Toxicol.*, 1992, **5**, 834.
5. J. S. Beckman and W. H. Koppenol, *Am. J. Physiol.-Cell Physiol.*, 1996, **271**, C1424.
6. C. Szabo, H. Ischiropoulos and R. Radi, *Nature Rev. Drug Discov.*, 2007, **6**, 662.
7. P. Pacher, J. S. Beckman and L. Liaudet, *Physiol. Rev.*, 2007, **87**, 315.

8. S. F. Peteu, S. Banihani, M. M. Gunesekera, P. Peiris, O. A. Sicuia, and M. Bayachou, *Oxidative Stress: Diagnostics, Prevention, and Therapy*, S. Andreescu and M. Hepel, Am. Chem. Soc., Washington DC, 1st edn, 2011, *ACS Symposium Series* 1083, 311–339.
9. H. Cai and D. G. Harrison, *Circulat. Res.*, 2000, **87**, 840.
10. G. Ferrer-Sueta and R. Radi, *ACS Chem. Biol.*, 2009, **4**, 161.
11. C. Amatore, S. Arbault, C. Bouton, J. C. Drapier, H. Ghandour and A. C. W. Koh, *Chem. Biol. Chem.*, 2008, **9**, 1472.
12. S. Amemiya, J. D. Guo, H. Xiong and D. A. Gross, *Anal. Bioanal. Chem.*, 2006, **386**, 458.
13. T. Finkel, *Curr. Opin. Cell Biol.*, 2003, **15**, 247.
14. B. Halliwell and J. M. C. Gutteridge, *Free Radicals in Biology and Medicine*, 3rd edn, Oxford University Press, Oxford, 1999.
15. R. F. Furchgott, *Biosci. Rep.*, 1999, **19**, 235.
16. M. A. Arstall, D. B. Sawyer, R. Fukazawa and R. A. Kelly., *Circ. Res.*, 1999, **85**, 829.
17. V. Grishko, V. Pastukh, V. Solodushko, M. Gillespie, J. Azuma and S. Schaffer, *Am. J. Physiol. Heart. Circ. Physiol.*, 2003, **285**, H2364.
18. J. G. Dickhout, G. S. Hossain, L. M. Pozza, J. Zhou, S. Lhotak and R. C. Austin, *Arterioscler. Thromb. Vasc. Biol.*, 2005, **25**, 2623.
19. J. Li, W. Li, J. Su, W. Liu, B. T. Altura and B. M. Altura, *Exp. Biol. Med.*, 2004, **229**, 264.
20. J. Li, W. Li, J. Su, W. Liu, B. T. Altura and B. M. Altura, *Neurosci. Lett.*, 2003, **350**, 173.
21. P. Ferdinandy and R. Schulz, *Br. J. Pharmacol.*, 2003, **138**, 532.
22. P. Pacher, R. Schulz, L. Liaudet and C. Szabo, *Trends Pharmacol. Sci.*, 2005, **26**, 302.
23. P. Wang and J. L. Zweier, *J. Biol. Chem.*, 1996, **271**, 29223.
24. W. Yasmin, K. D. Strynadka and R. Schulz, *Cardiovasc. Res.*, 1997, **33**, 422.
25. A. A. Chaves, M. J. Mihm, B. L. Schanbacher, A. Basuray, C. Liu, L. W. Ayers and J. A. Bauer, *Cardiovasc. Res.*, 2003, **60**, 108.
26. N. W. Kooy, S. J. Lewis, J. A. Royall, Y. Z. Ye, D. R. Kelly and J. S. Beckman, *Crit. Care Med.*, 1997, **25**, 812.
27. J. S. Beckman, Y. Z. Ye, P. G. Anderson, J. Chen, M. A. Accavitti, M. M. Tarpey and C. R. White, *Biol. Chem. Hoppe-Seyler*, 1994, **375**, 81.
28. N. Escobales and M. J. Crespo, *Curr. Vasc. Pharmacol.*, 2005, **3**, 231.
29. A. Csiszar, Z. Ungvari, J. G. Edwards, P. Kaminski, M. S. Wolin, A. Koller and G. Kaley, *Circ. Res.*, 2002, **90**, 1159.
30. A. Denicola and R. Radi, *Toxicol.*, 2005, **208**, 273.
31. P. Sarchielli, F. Galli, A. Floridi and V. Gallai, *Amino Acids*, 2003, **25**, 427–436.
32. M. J. Stevens, *Curr. Vasc. Pharmacol.*, 2005, **3**, 253.
33. W. L. Suarez-Pinzon, C. Szabo and A. Rabinovitch, *Diabetes*, 1997, **46**, 907.

34. C. Szabo, J. G. Mabley, S. M. Moeller, R. Shimanovich, P. Pacher, L. Virag, F. G. Soriano, J. H. Van Duzer, W. Williams, A. L. Salzman and J. T. Groves, *Mol. Med.*, 2002, **8**, 571.
35. M. Thiyagarajan, C. L. Kaul and S. S. Sharma, *Br. J. Pharmacol.*, 2004, **142**, 899.
36. S. S. Sharma, S. Munusamy, M. Thiyagarajan and C. L. Kaul, *J. Neurosurg.*, 2004, **101**, 669.
37. D. Liu, F. Bao, D. S. Prough and D. S. Dewitt, *J. Neurotrauma*, 2005, **22**, 1123.
38. C. Bianchi, H. Wakiyama, R. Faro, T. Khan, J. D. McCully, S. Levitsky, C. Szaboœ and F. W. Sellke, *Ann. Thorac. Surg.*, **74**, 1201.
39. S. Cuzzocrea, E. Mazzon, R. Di Paola, E. Esposito, H. Macarthur, G. M. Matuschak and D. Salvemini, *J. Pharmacol. Exp. Ther.*, 2006, **319**, 73.
40. D. Salvemini, Z. Q. Wang, D. M. Bourdon, M. K. Stern, M. G. Currie and P. T. Manning, *Eur. J. Pharmacol.*, 1996, **303**, 217.
41. D. Salvemini, Z. Q. Wang, M. K. Stern, M. G. Currie and T. P. Misko, *Proc. Natl. Acad. Sci. USA*, 1998, **95**, 2659.
42. J. G. Mabley, J. G. L. Liaudet, P. Pacher, G. Southan, J. T. Groves, A. L. Salzman and C. Szabo, *Mol. Med.*, 2002, **8**, 581–590.
43. G. S. Scott, S. Spitsin, R. Kean, T. Mikheeva, H. Koprowski and D. Hooper, Proc. Natl Acad. Sci. *USA.*, 2002, **99**, 16303–16308.
44. J. P. Crow, N. Y. Calingasan, J. Chen, J. L. Hill and M. F. Beal, *Ann. Neurol.*, 2005, **58**, 258.
45. A. H. Cross, M. San, M. K. Stern and T. P. Misko, *J. Neuroimmunol.*, 2001, **107**, 21.
46. R. Sugawara, T. Hikichi, T. N. Kitaya, F. Mori, I. Nagaoka, A. Yoshida and C. Szabo, *Curr. Eye Res*, 2004, **29**, 11–16.
47. S. Borgmann, *Analyt. Bioanalyt. Chem.*, 2009, **394**, 95.
48. F. Bedioui, D. Quinton, S. Griveau and T. Nyokong, *Phys. Chem. Chem. Phys.*, 2010, **12**, 9976.
49. M. M. Tarpey and I. Fridovich, *Circulation Res.*, 2001, **89**, 224.
50. C. Amatore, S. Arbault, M. Guille and F. Lemaitre, *Chem. Rev.*, 2008, **108**, 2585.
51. R. M. Uppu, G. L. Squadrito, R. Cueto and W. A. Pryor, *Nitric Oxide, Pt. B*, 1996, **269**, 285.
52. M. N. Hughes, H. G. Nicklin and W. A. Sackrule, *J. Chem. Soc.-Inorg. Phys. Theor.*, 1971, **23**, 3722.
53. J. W. H. Reed and W. L. Jolly, *J. Am. Chem. Soc.*, 1974, **96**, 1248.
54. J. R. Leis, M. E. Pena and A. Rios, *J. Chem. Soc.-Chem. Commun.*, 1993, **16**, 1298.
55. R. C. Plumb and J. O. Edwards, *J. Phys. Chem.*, 1992, **96**, 3245.
56. R. M. Uppu and W. A. Pryor, *Analyt. Biochem.*, 1996, **236**, 242.
57. S. F. Peteu, P. Peiris, E. Gebremichael and M. Bayachou, *Biosens. Bioelectr.*, 2010, **25**, 1914.
58. E. de la Fuente, G. Villagra and S. Bollo, *Electroanalysis*, 2007, **19**, 1518.

59. A. Chatterjee, A. Smith, J. D. Catravas and S. M. Black, *Free Rad. Biol. Med.*, 2008, **45**, S44.
60. L. K. Cuddy, A. C. Gordon, S. A. G. Black, E. Jaworski, S. S. G. Ferguson and R. J. Rylett, *J. Neurosci.*, 2012, **32**, 5573.
61. J. B. Gramsbergen, T. R. Larsen, S. Rossen and P. Roepstorff, *J. Neurochem.*, 2007, **101**, 31.
62. M. J. Kohr, C. J. Traynham, S. R. Roof, J. P. Davis and M. T. Ziolo, *J. Mol. Cell. Cardiol.*, 2010, **48**, 645.
63. J. L. Trackey, T. F. Uliasz and S. J. Hewett, *J. Neurochem.*, 2001, **79**, 445.
64. P.T. Liu, C.E. Hock, R. Nagele, and P.Y.K. Wong, *Am. J. Physiol.-Heart Circulat. Physiol.*, 1997, **272**, H2327.
65. Z. H. Taha, *Talanta*, 2003, **61**, 3–10.
66. A. Daiber, M. Mehl and V. Ullrich, *Nitric Oxide*, 1998, **2**, 259.
67. F. L. Kiechle and T. Malinski, *Ann. Clin. Lab. Sci.*, 1996, **26**, 501.
68. C. Amatore, S. Arbault and D. Bruce, P. de Oliveira, M. Erard, and M. Vuillaume, *Chem. Eur. J.*, 2001, **7**, 4171.
69. C. Amatore, S. Arbault and A. C. W. Koh, *Analyt. Chem.*, 2010, **82**, 1411.
70. J. S. Cortes, S. G. Granados, A. A. Ordaz, J. A. L. Jimenez, S. Griveau and F. Bedioui, *Electroanal.*, 2007, **19**, 61.
71. J. Xue, X. Y. Ying, J. S. Chen, Y. H. Xian, L. T. Jin and J. Jin, *Analyt. Chem.*, 2000, **72**, 5313.
72. Y. Wang and Z.-Z. Chen, *Talanta*, 2010, **82**, 534.
73. W. C. A. Koh, J. I. Son, E. S. Choe and Y.-B. Shim, *Analyt. Chem.*, 2010, **82**, 10075.
74. R. Kubant, C. Malinski, A. Burewics and T. Malinski, *Electroanal.*, 2006, **18**, 410.
75. F. Dumoulin, M. Durmus, V. Ahsen and T. Nyokong, *Coord. Chem. Rev.*, 2010, **254**, 2792.
76. N. Sehlotho, T. Nyokong, J. H. Zagal and F. Bedioui, *Electrochim. Acta*, 2006, **51**, 5125.
77. X. H. Qi and R. P. Baldwin, *Electroanal.*, 1994, **6**, 353.
78. X. H. Qi and R. P. Baldwin, *J. Electrochem. Soc.*, 1996, **143**, 1283.
79. M. Gulppi, S. Griveau, F. Bedioui and J. H. Zagal, *Electrochim. Acta*, 2001, **46**, 3397.
80. T. Mugadza and T. Nyokong, *J. Colloid Interface Sci.*, 2011, **354**, 437.
81. K. S. Lokesh, K. De Wael and A. Adriaens, *Langmuir*, 2010, **26**, 17665.
82. J. Arguello, H. A. Magosso, R. Landers and Y. Gushikem, *J. Electroan. Chem.*, 2008, **617**, 45.
83. J. H. Zagal, S. Griveau, K. I. Ozoemena, T. Nyokong and F. Bedioui, *J. Nanosci. Nanotechnol.*, 2009, **9**, 2201.
84. A. B. P. Lever, *J. Porphyrins Phthalocyanines*, 1999, **3**, 488–499.
85. G. G. Wallace and L. A. P. Kane-Maguire, *Adv. Mater.*, 2002, **14**, 953.
86. G. G. Wallace, M. Smyth and H. Zhao, *TRAC-Trends Analyt. Chem.*, 1999, **18**, 245.

87. C. Zanardi, F. Terzi and R. Seeber, *Analyt. Bioanalyt. Chem.*, 2012, **405**, 509.
88. H. Shirakawa, E. J. Louis, A. G. MacDiarmid, C. K. Chiang and A. J. Heeger, *J. Chem. Soc., Chem. Commun.*, 1977, **16**, 578.
89. R. W. Murray, *Acc. Chem. Res.*, 1980, **13**, 135.
90. C. Dhand, M. Das, M. Datta and B. D. Malhotra, *Biosens. Bioelectron.*, 011, **26**, 2811.
91. A. Ramanavčius, A. Ramanavičlienė and A. Malinauskas, *Electrochim. Acta*, 2006, **51**, 6025.
92. H. J. Ahonen, J. Lukkari and J. Kankare, *Macromolecules*, 2000, **33**, 6787.
93. M. Lapkowski and A. Prón, *Synth. Met.*, 2000, **110**, 79.
94. L. Groenendaal, G. Zotti, P. H. Aubert, M. S. Waybright and J. R. Reynolds, *Adv. Mater.*, 2003, **15**, 855.
95. D. W. Hatchett and M. Josowicz, *Chem. Rev.*, 2008, **108**, 746.
96. C. Zanardi, F. Terzi, L. Pigani, and R. Seeber, *Encyclopedia of Polymer Composites: Properties, Performance and Applications*, M. Lechkov and S. Prandzheva, Nova, New York, 2011, pp 1–74.
97. A. Bhattacharya and A. De, *Prog. Solid. State. Chem.*, 1996, **24**, 141.
98. R. Gangopadhyay and A. De, *Chem. Mater.*, 2000, **12**, 608.
99. S. Bahaduri and D. Mukesh, *Homogeneous Catalysis: Mechanisms and Industrial Applications*, S. Bahaduri and D. Mukesh, 2000, Wiley, New York.
100. S. F. Peteu and M. Bayachou, unpublished results.
101. F. Terzi, L. Pasquali, M. Montecchi, S. Nannarone, A. Viinikanoja, T. Ääritalo, M. Salomäki, J. Lukkari, B. P. Doyle and R. Seeber, *J. Phys. Chem. C*, 2011, **115**, 17836.
102. M. Innocenti, F. Loglio, L. Pigani, R. Seeber, F. Terzi and R. Udisti, *Electrochim. Acta*, 2010, **50**, 1497.
103. A. R. Gonçalves, M. E. Ghica and C. M. A. Brett, *Electrochim. Acta*, 2011, **56**, 3685.
104. V. Ruiz, A. Colina, A. Heras, J. Lopez-Palacios and R. Seeber, *Helv. Chim. Acta*, 2001, **84**, 3628.
105. L. Pigani, A. Heras, A. Colina, R. Seeber and J. Lopez-Palacios, *Electrochem. Commun.*, 2004, **6**, 1192.
106. R. L. McCrerry, *Chem. Rev.*, 2008, **108**, 2646.
107. D. Tsasis, N. Tagmatarchis, A. Bianco and M. Prato, *Chem. Rev.*, 2002, **106**, 1105.
108. A. H. Castro Neto, F. Guinea, N. M. R. Peres, K. S. Novoselov and A. K. Geim, *Rev. Mod. Phys.*, 2009, **81**, 109.
109. A. Fujishima, Y. Einaga, T. N. Rao, and D. A. Tryk, *Diamond Electrochemistry*, Elsevier, co-publication with BKC Inc. Japan, 2005.
110. E. Mercey, R. Sadir, E. Maillart, A. Roget, F. Baleux, H. Lortat-Jacob and T. Livache, *Analyt. Chem.*, 2008, **80**, 3476.
111. M. Pumera, *Langmuir*, 2007, **23**, 6453.

112. C. E. Banks, A. Crossley, C. Salter, S. J. Wilkins and R. G. Compton, *Angew. Chem. Int. Ed.*, 2006, **45**, 2533.

113. M. Pumera and Y. Miyahara, *Nanoscale*, 2009.

114. K. S. Novoselov, A. K. Geim, S. V. Morozov, D. Jiang, Y. Zhang, S. V. Dubonos, I. V. Grigorieva and A. A. Firsov, *Science*, 2004, **306**, 666–669.

115. K. S. Novoselov, D. Jiang, F. Schedin, T. J. Booth, S. V. Khotkevich, S. V. Morozov and A. K. Geim, *Proc. Natl. Acad. Sci. USA*, 2005, **102**, 10451.

116. R. Oprea, S. F. Peteu, P. Subramanian, W. Qi, E. Pichonat, H. Happy, M. Bayachou, R. Boukherroub, and S. Szunerits, *Analyst*, 2013, **138**, 4345.

117. A. A. Vernekar and G. Mugesh, *Chem. Eur. J.*, 2012, **18**, 15122.

118. T. Xue, S. Jiang, Y. Qu, Q. Su, R. Cheng, S. Dubin, C.-Y. Chiu, R. B. Kaner, Y. Huang and X. Duan, *Angew. Chem. Int. Ed.*, 2012, **51**, 3822.

119. Y. Guo, J. Li and S. Dong, *Sens. Actuators B Chem.*, 2011, **160**, 295.

120. C. X. Guo, Y. Lei and C. M. Li, *Electroanalysis*, 2011, **23**, 885.

121. C. Xu, J. Li, X. Wang, J. Wang, L. Wan, Y. Li, M. Zhang, X. Shang and Y. Yang, *Mater. Chem. Phys.*, 2012, **132**, 858.

122. Y. Guo, L. Deng, J. Li, S. Guo, E. Wang and S. Dong, *ACS Nano*, 2011, **5**, 1282.

123. J. Wei, J. Qiu, L. Li, L. Ren, X. Zhang, J. Chaudhuri and S. Wang, *Nanotechnol*, 2012, **23**, 335707.

124. J. Chen, U. Wollenberger, F. Lisdat, B. Ge and F. W. Scheller, *Sens. Actuators B*, 2000, **70**, 115.

125. J. S. Ye, Y. Wen, W. DeZhang, H. F. Cui, L. M. Gan, G. Q. Xu and F. S. Sheu, *J. Electroanal. Chem.*, 2004, **562**, 241.

126. H. Cao, X. Sun, Y. Zhang, C. Hu and N. Jia, *Analyt. Meth.*, 2012, **4**, 2412.

127. S. F. Peteu and M. Bayachou, *Analyt. Chim. Acta*, 2013, **780**, 81.

128. D. Emerson, S. F. Peteu and R. M. Woden, *Biotechnol. Tech.*, 1996, **10**, 673.

129. S. F. Peteu, M. T. Widman and R. M. Worden, *Biosens. Bioelectr.*, 1998, **780**, 81.

130. R. Gutierez-Osuna and A. Hierlmann, *Ann. Rev. Bioanalyt. Chem.*, 2010, **3**, 250.

CHAPTER 7

The Diagnosis of Myelodysplastic Syndromes

ALISON S. THOMAS AND CHRISTOPHER MCNAMARA*

Department of Haematology, Royal Free Hospital, London NW3 2QG,
United Kingdom
*Email: cmcnamara1@nhs.net

Myelodysplastic syndromes (MDS) comprise a heterogeneous group of clonal stem-cell disorders characterised by dysplasia, ineffective haematopoiesis and a risk of transformation to acute myeloid leukaemia. MDS may arise *de novo* or secondary to exposure to mutagenic agents. MDS is usually first suspected due to the presence of peripheral blood cytopenias or evidence of dysplasia on a blood film and the diagnosis is confirmed by the presence of otherwise unexplained significant dysplasia in the bone marrow, often in the presence of cytogenetic abnormalities. Originally described as a preleukaemic condition in the early 20th century, the cytopenias that form part of the clinical picture of MDS can give rise to transfusion dependence, recurrent infections and a bleeding diathesis and the complications of bone marrow failure are the main cause of mortality in patients, rather than leukaemic transformation. The definition of MDS has evolved since the initial proposal of a French–American–British classification in 1976[1] as our understanding of the pathophysiology and natural history of the syndrome has advanced. Alongside this has been the development of subclassifications, aimed at defining groups with differing diagnostic and prognostic features within this highly heterogeneous disorder. Despite many scientific advances there remains, however, no single biological or genetic marker that reliably identifies all cases of MDS and morphology remains the cornerstone of diagnosis. Accurate diagnosis of MDS

RSC Detection Science Series No. 2
Detection Challenges in Clinical Diagnostics
Edited by Pankaj Vadgama and Serban Peteu
© The Royal Society of Chemistry 2013
Published by the Royal Society of Chemistry, www.rsc.org

relies on the ability to differentiate MDS from other causes of myelodysplasia and a willingness to revisit the diagnosis in cases of uncertainty. There is a pressing need for increased engagement in research programs that seek to further our understanding of the molecular and biological mechanisms underpinning the pathogenesis of this disorder.

7.1 Morphological Features of Myelodysplasia

Dysplasia, derived from the Greek words δυσ (dys-, abnormal) and πλασις (plasis, growth or formation) describes the aberrant maturation of cells. Myeloid dysplasia manifests morphologically as an excess of immature myeloid precursors with cytoplasmic and nuclear abnormalities. These precursors may be abnormally located within the bone marrow compartment, as visible on trephine biopsy, denoted as abnormal location of immature precursors (ALIP). Although patients have peripheral blood cytopenias, the bone marrow is commonly hypercellular with an increased proportion of immature cells ("left-shifted").

Myeloid dysplasia can affect any of the myeloid lineages (erythroid, granulocytic or megakaryocytic) and commonly affects more than one lineage even if there is only a cytopenia in one lineage. Detection of dysplasia requires examination of blood films and bone marrow aspirates that are stained to a high standard. Inadequately stained slides may lead to erroneous identification of abnormalities of granularity. Diagnosis of MDS requires the presence of unequivocal dysplasia in ≥10% of at least one of the myeloid series. Accurate determination of the percentage of blast cells is required for sub-classification and at least 200 cells should be counted in the peripheral blood and 500 in the bone marrow. Common dysplastic features in each lineage are summarised in Table 7.1 and shown in Figure 7.1.

7.1.1 Erythroid Dysplasia

In MDS, dysplasia is most commonly seen in the erythroid series. The most reliable dysplastic features are nuclear abnormalities including: nuclear budding,[2,3] internuclear bridging (connections between the nuclei of two different cells), multinuclearity and megaloblastic features. Cytoplasmic features include patchy haemoglobinisation, intercytoplasmic bridging and vacuolisation. An iron stain (*e.g.* Perl's stain) should be performed on all bone-marrow aspirates of patients with suspected MDS, both to assess iron stores, as iron deficiency can cause dysplasia, and to look for the presence of ringed sideroblasts, the percentage of which is important in the sub-classification of MDS. Ringed sideroblasts are erythroid precursors which have at least five siderotic granules in a perinucelar distribution, covering at least a third of the nuclear circumference.[4] An increased number of ringed sideroblasts is a feature of a specific subclass of MDS, refractory anaemia with ringed sideroblasts (RARS).

Table 7.1   Common morphological features of dysplasia.

Lineage	Peripheral blood	Bone-marrow aspirate	Bone-marrow trephine
Erythroid	Poikilocytosis:	Megaloblastoid features	Defective maturation in erythroid islands – erythroblasts at same stage of maturation
	Ovalomacrocytes	Nuclear budding and multinucleation	
	Dacrocytes	Internuclear and intercytoplasmic bridging	
	Elliptocytes	Cytoplasmic vacuolisation	
	Howell–Jolly bodies	Patchy haemoglobinisation	
	Nucleated RBC	Nuclear-cytoplasmic asynchrony	
	Basophilic stippling Acquired HbH features, including H inclusions	Ringed sideroblasts	
Granulocytic	Hyposegmentation (pseudo-Pelger–Huet) or hypersegmentation Hypogranulation Ring-shaped nuclei	Increased blasts Maturation arrest Nuclear-cytoplasmic asynchrony	Abnormal location of precursors (ALIP): central rather than periosteal location of blasts
Megakaryocytic	Platelet anisocytosis	Hypolobated or nonlobated nuclei Multiple separated nuclei Micromegakaryocytes	Megakaryocyte clusters Monolobated megakaryocytes

7.1.2   Granulocytic Dysplasia

Dysgranulopoiesis is characterised by nuclear abnormalities with both hypolobation (pseudo-Pelger–Huet) and hyperlobation.[5] Cytoplasmic granulation may be decreased. There is often an increase in immature myeloid precursors (*e.g.* blasts, promyelocytes). Accurate enumeration of the blast percentage is important for diagnosis of acute myeloid leukaemia (myeloid blasts ≥20%) and for subclassification and prognosis of MDS. Myeloid blasts comprise: myeloblasts, monoblasts and megakaryoblasts; erythroid precursors are not counted as blasts except in cases of erythroleukaemia. To ensure consistency in identification of blast cells in MDS a consensus statement containing detailed descriptions and diagrams of blast cells has been published.[4]

7.1.3   Megakaryocytic Dysplasia

It is recommended that at least 30 megakaryocytes are evaluated; megakaryocytic dysplasia is more easily identified on trephine sections. Micromegakaryocytes

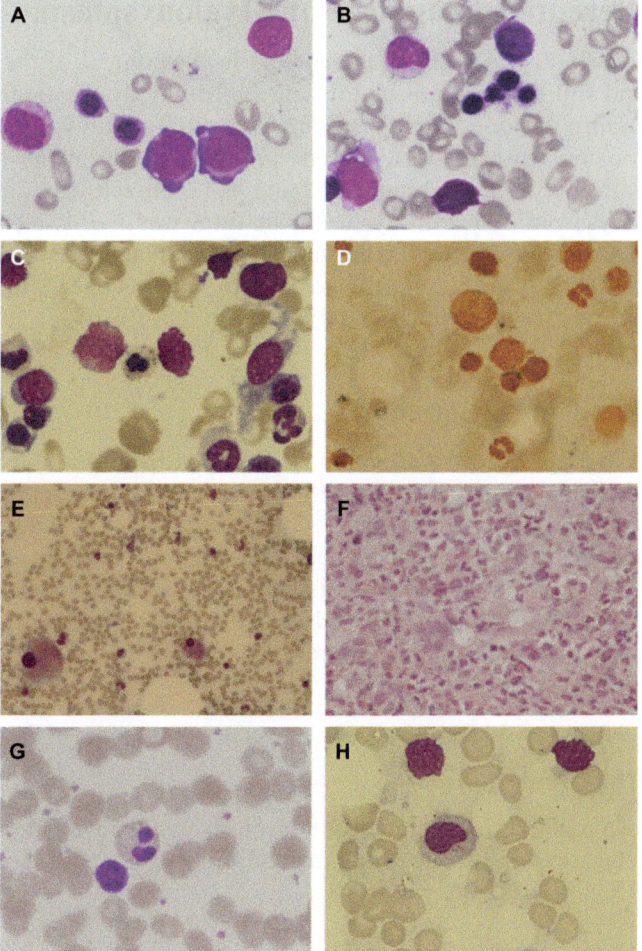

Figure 7.1 Morphological features of myelodysplasia. (A) erythroid intercytoplasmic bridging and red blood cell poikilocytosis, MGG×100. (B) erythroid dysplasia and hypogranular myelocyte, MGG×100. (C) multinucleate late erythroblast with megaloblastic features, erythroblast with patchy haemoglobinisation and abnormally lobated neutrophil, MGG×100. (D) ringed sideroblasts, Perls' stain, ×100. (E) monolobated megakaryocytes in 5q- syndrome, MGG×10. (F) trephine showing dysplastic megakaryocytes, (H) and (E), ×10. (G) pseudo-Pelger-Huët abnormality, MGG×100 (H) hypogranular myelocyte, MGG×100.

and multinucleate megakaryocytes are considered most reliable for the diagnosis of MDS.[6] Micromegakaryocytes are small megakaryocytes with hypolobated nuclei. Multinucleate megakaryocytes have multiple, widely separated nuclei in contrast to the polylobated nucleus of normal megakaryocytes and there may be increased variation in size.

7.2   Differential Diagnosis of Myelodysplasia

Bone marrow findings should be interpreted in the context of the clinical presentation and full blood count abnormalities. Minor dysplastic features are a common finding in bone marrow specimens from healthy volunteers:[7] minor dyserythropoietic features were present in 84% of specimens (patchy haemoglobinisation 62%, cytoplasmic bridging 42%, binuclearity 24%); the percentage of erythroid cells with dysplastic features was 1–7%; 38% of samples had a small number of dysplastic megakaryocytes. Diagnosis of MDS requires the presence of significant (*i.e.* ≥10% of cells in any one lineage) dysplasia and the exclusion of causes of nonclonal dysplasia (see Table 7.2).

7.2.1   Metabolic Abnormalities

Nutritional deficiencies are a common cause of cytopenias and dysplasia. B12 and folate deficiency both cause dyserythropoiesis and, if severe, pancytopaenia and multilineage dysplasia. Deficiencies of other B vitamins may also result in similar abnormalities.[8] Copper deficiency, whether primary or secondary to zinc excess, is increasingly recognised as a cause of reversible dysplasia.[9] Exposure to heavy metals, particularly arsenic, which can be found in contaminated water supplies in certain geographic regions, may result in bone-marrow failure and dysplasia.[10,11]

7.2.2   Drugs and Toxins

Cytotoxic drugs can result in significant dysplasia in any myeloid lineage. Many drugs have been reported as causing myelodysplastic changes including:

Table 7.2   Differential diagnosis of myelodysplasia.

Vitamin/minerals	B12 deficiency
	Folate deficiency
	Copper deficiency (may be secondary to zinc excess)
	Arsenic poisoning
Infections	HIV
	Parvovirus
	Any severe infection
Drugs and toxins	Alcohol
	Chemotherapeutic agents
	Mycophenolate mofetil
	Ganciclovir
	Sodium valproate
	G-CSF, erythropoietin
Inherited disorders	Congenital dyserythropoietic anaemia
	Inherited bone-marrow-failure syndromes
Other myeloid neoplasms	Acute myeloid leukaemia
	Myeloproliferative syndromes

- growth factor therapy: granulocyte colony stimulating factor results in hypergranulation, abnormal granulocytic nuclear lobation and an increased proportion of myeloblasts, particularly in the peripheral blood;
- ganciclovir;[12,13]
- sodium valproate;
- mycophenolate mofetil;[12,13]
- co-trimoxazole;
- alemtuzumab.[14]

7.2.3 Infectious Diseases

Myelodysplasia in very common in patients with HIV[15] as a consequence of the direct effects of the virus on haematopoiesis as well as opportunistic infections and effects of antiretrovirals. Acute episodes of sepsis can stimulate a reactive bone marrow response that may exhibit dysplastic features. Less commonly, parvovirus infection can cause transient red cell aplasia with dyserythropoiesis.[16]

7.2.4 Congenital Disorders

In isolated erythroid dysplasia, the diagnosis of congenital dyserythropoietic anaemia should be considered. In younger patients, congenital bone-marrow failure syndromes, which often have a myelodysplastic phase prior to leukaemic transformation, should be included in the differential. These include: Fanconi anaemia, dyskeratosis congenita, Diamond–Blackfan syndrome and Schwachmann–Diamond syndrome. Although the dysplasia in these inherited bone marrow failure syndromes is clonal, they are due to an inherited rather than acquired genetic defect and as such represent distinct disorders which require specialised management.

7.2.5 Acute Myeloid Leukaemia and Other Haematopoietic Neoplasms

The presence of $\geq 20\%$ myeloid blasts in the peripheral blood or bone marrow leads to a diagnosis of acute myeloid leukaemia (AML) regardless of the presence or absence of dysplasia. Patients with a preceding history of cytotoxic chemotherapy or radiotherapy who develop myelodysplastic changes as a late complication are classified as having a therapy-related myeloid neoplasm, a subclassification of AML, regardless of blast percentage and have a generally poor prognosis. The most commonly implicated cytotoxic agents are alkylating agents and topoisomerase II inhibitors. As discussed below, some patients may present with features of a myeloproliferative disorder (*e.g.* thrombocytosis or monocytosis) but also have evidence of myelodysplasia. The final diagnosis of many of these patients is within the category of myelodysplastic/myeloproliferative neoplasms, which includes chronic myelomonocytic leukaemia.

7.3 Initial Assessment of a Patient with Possible MDS

Initial assessment of a patient with suspected MDS starts with a full clinical history, including a detailed drug history and history of any malignancy. Clinical examination should be unremarkable in MDS, except for occasional bruising in thrombocytopenic patients. Lymphadenopathy and hepatosplenomegaly are rare in MDS and their presence should prompt consideration of alternative diagnoses.

Initial investigations should include assessment of peripheral blood cytopenias and morphology and detection of causes of nonclonal dysplasia (see Figure 7.2). Anaemia is usually normocytic or macrocytic. An increased red cell distribution width (RDW) reflects variability in red cell size, but a normal mean cell haemoglobin concentration (MCHC), indicates a normal amount of haemoglobin for the cell size. Neutropenia is common and circulating immature cells may be seen on the blood film. Whilst thrombocytopenia is more common, occasionally thrombocytosis may be present. The importance of the haematologist examining the blood film for themselves prior to deciding to perform a bone-marrow biopsy cannot be overemphasised.

Any metabolic deficiencies should be corrected prior to further assessment of the patient. Acute illness results in bone marrow stress and apparent dysplasia making the diagnosis of MDS challenging in these circumstances; further investigation should be deferred until recovery if possible. If cytopenias remain unexplained, then a bone marrow aspirate and trephine should be performed, with a bone marrow sample being taken for cytogenetic analysis. The results of the aspirate, trephine and cytogenetic analysis should be reviewed together in conjunction with the clinical history and a summative, integrated report agreed. In some cases a concrete diagnosis may not be possible and a period of observation followed by repeat evaluation is required.

7.4 Additional Diagnostic Tools for MDS

7.4.1 Bone Marrow Aspirate Cytochemistry

An iron stain (*e.g.* Perl's stain) should be performed on all aspirates to enable enumeration of ringed sideroblasts. Cytochemical staining can be performed to confirm myeloid lineage of blasts (*e.g.* Sudan Black B) or to differentiate granulocytic from monocytic cells (*e.g.* dual esterase) but in most laboratories these stains have been superseded by immunological techniques.

7.4.2 Bone Marrow Trephine Biopsy

A bone marrow trephine complements an aspirate, with different morphological information available from each.[17] Whilst the aspirate provides detailed cytological information, the trephine provides an overview of bone marrow architecture and maturation. Dysplastic features may been evident as disruption of normal bone marrow architecture[18,19] (see Table 7.1). Cellularity is

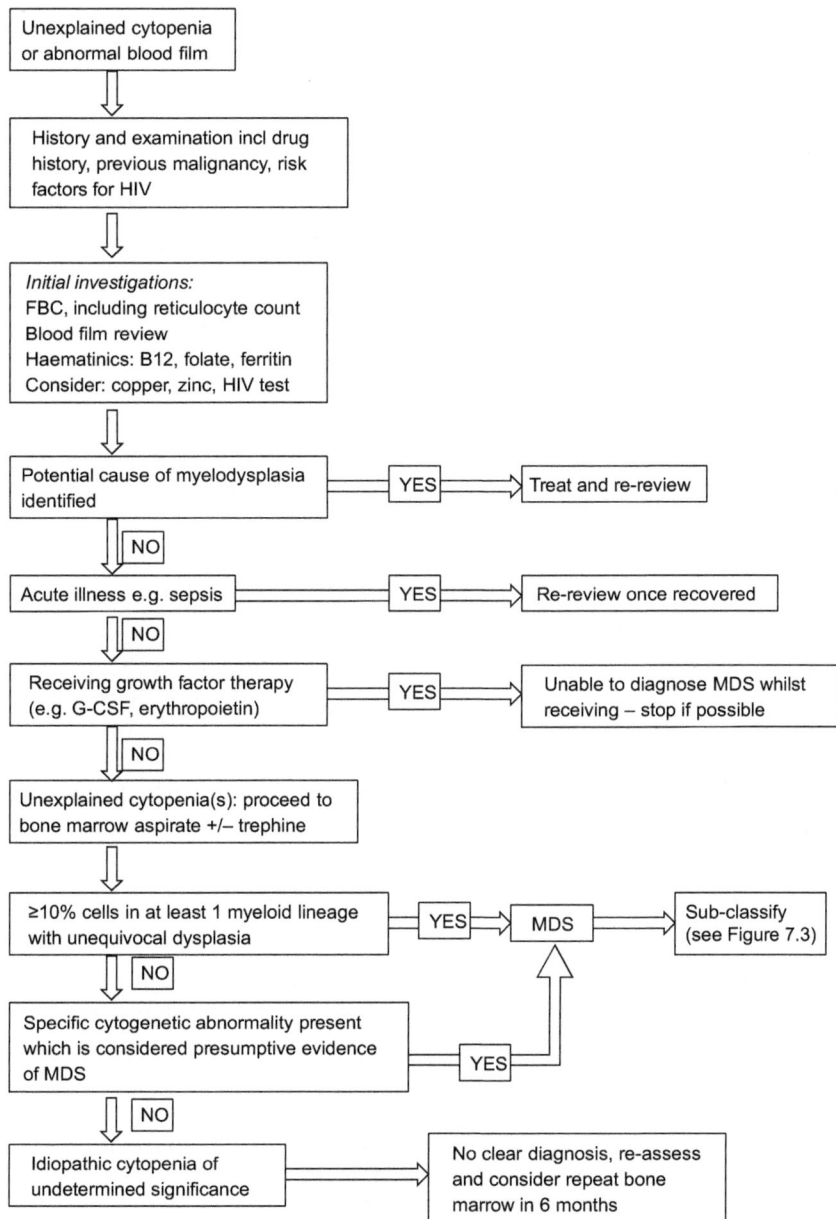

Figure 7.2   Approach to the patient with suspected MDS.

normally increased and reactive changes secondary to the disease process may also be seen, *e.g.* mastocytosis, fibrosis and increased histiocytes.[20] Immuno-histochemistry for lineage confirmation and enumeration of blast cells can be performed on trephine biopsies and are of particular benefit if the aspirate has been suboptimal. Importantly, a trephine biopsy may also provide evidence of

an alternative disease process as explanation of cytopenias or dysplasia, *e.g.*
hairy cell leukaemia, metastatic carcinoma.

7.4.3 Cytogenetics

Cytogenetic abnormalities are seen in 40–70% patients with MDS[21,22] and their
detection has an important role both in the diagnosis and prognosis of MDS.
A wide range of chromosomal abnormalities have been described in MDS including numerical abnormalities (monosomy or trisomy), inversions and interstitial deletions within one chromosome and balanced translocations between two
chromosomes. Cytogenetic studies should be performed on bone marrow samples
rather than peripheral blood. Karyotyping can be undertaken using conventional
metaphase analysis. This has the advantage of detecting a wide range of abnormalities but can be difficult in the presence of complex abnormalities and can
be hampered by poor sample quality. Fluorescent *in situ* hybridisation (FISH)
can be performed on metaphase chromosomes or interphase nuclei but will only
detect the abnormalities specific to the panel of probes utilised.[23]

7.4.3.1   Role of Cytogenetics in Diagnosis of MDS

In cases of suspected MDS with cytopenias but minimal dysplasia, the presence
of specific cytogenetic abnormalities is considered sufficient for the diagnosis of
MDS to be made in the absence of significant morphological dysplasia. These
abnormalities are listed in Table 7.3. MDS can precede AML; there are three
cytogenetic abnormalities which are considered diagnostic of AML, regardless
of blast percentage: t(8;21)(q22.q22), inv(16)(p13.1q22) or t(16;16)(p13.1;q22)
or t(15;17)(q22;q21.1)

7.4.3.2   Role of Cytogenetics in Subclassification and Prognosis
           of MDS

The finding of 5q- as the sole cytogenetic abnormality comprises a specific
subclass of MDS, 5q- syndrome. This is associated with characteristic morphological abnormalities of hypolobated and monolobated megakaryocytes; it

Table 7.3   Cytogenetic abnormalities, which if present, are diagnostic
            of MDS in the presence of peripheral blood cytopenias,
            even in the absence of significant dysplasia.

Deletions	−5 or del(5q) or −7/del(7q)
	Del(9q) or del(11q) or del(12q) or del(13q)
Translocations and inversions	t(1;3)(p36.3;q21)
	t(2;11)(p21;q23)
	inv(3)(q21;q26.2)
	t(3;21)(q26.2;q22.1)
	t(6;9)(p23;q34)
	t(11;16)(q23;p13.3)
	t(17p) or i(17q) with loss of 17p

is associated with a favourable prognosis. Currently, there are no other cytogenetic abnormalities associated with specific clinical and morphological features. Monosomy 7 or complex cytogenetic abnormalities (>5 cytogenetic abnormalities) are associated with a poorer prognosis[21] and are more common in RAEB than other subclasses of MDS.[22]

7.4.4 Immunophenotyping

As myeloid cells develop, different genes are transcribed and their proteins expressed intracellularly and on the cell surface membrane. Loss of coordination of the processes governing normal cell maturation occurs in MDS and may result in aberrant or asynchronous expression of maturation-associated antigens on myeloid cells. The aberrant expression or absence of a cell surface antigen may provide supporting evidence for the presence of an abnormal cell population. The role of immunophenotyping in the diagnosis of MDS remains, however, controversial. The specificity of abnormalities in antigen expression to MDS has not been extensively studied.[24] Aberrant expression has been described in reactive left-shift[25] and in nonhaematological disorders that can mimic MDS.[26] Further data is required to clarify the role of immunophenotyping in the diagnosis of MDS. This requires standardisation of sample processing, antibody panels and analysis protocols to enable comparisons between different data sets.[27] The current guidance is that immunophenotyping reports should be descriptive; the presence of ≥ 3 phenotypic abnormalities may be suggestive of MDS, but is not currently considered diagnostic.[2] Immunophenotyping should not be used in place of morphology to enumerate blast cells. Samples are often haemodilute, leading to underestimation and not all blast cells express CD34.

7.5 Diagnostic Difficulties in MDS

7.5.1 Unexplained Cytopenias in the Presence of Minimal Dysplasia

In some cases, patients will have unexplained cytopenias but will not meet the "minimum" morphological criteria for significant dysplasia of at least 10% of cells of at least one myeloid lineage showing unequivocal dysplasia. In these cases, additional effort should be made to ensure that the diagnostic samples are of adequate quality and that other causes of cytopenias and dysplasia have been excluded. There are some specific cytogenetic abnormalities (see Table 7.3), which if present, are considered sufficient for a presumptive diagnosis of MDS to be made in the absence of significant dysplasia.[2] Diagnosis can be difficult in the absence of cytogenetic abnormalities; virtually all bone marrow specimens will have at least some mild dysplasia. In a study of bone marrows of patients with different causes of anaemia,[28] dysplasia was seen in at least one lineage in 93% of iron-deficiency anaemia, 92% of megaloblastic anaemia and 84% of haemolytic anaemia. The percentage of dysplastic cells was similar in these anaemias to patients given a diagnosis of refractory anaemia with normal cytogenetics. In patients with unexplained significant cytopenias (haemoglobin <10 g/dL, neutrophils <1.8×10^9/L or platelets <100×10^9/L) the term "idiopathic cytopenia of undetermined significance" has been suggested, emphasising the lack of diagnostic clarity. These patients should be monitored and a repeat bone marrow evaluation after six months considered.

7.5.2 Myelioproliferative/Myelodysplastic Overlap Syndromes

Persistent monocytosis (>1×10^9/L) accompanied by the presence of dysplasia in one or myeloid lineages but with a blast percentage of <20% is diagnostic for chronic myelomonocytic leukaemia (CMML). Although originally considered part of MDS, CMML also shares some features of myeloproliferative neoplasms (MPN) and in the WHO 3rd edition, a new category of myelodysplastic/myeloproliferative neoplasms was created, incorporating CMML.

Some patients have laboratory and morphological features consistent with a diagnosis of MDS, but have additional myeloproliferative features (e.g. platelets ≥450×10^9/L or WBC >13×10^9/L) without a preceding history of myeloproliferative disorders. These patients are also considered to have a myelodysplastic/myeloproliferative neoplasm. Within this group are patients who have anaemia accompanied by thrombocytosis, with ringed sideroblasts present in excess on the bone marrow. Up to 2/3 of these patients will have a JAK2 V617F mutation;[29,30] in patients initially diagnosed with refractory anaemia with ringed sideroblasts who develop a thrombocytosis, acquisition of the JAK2 V617F mutation at the time of development of thrombocytosis has been demonstrated.[30] These patients fall into a provisional entity of "refractory anaemia with ring sideroblasts associated with marked thrombocytosis".

7.5.3 Cytopenias in the Presence of a Hypocellular Bone Marrow

In approximately 10% of cases of MDS the bone marrow is hypocellular and this can raise a diagnostic challenge in differentiating MDS from aplastic anaemia. In aplastic anaemia, cellular elements will be reduced but should be morphologically normal. The peripheral blood film should be carefully examined for evidence of dysplasia – if leukocyte counts are significantly reduced a film may need to be prepared from the buffy coat. If the aspirate is paucicellular, blast percentage can be estimated by immunohistochemistry on the trephine specimen. Cytogenetic abnormalities are far commoner in MDS than aplastic anaemia and the presence of clonal cytogenetic abnormalities would favour a diagnosis of MDS.[31]

7.5.4 Dysplasia in the Presence of Bone Marrow Fibrosis

Bone marrow fibrosis occurs in approximately 10% MDS. Difficulty in obtaining an aspirate, which is often haemodilute, reduces the quantity of diagnostic material available and a good quality trephine specimen of adequate length is essential. The differential diagnosis includes: acute panmyelosis with myelofibrosis, acute megakaryoblastic leukaemia and chronic myeloproliferative neoplasms. Immunohistochemistry should be performed to enumerate CD34 positive blast cells and to identify megakaryocytic cells. Evaluation for the JAK2 V617F mutation may be helpful in distinguishing fibrotic MDS from myelofibrosis; the mutation is present in approximately 50% of myelofibrosis but is infrequent in MDS.[32]

7.6 Subclassification of MDS

Once the diagnosis of MDS has been established, classification into a specific subtype of MDS is required. Classifications are consensus definitions of specific disease entities and currently the most widely used classification of MDS is the WHO 4th Edition Classification.[2] This incorporates clinical and laboratory data to arrive at diagnostic criteria and subclassifications of MDS. Subclassification is important given the heterogeneous features and variable prognosis of MDS. The classification of MDS is designed to be used in routine clinical practice, as well as to provide common definitions for clinical trials and other studies. It incorporates data available at the time it was written; in this rapidly advancing field new data is constantly being published as our understanding of MDS evolves. This is recognised in the 4th edition of the WHO classification where provisional entities are described based on new findings where further data is needed to clarify and confirm the clinical significance.

7.6.1 Original FAB Classification

The original classification of MDS was the 1976 French–American–British (FAB) classification,[1] where the category of dysmyelopoietic syndromes was designed to help distinguish acute leukaemia from less acute disorders that carried a risk of progression to acute leukaemia. Two categories were described: CMML and refractory anaemia with excess blasts (RAEB). Diagnosis was based entirely on the morphological appearances of peripheral blood and bone marrow. RAEB was defined as bone marrow blasts 10–30%, with 30–50% representing probable progression to AML. AML was diagnosed if blasts were >50%.

The FAB classification was revised in 1982[33] in recognition that there was a wide range of dysplastic morphological appearances and that the risk of transformation to AML was variable and, to a certain extent, correlated with morphological features. The expansion to five categories aimed to enable better assessment of the prognostic significance of these morphological features:

- Refractory anaemia (RA): <1% blasts in peripheral blood (PB), <5% blasts in bone marrow (BM); granulocytic and megakaryocytic series usually normal
- Refractory anaemia with ringed sideroblasts (RARS): <1% blasts in PB, <5% blasts in BM, ringed sideroblasts >15% all nucleated cells in BM

- RA with excess blasts (RAEB): 5–20% blasts in BM and <5% in PB
- CMML: monocytes $>1 \times 10^9/L$, <5% blasts in PB
- RAEB in transformation (RAEB-T): >5% blasts in PB, 20–30% blasts in BM or Auer rods.

7.6.2 WHO Classification 3rd Edition and 4th Edition

Whilst the FAB classifications were based solely on morphological appearances, the World Health Organisation (WHO) classifications incorporate morphological, genetic, immunophenotypic and clinical features in an attempt to identify distinct molecular pathways underlying neoplastic disorders. Haematological malignancies, including MDS, were first included in the 3rd edition of the WHO Classification of tumours in 2001. These have been subsequently revised and a 4th edition published in 2008.[24] The WHO classification introduced a new category, MDS/MPN overlap syndromes, recognising the dysplastic and proliferative features present in some patients such as those with CMML. Building on the FAB classification, the WHO classifications have continued to refine subclassifications of MDS, aiming to be prognostically relevant. A diagnostic algorithm for subclassifying MDS is presented in Figure 7.3. As the availability of therapies for MDS increases, it is vital to correctly subclassify patients and all diagnostic material should be reviewed by haematopathologists experienced in the diagnosis of MDS. A review of the diagnosis of patients referred to a tertiary cancer centre in the USA revealed a discordant diagnosis in 12% with over two-thirds of these having higher risk disease (*i.e.* RAEB) than initially diagnosed.[34] A number of prognostic scoring systems are available (*e.g.* revised IPSS,[35] WPSS[36]), aimed at stratifying risk of overall survival and evolution to AML. These incorporate information on blast percentages, cytopenias and cytogenetic abnormalities alongside other clinical data (*e.g.* transfusion dependence); sufficient information should be provided on the integrated diagnostic report to enable clinicians to utilise these scores.

7.7 Evolving Understanding of the Pathophysiology of MDS

Much is still unknown about the molecular mechanisms underlying the pathogenesis of MDS. There are many similarities in the gene abnormalities identified in MDS and AML, including abnormalities in tumour suppressor genes and oncogenes, but progression from MDS to AML is highly variable. Whilst metaphase analysis and FISH-based approaches allow detection of translocations and deletions, array-based approaches enable identification of gene over/underexpression, loss of heterozygosity and copy-number variations. In addition to abnormalities in genes involved in regulation of the cell cycle, epigenetic mechanisms of gene expression regulation and mRNA splicing may be important in the pathogenesis of MDS.

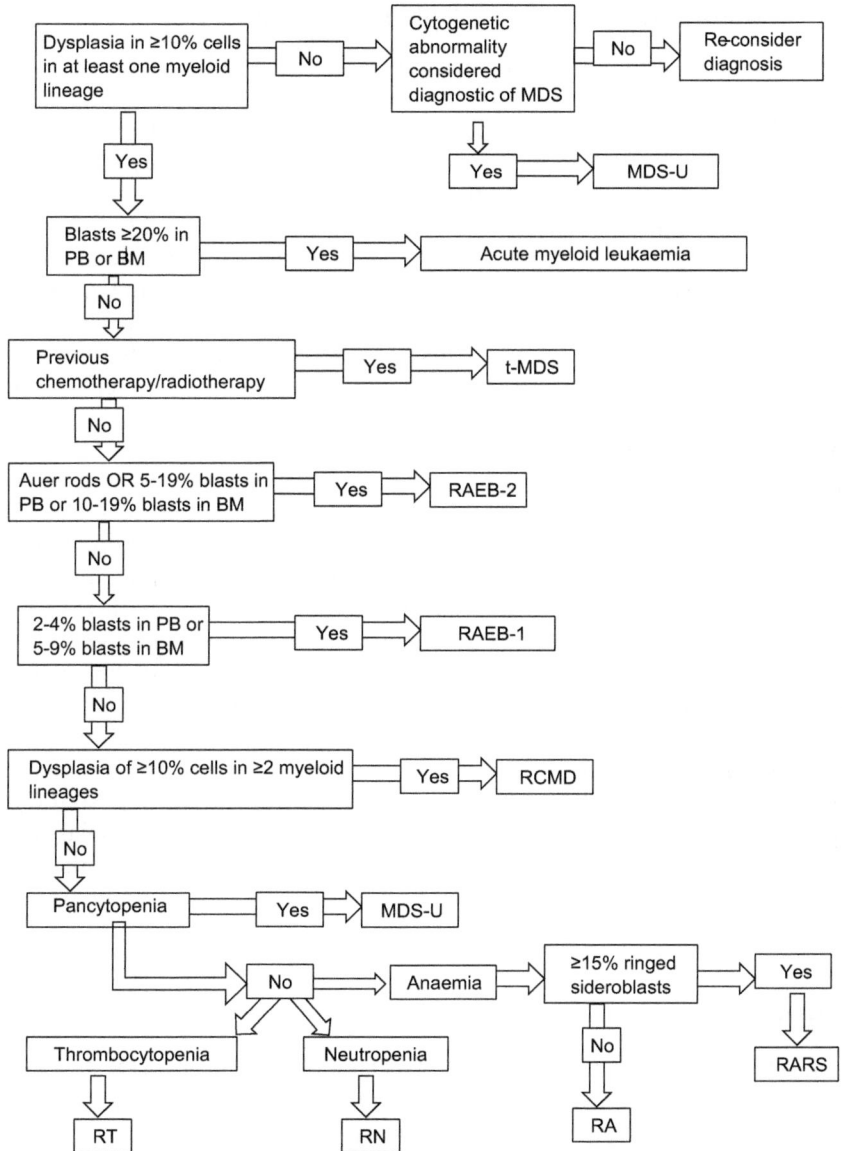

Figure 7.3 Subclassification of MDS.

7.7.1 Altered Methylation

Altered methylation is an important epigenetic mechanism of gene expression regulation and mutations have been found in the TET2 (ten-eleven-translocation) gene family in up to 23% patients with MDS[37] and isocitrate dehydrogenase (IDH) oncogenes in 3.6% patients,[38] both resulting in hypermethylation.

7.7.2   Abnormalities in RNA Splicing Machinery

RNA splicing is accomplished by a complex of five small nuclear ribonucleo-proteins (snRNPs) and other proteins on pre-mRNAs that form the spliceo-some. Mutations within any of these proteins can result in aberrantly spliced mRNA. Whole exome sequencing of MDS specimens has revealed novel mu-tations involving components of the spliceosome, particularly those involved in 3' splice site recognition[39] in up to 40% of MDS patients.[39,40] Mutations have also been found in patients with other haematological and nonhaematological malignancies, but at lower levels (1–5%). Mutations in SRSF2 and ZRSR2 are more common in those with RAEB and associated with poorer survival.[40] Mutations in SF3B1 are found in up to 75% of patients with ringed side-roblasts;[39] knockdown of SF3B1 in normal cells leads to development of ringed sideroblasts.[41]

MDS represents a heterogeneous group of clonal disorders of haemato-poietic stem cell maturation. First identified by the presence of bone marrow dysplasia and a propensity to develop into acute leukaemia, it is clear that current diagnostic criteria for MDS incorporate a number of different entities with vastly different natural histories, ranging from RAEB-2 with a high risk of leukaemia and short life expectancy to refractory anaemia, which may run a more indolent course. Care should be given in arriving at the diagnosis, par-ticularly in the absence of significant dysplasia and cytogenetic abnormalities; observation and re-evaluation are prudent in these circumstances. Common nonclonal causes of MDS must be excluded prior to the initiation of any specific (*e.g.* chemotherapeutic) therapy for MDS. As molecular diagnostic techniques advance, coupled with large cohort analyses of patients, so too may our understanding of the underlying pathogenesis of the myelodysplastic syn-dromes. In future it may be possible to diagnose and subclassify these syn-dromes on the basis of molecular and genetic, rather than primarily morphological abnormalities.

References

1. J. M. Bennett, D. Catovsky, M. T. Daniel, G. Flandrin, D. A. Galton, H. R. Gralnick and C. Sultan, Proposals for the classification of the acute leukaemias. French-American-British (FAB) co-operative group, *Br. J. Haematol.*, 1976, **33**(4), 451–458.
2. S. H. Swerdlow, E. Campo, N. L. Harris, E. S. Jaffe, S. A. Pileri, H. Stein, J. Thiele and J. W. Vardiman, *WHO Classification of Tumours of Hae-matopoietic and Lymphoid Tissues*, 4th edn. World Health Organisation, 2008.
3. J. Fradera, Erythroid karyorrhexis in myelodysplasia: bone marrow as-pirate. *Blood*, 2008, **112** (3), 479.
4. G. J. Mufti, J. M. Bennett, J. Goasguen, B. J. Bain, I. Baumann, R. Brunning, M. Cazzola, P. Fenaux, U. Germing, E. Hellstrom-Lindberg, I. Jinnai, A. Manabe, A. Matsuda, C. M. Niemeyer, G. Sanz,

M. Tomonaga, T. Vallespi and A. Yoshimi, Diagnosis and classification of myelodysplastic syndrome: International Working Group on Morphology of myelodysplastic syndrome (IWGM-MDS) consensus proposals for the definition and enumeration of myeloblasts and ring sideroblasts, *Haematologica*, 2008, **93**(11), 1712–1717.

5. B. J. Bain, Dysplastic macropolycytes in myelodysplasia-related acute myeloid leukemia, *Am. J. Hematol.*, 2011, **86**(9), 776.

6. E. Verburgh, R. Achten, V. J. Louw, C. Brusselmans, M. Delforge, M. Boogaerts, A. Hagemeijer, P. Vandenberghe and G. Verhoef, A new disease categorization of low-grade myelodysplastic syndromes based on the expression of cytopenia and dysplasia in one versus more than one lineage improves on the WHO classification, *Leukemia*, 2007, **21**(4), 668–677.

7. B. J. Bain, The bone marrow aspirate of healthy subjects, *Br. J. Haematol.*, 1996, **94**(1), 206–209.

8. A. Bazarbachi, S. Muakkit, M. Ayas, A. Taher, Z. Salem, H. Solh and J. H. Haidar, Thiamine-responsive myelodysplasia, *Br. J. Haematol.*, 1998, **102**(4), 1098–1100.

9. J. D. Huff, Y. K. Keung, M. Thakuri, M. W. Beaty, D. D. Hurd, J. Owen and I. Molnar, Copper deficiency causes reversible myelodysplasia, *Am. J. Hematol.*, 2007, **82**(7), 625–630.

10. W. N. Rezuke, C. Anderson, W. T. Pastuszak, S. R. Conway and S. I. Firshein, Arsenic intoxication presenting as a myelodysplastic syndrome: a case report, *Am. J. Hematol.*, 1991, **36**(4), 291–293.

11. D. D. Westhoff, R. J. Samaha and A. Barnes, Arsenic intoxication as a cause of megaloblastic anemia, *Blood*, 1975, **45**(2), 241–246.

12. G. A. Kennedy, T. D. Kay, D. W. Johnson, C. M. Hawley, S. B. Campbell, N. M. Isbel, P. Marlton, R. Cobcroft, D. Gill and G. Cull, Neutrophil dysplasia characterised by a pseudo-Pelger-Huet anomaly occurring with the use of mycophenolate mofetil and ganciclovir following renal transplantation: a report of five cases, *Pathology (Phila.)*, 2002, **34**(3), 263–266.

13. A. B. Taegtmeyer, O. Halil, A. D. Bell, M. Carby, D. Cummins and N. R. Banner, Neutrophil dysplasia (acquired pseudo-Pelger anomaly) caused by ganciclovir, *Transplantation*, 2005, **80**(1), 127–130.

14. S. D. Gibbs, D. A. Westerman, C. McCormack, J. F. Seymour and P. H. Miles, Severe and prolonged myeloid haematopoietic toxicity with myelodysplastic features following alemtuzumab therapy in patients with peripheral T-cell lymphoproliferative disorders, *Br. J. Haematol.*, 2005, **130**(1), 87–91.

15. D. S. Karcher and A. R. Frost, The bone marrow in human immuno-deficiency virus (HIV)-related disease. Morphology and clinical correlation, *Am. J. Clin. Pathol.*, 1991, **95**(1), 63–71.

16. S. L. Carpenter, S. A. Zimmerman and R. E. Ware, Acute parvovirus B19 infection mimicking congenital dyserythropoietic anemia, *J. Pediatr. Hematol. Oncol.*, 2004, **26**(2), 133–135.

17. N. Ngo, I. A. Lampert and K. N. Naresh, Bone marrow trephine findings in acute myeloid leukaemia with multilineage dysplasia, *Br. J. Haematol.*, 2008, **140**(3), 279–286.

18. G. Tricot, C. De Wolf-Peeters, R. Vlietinck and R. L. Verwilghen, Bone marrow histology in myelodysplastic syndromes. II. Prognostic value of abnormal localization of immature precursors in MDS, *Br. J. Haematol.*, 1984, **58**(2), 217–225.

19. G. Tricot, C. De Wolf-Peeters, B. Hendrickx and R. L. Verwilghen, Bone marrow histology in myelodysplastic syndromes. I. Histological findings in myelodysplastic syndromes and comparison with bone marrow smears, *Br. J. Haematol.*, 1984, **57**(3), 423–430.

20. A. J. Stewart and R. D. Jones, Pseudo-Gaucher cells in myelodysplasia, *J. Clin. Pathol.*, 1999, **52**(12), 917–918.

21. D. Haase, U. Germing, J. Schanz, M. Pfeilstocker, T. Nosslinger, B. Hildebrandt, A. Kundgen, M. Lubbert, R. Kunzmann, A. A. Giagounidis, C. Aul, L. Trumper, O. Krieger, R. Stauder, T. H. Muller, F. Wimazal, P. Valent, C. Fonatsch and C. Steidl, New insights into the prognostic impact of the karyotype in MDS and correlation with subtypes: evidence from a core dataset of 2124 patients, *Blood*, 2007, **110**(13), 4385–4395.

22. O. Pozdnyakova, P. M. Miron, G. Tang, O. Walter, A. Raza, B. Woda and S. A. Wang, Cytogenetic abnormalities in a series of 1,029 patients with primary myelodysplastic syndromes: a report from the US with a focus on some undefined single chromosomal abnormalities, *Cancer*, 2008, **113**(12), 3331–3340.

23. A. M. Cherry, M. L. Slovak, L. J. Campbell, K. Chun, V. Eclache, D. Haase, C. Haferlach, B. Hildebrandt, A. M. Iqbal, S. C. Jhanwar, K. Ohyashiki, F. Sole, P. Vandenberghe, D. L. VanDyke, Y. Zhang and G. W. Dewald, Will a peripheral blood (PB) sample yield the same diagnostic and prognostic cytogenetic data as the concomitant bone marrow (BM) in myelodysplasia?, *Leuk. Res.*, 2012, **36**(7), 832–840.

24. J. W. Vardiman, J. Thiele, D. A. Arber, R. D. Brunning, M. J. Borowitz, A. Porwit, N. L. Harris, M. M. Le Beau, E. Hellstrom-Lindberg, A. Tefferi and C. D. Bloomfield, The 2008 revision of the World Health Organization (WHO) classification of myeloid neoplasms and acute leukemia: rationale and important changes. *Blood*, 2009, **114** (5), 937–951.

25. W. G. Finn, A. M. Harrington, K. M. Carter, R. Raich, S. H. Kroft and A. O. Hero, Immunophenotypic signatures of benign and dysplastic granulopoiesis by cytomic profiling, *Cytom. B Clin. Cytom.*, 2011, **80**(5), 282–290.

26. F. Steinbach, F. Henke, B. Krause, B. Thiele, G. R. Burmester and F. Hiepe, Monocytes from systemic lupus erythematous patients are severely altered in phenotype and lineage flexibility, *Ann. Rheum. Dis.*, 2000, **59**(4), 283–288.

27. A. A. van de Loosdrecht, C. Alhan, M. C. Bene, M. G. Della Porta, A. M. Drager, J. Feuillard, P. Font, U. Germing, D. Haase,

C. H. Homburg, R. Ireland, J. H. Jansen, W. Kern, L. Malcovati, J. G. Te Marvelde, G. J. Mufti, K. Ogata, A. Orfao, G. J. Ossenkoppele, A. Porwit, F. W. Preijers, S. J. Richards, G. J. Schuurhuis, D. Subira, P. Valent, V. H. van der Velden, P. Vyas, A. H. Westra, T. M. de Witte, D. A. Wells, M. R. Loken and T. M. Westers, Standardization of flow cytometry in myelodysplastic syndromes: report from the first European LeukemiaNet working conference on flow cytometry in myelodysplastic syndromes, *Haematologica*, 2009, **94**(8), 1124–1134.

28. D. Liu, Z. Chen, Y. Xue, D. Lu, Y. Zhou, J. Gong, W. Wu, J. Liang, Q. Ma, J. Pan, Y. Wu, Y. Wang, J. Zhang and J. Shen, The significance of bone marrow cell morphology and its correlation with cytogenetic features in the diagnosis of MDS-RA patients, *Leuk. Res.*, 2009, **33**(8), 1029–1038.

29. H. Szpurka, R. Tiu, G. Murugesan, S. Aboudola, E. D. Hsi, K. S. Theil, M. A. Sekeres and J. P. Maciejewski, Refractory anemia with ringed sideroblasts associated with marked thrombocytosis (RARS-T), another myeloproliferative condition characterized by JAK2 V617F mutation, *Blood*, 2006, **108**(7), 2173–2181.

30. L. Malcovati, M. G. Della Porta, D. Pietra, E. Boveri, A. Pellagatti, A. Galli, E. Travaglino, A. Brisci, E. Rumi, F. Passamonti, R. Invernizzi, L. Cremonesi, J. Boultwood, J. S. Wainscoat, E. Hellstrom-Lindberg and M. Cazzola, Molecular and clinical features of refractory anemia with ringed sideroblasts associated with marked thrombocytosis, *Blood*, 2009, **114**(17), 3538–3545.

31. F. R. Appelbaum, J. Barrall, R. Storb, R. Ramberg, K. Doney, G. E. Sale and E. D. Thomas, Clonal cytogenetic abnormalities in patients with otherwise typical aplastic anemia, *Exp. Hematol.*, 1987, **15**(11), 1134–1139.

32. D. P. Steensma, G. W. Dewald, T. L. Lasho, H. L. Powell, R. F. McClure, R. L. Levine, D. G. Gilliland and A. Tefferi, The JAK2 V617F activating tyrosine kinase mutation is an infrequent event in both "atypical" mye-loproliferative disorders and myelodysplastic syndromes, *Blood*, 2005, **106**(4), 1207–1209.

33. J. M. Bennett, D. Catovsky, M. T. Daniel, G. Flandrin, D. A. Galton, H. R. Gralnick and C. Sultan, Proposals for the classification of the myelodysplastic syndromes, *Br. J. Haematol.*, 1982, **51**(2), 189–199.

34. K. Naqvi, E. Jabbour, C. Bueso-Ramos, S. Pierce, G. Borthakur, Z. Estrov, F. Ravandi, S. Faderl, H. Kantarjian and G. Garcia-Manero, Implications of discrepancy in morphologic diagnosis of myelodysplastic syndrome between referral and tertiary care centers, *Blood*, 2011, **118**(17), 4690–4693.

35. P. L. Greenberg, H. Tuechler, J. Schanz, G. Sanz, G. Garcia-Manero, F. Sole, J. M. Bennett, D. Bowen, P. Fenaux, F. Dreyfus, H. Kantarjian, A. Kuendgen, A. Levis, L. Malcovati, M. Cazzola, J. Cermak, C. Fonatsch, M. M. Le Beau, M. L. Slovak, O. Krieger, M. Luebbert, J. Maciejewski, S. M. Magalhaes, Y. Miyazaki, M. Pfeilstocker, M. Sekeres, W. R. Sperr, R. Stauder, S. Tauro, P. Valent, T. Vallespi, A. A. van de Loosdrecht, U. Germing and D. Haase, Revised international

prognostic scoring system for myelodysplastic syndromes, *Blood*, 2012, **120**(12), 2454–2465.

36. L. Malcovati, M. G. Porta, C. Pascutto, R. Invernizzi, M. Boni, E. Travaglino, F. Passamonti, L. Arcaini, M. Maffioli, P. Bernasconi, M. Lazzarino and M. Cazzola, Prognostic factors and life expectancy in myelodysplastic syndromes classified according to WHO criteria: a basis for clinical decision making, *J. Clin. Oncol.*, 2005, **23**(30), 7594–7603.

37. O. Kosmider, V. Gelsi-Boyer, M. Cheok, S. Grabar, V. la-Valle, F. Picard, F. Viguie, B. Quesnel, O. Beyne-Rauzy, E. Solary, N. Vey, M. Hunault-Berger, P. Fenaux, M. Mansat-De, E. V. Delabesse, P. Guardiola, C. Lacombe, W. Vainchenker, C. Preudhomme, F. Dreyfus, O. A. Bernard, D. Birnbaum and M. Fontenay, TET2 mutation is an independent favorable prognostic factor in myelodysplastic syndromes (MDSs), *Blood*, 2009, **114**(15), 3285–3291.

38. F. Thol, E. M. Weissinger, J. Krauter, K. Wagner, F. Damm, M. Wichmann, G. Gohring, C. Schumann, G. Bug, O. Ottmann, W. K. Hofmann, B. Schlegelberger, A. Ganser and M. Heuser, IDH1 mutations in patients with myelodysplastic syndromes are associated with an unfavorable prognosis, *Haematologica*, 2010, **95**(10), 1668–1674.

39. K. Yoshida, M. Sanada, Y. Shiraishi, D. Nowak, Y. Nagata, R. Yamamoto, Y. Sato, A. Sato-Otsubo, A. Kon, M. Nagasaki, G. Chalkidis, Y. Suzuki, M. Shiosaka, R. Kawahata, T. Yamaguchi, M. Otsu, N. Obara, M. Sakata-Yanagimoto, K. Ishiyama, H. Mori, F. Nolte, W. K. Hofmann, S. Miyawaki, S. Sugano, C. Haferlach, H. P. Koeffler, L. Y. Shih, T. Haferlach, S. Chiba, H. Nakauchi, S. Miyano and S. Ogawa, Frequent pathway mutations of splicing machinery in myelodysplasia, *Nature*, 2011, **478**(7367), 64–69.

40. F. Damm, O. Kosmider, V. Gelsi-Boyer, A. Renneville, N. Carbuccia, C. Hidalgo-Curtis, V. Della, L. V. Couronne, L. Scourzic, V. Chesnais, A. Guerci-Bresler, B. Slama, O. Beyne-Rauzy, A. Schmidt-Tanguy, A. Stamatoullas-Bastard, F. Dreyfus, T. Prebet, B. S. de, N. Vey, M. A. Morgan, N. C. Cross, C. Preudhomme, D. Birnbaum, O. A. Bernard and M. Fontenay, Mutations affecting mRNA splicing define distinct clinical phenotypes and correlate with patient outcome in myelodysplastic syndromes, *Blood*, 2012, **119**(14), 3211–3218.

41. V. Visconte, H. J. Rogers, J. Singh, J. Barnard, M. Bupathi, F. Traina, J. McMahon, H. Makishima, H. Szpurka, A. Jankowska, A. Jerez, M. A. Sekeres, Y. Saunthararajah, A. S. Advani, E. Copelan, H. Koseki, K. Isono, R. A. Padgett, S. Osman, K. Koide, C. O'Keefe, J. P. Maciejewski and R. V. Tiu, SF3B1 haploinsufficiency leads to formation of ring sideroblasts in myelodysplastic syndromes, *Blood*, 2012, **120**(16), 3173–3186.

CHAPTER 8

The Prospects for Real-Time Raman Spectroscopy for Oesophageal Neoplasia

MAX ALMOND,[a] GAVIN RHYS-LLOYD,[a]
JO HUTCHINGS,[a] GEETA SHETTY,[a] NEIL SHEPHERD,[a,b,e]
CATHERINE KENDALL,[a,e] NICHOLAS STONE[a,c] AND
HUGH BARR*[a,d,e]

[a] Biophotonics Research Unit, Leadon House, Gloucestershire Royal
Hospital, Great Western Road, Gloucester GL13NN, United Kingdom;
[b] Department of Histopathology, Gloucestershire Hospitals NHS Foundation
Trust, United Kingdom; [c] Biomedical Imaging and Bio Sensing, School of
Physics, University of Exeter, EX4 4QL, United Kingdom; [d] Three Counties
Cancer Resection Unit & Upper Gastrointestinal Surgical Unit,
Gloucestershire Hospitals NHS Trust, Gloucester GL13NN, United
Kingdom; [e] Faculty of Bioscience and Medicine, Cranfield Health, Cranfield
University, Bedfordshire, United Kingdom
*Email: hugh.barr@glos.nhs.uk

8.1 The Clinical Problem

Over the past 40 years the incidence of oesophageal adenocarcinoma has increased more rapidly than any other solid tumour in the Western world and now represents a major public health problem. Recently, the National United Kingdom Oesophagogastric Cancer Audit reported a 5-year age-adjusted

RSC Detection Science Series No. 2
Detection Challenges in Clinical Diagnostics
Edited by Pankaj Vadgama and Serban Peteu
© The Royal Society of Chemistry 2013
Published by the Royal Society of Chemistry, www.rsc.org

survival rate of 10%.[1] There is a clear imperative to bring forward the diagnosis of this cancer and to identify and treat precancerous changes.

Barrett's oesophagus has been identified as a major risk factor. It is defined as "a change in the distal oesophageal epithelium of any length that can be recognised as columnar type mucosa at endoscopy and is confirmed by biopsy of the tubular oesophagus". This can be recognised endoscopically but often the subtle molecular degeneration to cancer cannot be seen.[2]

The major healthcare problem is that the population prevalence estimates of Barrett's oesophagus in the UK and US populations is approximately 1 million and 4 million individuals, respectively. There is also evidence that the incidence of Barrett's oesophagus is increasing by around 2% per year.[3–6]

Although Barrett's oesophagus is accepted as a significant risk factor for the development of adenocarcinoma, the overall risk of progression and the risk of disease specific mortality are low. Two large population-based cohort studies reported annual rates of progression to be 0.12–0.13% per year.[7,8]

The natural history of preneoplastic/dysplastic Barrett's oesophagus has been difficult to ascertain. Studies have suffered from inadequate power, insufficient follow-up durations, sampling errors at endoscopy, and considerable diagnostic variability due to histological subjectivity. Two phenotypes, low-grade dysplasia (LGD) and high-grade dysplasia (HGD), indicate increased risk of cancer development. A cohort study of 111 patients with LGD showed a relative risk of 9.7 (95% confidence intervals of 4.4–21.5) for progression to HGD or adenocarcinoma.[9] If detected at the baseline endoscopy the risk was 4.8 (95% CI 2.6-8.8).[7] LGD is hard to diagnose from biopsy samples and may be overdiagnosed. Of 1198 Barrett's patients undergoing surveillance 12.5% were diagnosed as LGD. A review of the histology by two external expert pathologists showed only 0.7% were considered to have "true LGD"; 42% of these patients progressed to cancer or HGD, after 51 months.[10] Patients with HGD have a 59% change of invasive cancer after 5 years.[11]

8.2   Endoscopic Recognition of Dysplasia and Cancer

Barrett's oesophagus is usually identified in patients who are undergoing an endoscopy for investigation of symptoms of acid reflux, see Figure 8.1. It can be easily recognised by due to its salmon pink appearance. The detection of early dysplasia and early cancer using standard white-light endoscopy is very difficult and highly subjective, even for experts. Therefore guidelines recommend that quadrantic biopsies are taken every 2 cm of Barrett's oesophagus with additional targeted biopsies from any visible areas of mucosal abnormality. These are poorly adhered to outside of clinical trials, sample less than 5% of the mucosa and may miss up to 57% of dysplasia.[12,13]

Several advanced endoscopic imaging techniques have been developed that may aid targeted biopsies and identify high-risk areas,[14–18] The latest endoscopes are high resolution (one million pixels) and do improve detection of early Barrett's neoplasia in expert hands.[19] In addition, spraying dyes or acetic acid onto the surface of the oesophagus can allow more accurate lesion

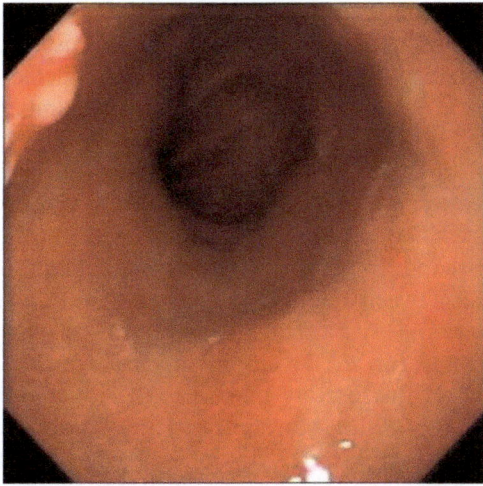

Figure 8.1 Endoscopic picture of Barrett's oesophagus. The nodular area with white areas in the corner is clearly abnormal and is an early cancer. The patchy reddened areas may simply be inflammation, Barrett's oesophagus or invisible early cancer or dysplasia. These are only clarified by biopsy. However, the biopsy may miss an area of dysplasia and full mapping is necessary. Every biopsy will produce bleeding further obscuring the view.

recognition. There are several methods for this "chromoendoscopy". Absorptive stains (such as methylene blue) cross epithelial membranes selectively, whereas contrast stains (such as indigo carmine) permeate into mucosal crevices highlighting surface topography and mucosal irregularities. Success is varied and acetic acid is currently favoured.

Specific targeted biomarkers do hold promise. The Fitzgerald group have demonstrated that alterations in cell-surface glycans occur in Barrett's oesophagus during progression to adenocarcinoma. Selective binding of a candidate lectin (wheat germ agglutinin) sprayed into an *ex vivo* oesophagus enabled visualisation of high-grade dysplastic lesions, which were not detectable by conventional endoscopy.[20]

To avoid staining the surface narrow-band imaging (NBI) is an alternative method to provide contrast during endoscopic examination. The oesophageal surface is illuminated with blue and green light. Longer wavelengths penetrate deeper into tissue. By using shorter wavelengths in the blue part of the spectrum superficial capillary networks can be seen, and green light displays the sub-epithelial vessels. In combination, they produce an extremely high-definition image of the mucosal surface allowing visualisation of subtle mucosal irregularities and alterations in vascular patterns consistent with dysplasia and intra-mucosal cancer.[21,22] A recent review of NBI with magnification demonstrated high accuracy for the diagnosis of HGD in Barrett's oesophagus based on recognition of irregular mucosal pit patterns and/or irregular microvascularisation.[23] Overall, data on the accuracy of NBI in Barrett's oesophagus are inconclusive and results of multicentre randomised controlled trials are awaited.

As well as morphological changes there are fluorophore concentration differences between normal, metaplastic and neoplastic tissue. Autofluoresecence endoscopy (AFI) endoscopy has been used for the detection of high grade dysplasia in Barrett's oesophagus. Curvers *et al.* demonstrated an increased detection rate of HGD/IMC using AFI compared to WLE alone (53% *vs.* 90%) but this came at the expense of a high false-positive rate of 81%.[24]

Optical coherence tomography OCT produces cross-sectional images of the oesophageal surface. The probe has to be passed through the instrument channel of an endoscope. In 55 patients with Barrett's oesophagus, OCT was able to differentiate between HGD and adenocarcinoma with a sensitivity of 83% and a specificity of 75%.[25] The acquisition of images and interpretation can take some time and further developments are awaited.

Endoscopic confocal microscopy magnifies the mucosa 1000-fold enabling real-time visualisation of cellular structures. It has shown considerable potential for use in patients with Barrett's oesophagus with reported accuracy of up to 97.4% for the detection of dysplasia.[26] There is considerable interuser variation, it is time consuming, requires administration of an exogenous agent; and is expensive.[27]

Elastic scattering spectroscopy (ESS) was one of the first methods to be examined for the detection of dysplasia in Barrett's using a probe passed down the instrumentation channel of the endoscope. Lovat *et al.* measured spectra from 181 tissue sites from 81 patients that were correlated with consensus histopathology. ESS identified HGD with a sensitivity of 92% and a specificity of 60% and was able to differentiate these sites from inflammation with a sensitivity and specificity of 79%.[28] This current level of accuracy is probably not sufficient to support clinical uptake of ESS, unless it can be improved through refinement of the technology or the use of a concomitant imaging modality.

8.3   Raman Spectroscopy

The attraction of Raman spectroscopy is near-instantaneous acquisition of molecular information. Raman scattering was first described in 1928 by an Indian physicist of the same name who was very shortly afterwards awarded the Nobel Prize for Physics.[29] The great attraction of the technique is the highly specific biochemical/molecular information obtained. In the past decade, the potential of the technique for differentiating between normal, premalignant, and cancerous tissue has been extensively explored. Technological developments have occurred, which have allowed the possibility of an endoscopic Raman probe to provide accurate, noninvasive optical diagnosis in "real-time", without the need for biopsy.

Two types of Raman scattering have been described, dependent on the state of the molecule when the interaction with light occurs. Stokes scattering results when light interacts with a molecule in its ground (unexcited) state. Nuclear motion is initiated and the scattered light has a lower energy than the incident light. However, where the molecule exists in an already vibrationally excited state, as is seen at high temperatures, the scattered light is of a higher energy than the incident light. This is known as anti-Stokes scattering. At room temperature, most molecules are at the ground vibrational state and therefore Stokes Raman scattering is much more prevalent than anti-Stokes scattering. It is possible to calculate the proportion of molecules in a substance that are in the ground and excited states using the ratio of intensities of Stokes and anti-Stokes scattering.

As a molecule vibrates the surrounding electron cloud changes shape. This can induce a change of the dipole moment or a change in the polarisation of the molecule. For the vibrational change in a molecule to be Raman active there must be a change in the polarisation of the molecule. That is to say that in shifting from one vibrational state to another, the electron density relative to the position of the nucleus must change. However, not all changes in vibrational states can be detected by Raman spectroscopy. In some cases the change in vibrational mode represents a change in the dipole moment of the molecule and not its polarisation. This change is not detected by Raman spectroscopy but can be detected by infrared (IR) spectroscopy. An example of the differences in vibrational modes can be appreciated when the properties of a phenyl ring are compared to that of OH bond. The phenyl ring has a large electron cloud and is highly polarisable, giving a strong Raman signal, whereas an OH bond is less polarisable (so gives a weak Raman signal), but a dipole change is readily induced, causing a strong IR signal. These properties of the OH bond have important consequences for the application of spectroscopy to living-tissue diagnostics. As a result of the strong IR signal, all water must be excluded from tissue samples before using IR spectroscopy to prevent profound distortion of the spectra from the marked dipole effect of the OH bonds. This effect is not seen with Raman spectroscopy and so allows this technique to be used to analyse fresh tissue. Any molecule of sufficient complexity will usually exhibit a usable Raman signal.

A Raman spectrum is a plot of "Raman shift" measured in wave numbers (on the x-axis) against intensity (on the y-axis) (see Figure 8.2). The zero point on the x-axis corresponds to the energy of the incident light. Data to the right of this point therefore represents Stokes scattering, whereas data to the left represents anti-Stokes scattering. A Raman spectrum is always relative to the wavelength of the incident light. As a result, a specific wavelength of incident light can be selected in order to determine the wavelength region of the measured Raman spectra. The x-axis of a Raman spectrum is therefore rarely measured in wavelength as this would vary with the incident light chosen for excitation. Instead, wave number is used to represent the "Raman shift" and facilitate comparison between spectra. A larger Raman shift indicates that a greater energy is required to create vibrational motion in a given molecule and crucially, the "value" of the Raman shift for a given compound is independent of the frequency of the excitational light. The excitational light frequency does directly affect the intensity of the Raman signal.

Raman peaks are spectrally narrow and correspond precisely to the vibration of specific bonds within molecules. In addition, the intensity of a peak is proportional to the concentration of the specific molecular constituent that gives rise to it. This allows specific vibrations to be assigned to certain spectral peaks,

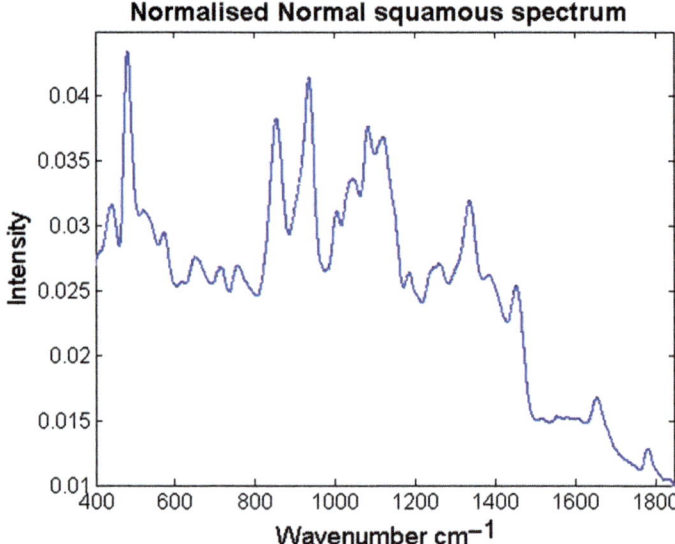

Figure 8.2   Raman spectrum acquired from normal squamous oesophageal mucosa.

enabling a precise qualitative and quantitative biochemical fingerprint to be obtained.[30] However, in complex molecules peak assignment can be difficult and unreliable as the various peaks merge to form bands. It is then the shape and distribution of these bands that can then be used as a chemically sensitive signature of the specimen.

Analysis of pure natural compounds such as glycogen and collagen has enabled scientists to assign peaks that are characteristic of these molecules. Glycogen molecules have characteristic peaks at 490, 853, 937, 1083 and 1123 cm^{-1}. Although potentially obscured by signals from other molecular constituents, modern multivariate analysis techniques can be used to extract these subtle spectral features and enable conclusions to be drawn regarding the nature of a tissue specimen. This highlights the potential of Raman spectroscopy to measure objective, quantitative information that can enable precise tissue analysis and medical diagnosis.

8.4   Instrumentation

Raman spectrometers must be very sensitive in order to detect Raman signals of extremely low intensity. Other signals, in particular elastically scattered light, have to be excluded. Otherwise, the weak Raman signal would be completely obscured. This is achieved using a series of filters within the instrument. As the positions of Raman peaks are relative to the wavelength of incident light, a monochromatic (single-wavelength) incident laser is used for excitation. When assessing biological tissue, incident laser light in the near-infrared (NIR) region is generally the preferred source for excitation. At this wavelength there is minimal tissue absorption and the risks of thermal or mutagenic effects are therefore avoided. In addition, the effect of fluorescence seen with visible and UV light is largely overcome.

A series of mirrors typically guide the laser onto the sample. Scattered light passes back into the spectrometer and is filtered by a notch or edge filter to remove elastically scattered light. A lens focuses the light onto a slit, which removes any stray light. The light is then split and dispersed by the diffraction grating and then reflected onto the charge-coupled device (CCD).

The CCD typically consists of a sectored length of silicon. This enables different frequencies of scattered light to be collected separately, and this information can then be relayed to a computer in order to construct a spectrum. The shape of the spectrum will depend on the number of photons striking each sector of the silicon. The two main types of CCD used for Raman spectroscopy are backthinned and deep-depletion devices. Backthinned CCDs suffer from etaloning and for this reason deep-depletion CCDs are commonly preferred for NIR applications.

8.5   Raman Spectroscopy for Medical Diagnosis

The ability of Raman spectroscopy to accurately identify pathological changes in tissues has been proven using *ex vivo* tissue specimens from breast, cervix, larynx, bladder, prostate, lymph nodes and oesophagus.[31–48] These studies have encouraged clinicians about the potential of Raman spectroscopy to bring forward cancer and precancer diagnosis. In particular, to understand the subtle molecular degeneration that occurs before an invasive cancer becomes morphologically visible. The aim is to provide real-time rapid diagnosis without the need for tissue removal and biopsy and to detect the earliest molecular changes that indicate a segment of Barrett's oesophagus has the potential to become "bad". A Raman probe could identify early neoplastic changes within the oesophagus, which are often invisible using standard endoscopy. In addition Raman spectroscopy could remove the subjectivity from histological assessment and even eliminate the need for a biopsy altogether (so called "optical biopsy"). Objective endoscopic diagnosis of neoplastic change in the oesophagus would enable immediate intervention to take place. HGD in Barrett's oesophagus could be diagnosed and then treated by immediate endoscopic resection during a single endoscopy session. Similarly, tumour margins could be assessed intra-operatively to ensure an adequate surgical resection. In addition, when rescoping a patient with biopsy-proven focal HGD or IMC, Raman spectroscopy could accurately identify the lesion so that endoscopic resection (of the correct area) could be performed. The challenge of developing *in vivo* Raman systems is considerable as several important factors must be addressed and overcome:

- Weak signal strength:
  - Raman signals are inherently weak (only 1 in 10^6–10^8 photons are inelastically scattered).
  - Biochemical changes in cells that have become dysplastic are very difficult to detect as they may be present in minute concentrations.

○ When used on living tissue, the intensity of the excitatory laser must be limited to prevent tissue damage. This limits the ability to generate strong scattering signals from the tissue.
- Signal interference:
  ○ Biological tissues are inhomogeneous in composition and thus highly scattering.
  ○ Tissue fluorescence.
  ○ Spectral contamination caused by the background Raman and fluorescence signals generated in the fibre-optic materials.

Previous studies have recognised that simple peak analysis comparing the intensities and wave numbers of individual peaks, fails to provide discrimination between pathology. Instead, one must consider diagnostic information to be spread throughout the spectrum and complex chemometric approaches such as principal component analysis (PCA) and linear discriminant analysis (LDA) have been utilised to tease out biochemical differences between tissue types.

When aiming to optimise signal strength we must remember that the intensity of a Raman peak is dependent on the polarisability of the molecule being excited, the concentration of that molecule in the tissue, and the intensity of the incident laser. The power of Raman scattering increases with the fourth power of the frequency of the excitation source. In addition, the signal-to-noise ratio (SNR) is proportional to the square of the intensity of the incident light multiplied by the acquisition time.

Therefore, increasing both the incident laser power and the acquisition time would be expected to generate a stronger Raman signal. However, in practice this could cause injury to human tissue and therefore does not offer a realistic solution. Moreover, prolonged acquisition times of more than a few seconds would be impractical for *in vivo* use.

Another problem when using Raman spectroscopy to examine biological tissue is that of tissue fluorescence that often obscures the Raman signal.

8.6 Raman Spectroscopy Assessment of Oesophageal Pathology

Several groups have explored the potential of Raman spectroscopy for diagnosis of oesophageal pathology *in vitro*.

In 2003 Kendall *et al.* analysed snap-frozen oesophageal biopsy specimens using Raman microspectroscopy.[49] Their subset analysis included normal squamous mucosa, Barrett's mucosa and adenocarcinoma, and showed a high level of agreement (kappa statistic, $k = 0.89$) between consensus pathological opinion (the gold standard) and Raman measurements. They also showed that the level of agreement between Raman spectroscopy and consensus histological diagnosis compared favourably to that between an independent pathologist and the consensus pathology opinion for the same biopsy specimens ($k = 0.76$).

This illustrated the importance of gaining a consensus pathology opinion in order to reduce interobserver variability, a finding supported by others.[11,49]

Shetty *et al.* used Raman spectroscopy to characterise some of the early biochemical changes that occur in oesophageal mucosal cells as they progress to Barrett's metaplasia and then to dysplasia and cancer.[43] They found a marked variation in the distribution of various biochemical constituents in normal, dysplastic and malignant tissue. The changes occurred in sugars, lipids, and amino acids. Raman spectroscopy identified increased levels of DNA, actin and oleic acid in dysplastic glandular tissue compared to normal Barrett's or squamous mucosa. Conversely, glycogen was particularly prominent in normal squamous tissue and reduced in neoplastic tissue. This study further supported the potential of Raman spectroscopy for early diagnosis of malignancy, and raised the possibility that Raman spectroscopy could perhaps enable detection of biochemical changes in tissue before morphological changes manifest. The proven ability of Raman spectroscopy to analyse and classify biopsied oesophageal tissue has demonstrated potential for the technique to be used to aid histological assessment of tissue samples. The ability to objectively characterise tissue based on its biochemical profile could help to overcome the inherent subjectivity associated with histopathological classification.

Recently, Hutchings *et al.* demonstrated that rapid Raman mapping of oesophageal tissue can identify histologically relevant biochemistry in order to distinguish between pathological subgroups in a clinically practicable timescale.[38] Hutchings *et al.* used a Renishaw StreamLine system coupled to a microscope fitted with a Leica ×50 long working distance objective (NA 0.5) to map 12 oesophageal biopsy samples that had been snap frozen and sectioned. A 830-nm diode laser was used for excitation which had a power of 70 mW at the sample. A variety of mapping parameters were evaluated (see Figure 8.3). The continuous readout of the CCD minimised time between sequential data measurements resulting in the acquisition of thousands of spectra in relatively short timescales. Mapping of a 2-mm oesophageal biopsy in 30–90 min (0.5 s acquisition time per 25.3 mm Raman pixels) in conjunction with principal component analysis was shown to be sufficient to generate spectral maps with clearly identifiable pathological regions.[34]

Rapid Raman mapping has the potential to provide prompt objective assessment of biopsied tissue samples without the need for tissue fixation or staining. This low signal-to-noise mapping lends itself to automation. Furthermore, objective diagnosis can be obtained by projecting measured spectral data onto pre-established multivariate algorithms in order to obtain a diagnosis based on the biochemical profile of the tissue, thus removing the subjectivity of histological reporting. Work is currently underway to further evaluate the diagnostic potential of rapid Raman mapping as an aid to oesophageal histopathology evaluation, particularly with regard to the grading of Barrett's-associated dysplasia. This ability to map out pathological regions within oesophageal biopsy specimens, which can be correlated with histology appearance, provides further evidence of the ability of Raman spectroscopy to

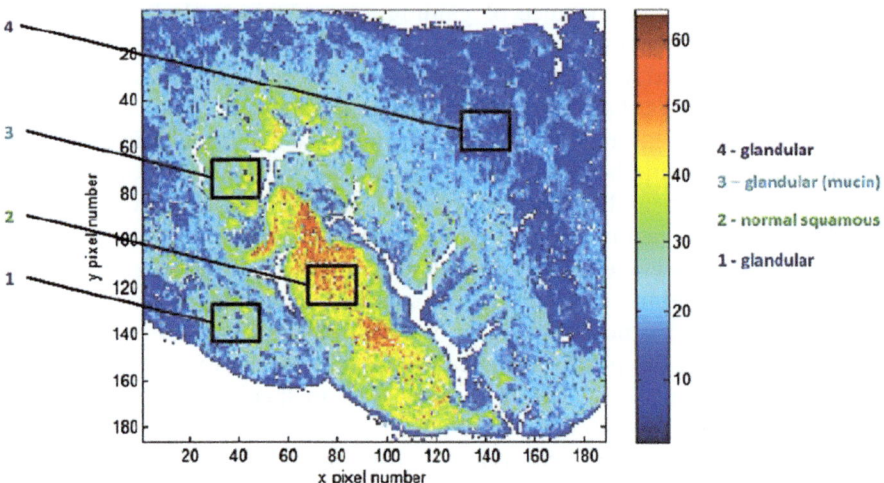

Figure 8.3   A high-quality Raman map is able to identify 4 distinct areas that have been discriminated through differences in their biochemical composition. Each has a characteristic spectral fingerprint which can be distinguished using PC-fed LDA.

objectively assess tissue and therefore gives additional support to the proposal of developing diagnostic Raman probes for *in vivo* use in the oesophagus.

8.7   Fibre-Optic Raman Spectroscopy

In order to develop a clinically applicable *in vivo* Raman probe for upper gastrointestinal endoscopy, there are several specific scientific and practical considerations that must be addressed and overcome.

- The probe must be narrow, long enough to fit down an endoscope channel – modern instruments have a 2.8-mm diameter port and are 90 cm in length. The design of the probe tip must not interfere with manipulation of the articulated distal end of the endoscope.
- Signal-to-noise ratio must be maximised, although acquisition times must remain clinically applicable (up to 2 s). The probe and spectrometer must be coupled to a data-processing unit so that short acquisition times are translated to meaningful clinical results in near "real time".
- Data measurement should also be close to "real time" (short acquisition times) to enable rapid assessment of large areas of oesophagus and to minimise inaccuracy caused by scope movement or oesophageal peristalsis.
- The probe must be safe for use *in vivo* (laser wavelength and power must not cause tissue destruction or long-term mutagenic effects).
- The probe must function effectively in the presence of endoscopic light – laboratory Raman spectroscopy requires reduced ambient light to minimise unwanted spectral noise.

- The sampling depth and volume must be optimised. Metaplasia and dysplasia develop in the oesophageal epithelium – the optimum depth from which to attain Raman signal is therefore 10–200 μm.[43] Sampling from large surface areas would provide more rapid assessment of clinical pathology but may decrease sensitivity and specificity for focal abnormalities.
- The probe must work in direct contact with the mucosal surface in order to facilitate endoscopic use.
- The probe must work safely and accurately in the presence of blood, saliva and gastric juice.
- The probe must be packaged so that it can be reused for multiple endoscopies and must be robust enough to withstand sterilisation procedures or must be simple and cheap enough to manufacture to enable single use.
- Measurements must be reproducible on different days, by different endoscopists and using different probes and spectrometers;
- High sensitivity and specificity must be achievable in the clinical setting (and this must be tested and reproduced *in vivo*).

As Raman scattering is weak and the power of the incident laser must be low (to minimise thermal injury), the design of a fibre-optic Raman probe must centre on maximising collection efficiency, whilst minimising intrinsic scattering and fluorescence produced by the probe.

One of the most potentially difficult problems to overcome is the propensity of probe fibres to generate their own Raman and fluorescence signals.[50,51] In some cases, the Raman signals may be greater than those arising from the tissue being studied. Fibre signal is proportional to fibre length and to the intensity of excitatory light. It is generated in the delivery fibre due to excitation by the laser light and in the collection fibres by elastically scattered light transmitting back along the collection fibre. Holographic notch filters or metal-oxide edge filters can be used to remove elastically scattered light from the collection fibre, and narrow-bandpass filters are commonly used to remove fibre-optic signal that has been generated in the delivery fibre.

In 1998, Mahadevan-Jansen *et al.* published *in vivo* results using a fibre-optic Raman probe to analyse dysplastic change in cervical tissue.[52] Silica–silica fibres were shown to cause less signal disruption than sapphire and liquid guide fibres despite their tendency to generate significant background signal at low wave numbers (below 900 cm^{-1}). The probe consisted of a single excitation fibre and 50 collection fibres and had a total diameter of 12 mm. The excitation fibre was positioned at the edge of the probe, and the laser output (at 789 nm) was reflected by a gold mirror at an angle to produce a 900-μm diameter excitation spot on the tissue surface. A bandpass filter on the distal tip of the excitation fibre blocked fluorescence and Raman scattering generated by the fibre, but transmitted the excitation light. Although demonstrating that Raman spectra could feasibly be measured *in vivo* using a fibre optic, the dimensions of this probe and the time taken for spectral acquisition (90 s) meant this design was impractical for use in the gastrointestinal tract.

The next important challenge was to develop an effective miniature fibre-optic probe that employed state-of-the-art probe technology but that was compatible for use with a medical endoscope. An endoscope is 90 cm in length and has a 2.8-mm diameter instrument channel through which a Raman probe could be delivered. This presents a considerable technical challenge; to package a probe with all its constituent parts (filters, collimating lenses, illumination and collection channels, and external casing) into a durable device with a diameter smaller than 2.8 mm.

In 1999 Shim *et al.* reported on utilising an endoscope compatible, fibre-optic Raman probe consisting of one central delivery fibre (400 µm diameter) surrounded by seven collection fibres each with a 300-µm diameter (Visionex, Inc, Sunnyvale, CA, USA, probe).[53] Filters were placed over the fibres approximately 2.5 cm from the probe tip and the collection fibres were part bevelled so that they predominantly collected signal from a zone in front of the delivery fibre and at a specified collection depth, determined by the bevelled angle. This design provided a means of controlling the sampling volume whilst optimising collection efficiency.

In 2000, Shim *et al.* reported the first *in vivo* Raman measurement of gastrointestinal tissue using the Visionex probe described above.[54] In this feasibility study the probe was passed through an endoscope into the oesophagus, stomach and colon. Twenty patients successfully underwent the endoscopic procedure following routine sedation and no complications were reported. The probe was positioned to touch an area of mucosa for 5 s during which time readings were taken. The probed site was then biopsied to allow comparison of Raman spectra with histopathology at that site. Subtle differences were reported in spectra from normal and diseased areas but the data acquired was not sufficient to discriminate between pathological groups. Whilst demonstrating feasibility, Shim *et al.* concluded that more work was needed to enhance the signal-to-noise ratio (SNR), to minimise acquisition times, and to improve data interpretation before reliable results could be generated. This study illustrated that endoscopic Raman spectroscopy could be performed safely in human patients, but was not large enough to enable comparison with results generated using laboratory-based Raman spectrometers.

Motz *et al.* used a novel probe to study dysplasia in rat palate tissue using 100 mW of laser light (at 830 nm) for excitation.[54,55] Multivariate data analysis of Raman spectra demonstrated differences in low- and high-grade dysplasia. They concluded that further modifications to the probe were necessary to optimise collection depth as a significant spectral contribution was collected from tissue that was deeper than 500 µm. More recently, a greater emphasis has been placed on targeted illumination of the epithelial lining of the gastrointestinal tract. Precancerous dysplasia develops in epithelial cells and, by definition, has not yet invaded deeper structures. For optimum diagnostic accuracy, probes must collect signal from the epithelial layer and minimise spectral contributions from scattering occurring in deeper tissues.

In 2005, Wilson and coworkers published results of an *in vivo* trial in the upper gastrointestinal tract of humans. They used a probe consisting of a central delivery fibre surrounded by six 300-μm diameter bevelled collection fibres that achieved an estimated sampling depth of 500 μm.[56] In conjunction with a diode laser (emitting at 785 nm) and a modern CCD detector they reported an accuracy of 87% for delineating between normal oesophageal tissue and dysplasia.

Huang and coworkers recently described the use of a ball lens at the distal tip of a novel Raman probe to focus excitation light onto the epithelium and help collimate scattered light into the collection fibres.[57] In 2010, following a number of preliminary *in vitro* and *in vivo* studies in rats, Huang *et al.* published a series of results from *in vivo* trials of endoscopic Raman spectroscopy for the detection of gastric neoplasia in humans. They used a 1.8-mm diameter Raman probe consisting of 32 collection fibres surrounding a single central delivery fibre. The probe had two stages of optical filters, at its proximal and distal ends, and used 785 nm excitation. Each spectrum was acquired in 0.5 s with a laser light irradiance of 1.5 W cm^{-2}. In December 2010 they reported a diagnostic sensitivity of 82.1% and a specificity of 95.3% for delineating malignant gastric ulcers from benign tissue.[58] In 2011, the same Raman probe system was used in the oesophagus demonstrating a sensitivity of 97.0% and a specificity of 95.2% for detecting oesophageal cancer.[59]

8.8 Novel Raman Probe

We have developed a novel probe for *in vivo* analysis of oesophageal tissue using a novel, custom-built, confocal Raman probe that has been designed specifically for compatibility with a standard medical endoscope. The Raman probe has been designed and built at the Interface Analysis Centre, at the University of Bristol.

A full description of the probe has been published by Day *et al.*[37] The laser is transmitted through the excitation fibre, collimated by a GRIN lens, and filtered to remove Raman and elastic scattering signals generated within the fibre. A monolith reflects the light onto a dichroic filter that reflects only light at the specified excitation wavelength onto a focusing GRIN lens. After interaction with the sample, scattered light is collimated by the GRIN lens and transmitted back to the dichroic filter that reflects the elastically scattered light but allows inelastically scattered wavelengths to pass through to a second dichroic filter. The signal is then collimated by a GRIN lens and transmitted back to the spectrometer through the collection fibre. The probe is confocal, that is to say, the exit aperture of the excitation fibre is equal to the entrance aperture of the collection fibre. It has been designed to focus to 150 μm in order to sample from the oesophageal epithelium and minimise signal interference from deeper tissues. In addition, unlike previous commercial fibre-optic Raman probes it has a short tip so that endoscopic manipulation will be possible when taking probe measurements *in vivo*. It is also designed for direct tissue contact in order to facilitate *in vivo* endoscopic measurements. Initial analysis was on

Figure 8.4 Photograph taken prior to tissue measurements after placement of the perspex mapping grid. The distal oesophagus and proximal stomach have been opened and the grid has been pinned in place so that it overlies the distal oesophagus and gastro-oesophageal junction.

oesophagectomy specimens using a fixed grid as shown (see Figure 8.4). Following spectral measurement, a biopsy was taken using standard endoscopy forceps at each of the specified grid positions and immediately fixed in formalin. All samples were paraffin embedded, microtomed and stained with Haematoxylin and Eosin prior to expert histological reporting. In addition, endoscopic resection (ER) and endoscopic biopsy samples were used to augment the spectral dataset. These specimens were placed on acetate and snap frozen in liquid nitrogen in the endoscopy suite. They were subsequently allowed to defrost at room temperature for 2 min (point biopsies) or 5 min (ERs) and mounted on calcium fluoride substrate with their mucosal surface facing upwards before measurement. All tissue samples were independently reported by two expert Consultant Pathologists based at different hospitals in order to gain consensus expert diagnosis.

Figure 8.5 shows the results for classification of neoplasia in Barrett's oesophagus. The results presented have demonstrated that a custom-built Raman probe designed for endoscopic use can accurately discriminate between a variety of oesophageal tissue types. Despite unavoidable background signal, an inherent problem when aiming to measure subtle spectral features *via* a fibre-optic, principal-component-fed linear discriminant analysis (PC-fed LDA) demonstrated sufficient ability to detect the specific biochemical information to enable tissue classification. Background subtraction/correction algorithms did not yield improvements in classification performance.

Results have been validated using independent test datasets and cross-validation techniques (*e.g.* leave-one-out). The independent test datasets, which were acquired over a different timescale to the training dataset, demonstrated

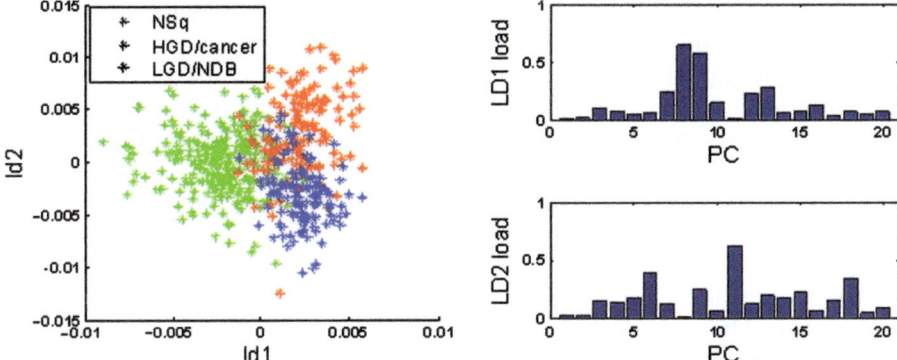

Figure 8.5 Linear discriminant analysis scatter plot of the three group classification model. All spectra were acquired in 1 s. NSq-normal squamous (normal oesophagus) HGD/cancer is high-grade dysplasia and cancer, LGD/ND8 is low-grade dysplasia and nondysplastic Barrett's Oesophagus. PC stands for principle component and ld (LD) is Linear Discriminant.

accurate classification performance. We have also demonstrated that reliable spectra can be acquired using different identically designed probes used by different operators and that these results can be accurately classified using an independent "training" algorithm.

8.9   Discussion

The data demonstrates that there is the possibility of real-time fibre-optic probe analysis of the oesophageal mucosa in clinically applicable timescales. In addition, spectra can feasibly be measured under direct vision using WLE and / or NBI. Shim *et al.* in 2000, demonstrated that *in vivo* Raman measurements were feasible and safe in the oesophagus, although diagnostically useful information was not obtained.[54] *In vivo* work by several groups has supported the conclusions that Raman spectroscopy is safe and feasible and has demonstrated more encouraging diagnostic data, although detection of early lesions using other probe systems has proved difficult.[55,56]

The results presented and the experience from previous pilot clinical trials support the need for an *in vivo* trial using this custom-built Raman probe system. There are several practical considerations that need to be addressed:

- Probe durability – Probes must be sufficiently durable to tolerate multiple passes down an endoscope without it affecting spectral recordings. Early in this project a probe was damaged during passage through a scope prompting a change to the design of the external rigid coating of the distal tip of the probe.
- Single or multiuse – Our data has shown that a multiuse probe could feasibly collect reliable data over at least a 2-year period.

- Manufacturing – There needs to be an established mechanism for reproducing identically built probes to high standards.
- CE marking – this is the mandatory legal conformity mark necessary for products placed on the European market. It confirms that the manufacturer has ensured that the product is not only safe but also functions in a way so as to achieve its intended purpose.
- Commercial support – There needs to be a mechanism for funding a future clinical trial. This will most likely be through a commercial collaboration.

After addressing these factors a carefully planned clinical trial will need to be undertaken. The aims of the trial would need to be clearly defined. Prior to a large, randomised trial a significant *in vivo* spectral dataset would be required to generate a robust classification model. The technique for ensuring that probe spectra can be correlated with the current gold standard (consensus expert histology) must be established. Huang *et al.* have described the use of an ER cap to ensure that probe and biopsy sites match. However, this requires that all patients are scoped using an ER cap, despite not necessarily needing therapeutic intervention. Alternatively, a double-lumen scope could be used so that the probe and the biopsy forceps are passed simultaneously through the scope. This would enable a biopsy to immediately be followed by a Raman measurement. However, slight adjustment by the endoscopists to allow for the different positions of the instrument channels will be required.

A Raman probe system could potentially have a role as a surgical adjunct during oesophagectomy. It could enable surgeons to assess resection margins intraoperatively to ensure they are clear of microscopic tumour deposits in order to minimise the risk of locally recurrent disease. Similarly, the probe could even act as an adjunct to oesohagogastric cancer staging, through laparoscopic assessment of suspicious peritoneal or lymphatic deposits, or potentially endoscopic assessment of the depth of tumour invasion through the oesophagus. Previously, the ability of RS to stage transitional cell carcinoma of the bladder based on urothelial measurements has been demonstrated.[44] There is potential for tumour depth "T-staging" in the oesophagus, however, modification to the confocal design of the probe described would be necessary if this were intended.

There are several methodological limitations that must be highlighted. These relate to the technique of spectral acquisition, the nature and design of the probe, the data preprocessing and analysis, and to assumptions made between the *ex vivo* environment and the real clinical environment at endoscopy.

- Point measurements – This is a recognised limitation of the technique. It is hoped that further developments may enable real-time imaging that could facilitate assessment of whole Barrett's segments in clinically applicable timescales. Thus, Raman point measurements may need to be combined with a wide field imaging system to identify areas of maximal concern.
- Fluorescence – The endogenous fluorophores in blood are well known to induce fluorescence that may mask weak Raman signals. Although water

lavage could be performed during endoscopy to remove excess blood, the fluorescence effects may interfere with a clinicians' ability to take multiple targeted biopsies. This will need further evaluation *in vivo*. As fluorescence was very rare *ex vivo* (following mucosal lavage) and these spectra were easily identifiable as visual outliers prior to data analysis, the problems caused by fluorescence were minimal.

- Probe background – The custom-built Raman probe has significant background signal due to Raman, fluorescence, and elastic scattering signals being generated within the fibre-optic cables. However, the probe has been shown to be sufficiently sensitive to detect the important spectral signals that enable tissue classification. Although not all visible above the probe background, PC-fed LDA was able to extract this data for tissue classification. Background subtraction/correction algorithms were not shown to improve classification performance.

- Exclusion of ambient light – All measurements presented and discussed were acquired following exclusion of ambient light as visible wavelengths can cause spectral disruption. However, both endoscopic white light and NBI have been shown to be compatible with Raman measurements.

- Pressure effects – Inadequate contact of the probe with the tissue may have affected the spectral signal. However, as explained previously, this was a translational project using a clinically applicable tool that was pressed against the tissue by hand, as would be the case *in vivo*. No attempt was taken to regulate probe pressure as this would be impractical *in vivo*.

- Inability to subclassify the grade of dysplasia – Despite measuring in excess of a thousand ex vivo tissue spectra, the numbers of homogeneous LGD and HGD samples that were confirmed by consensus histological diagnosis were too low to enable distinction between grades of dysplasia. Prior to widespread clinical use, an *in vivo* trial of sufficient power will need to generate a large spectral database in order to produce a robust classification model for *in vivo* spectra.

- Neoadjuvant chemotherapy – The majority of patients were treated with chemotherapy prior to surgical resection. This often led to shrinkage and fibrosis of the tumour, which may have affected its spectral signal.

- Previous ablation therapy – Some point biopsies/endoscopic resections were taken from patients who had undergone previous ablation therapy. It is unclear what effect, if any, this may have had on the Raman signal obtained from this tissue. However, in a clinical setting it will be essential that the probe can detect recurrent Barrett's/dysplasia in patients who have undergone previous endoscopic treatment. The ability to detect buried metaplasia has not been evaluated.

- Ischaemia effects – The effects of tissue ischemia have not been fully established. This may mean that the results cannot be generalised to *in vivo* results with complete accuracy. A large spectral dataset will need to be established following an *in vivo* trial and at that point comparisons with *ex vivo* data can be made.

- No assessment has been made as to whether the probe can identify macroscopically invisible areas of dysplasia/cancer. In order for RS to be a clinically applicable tool, it will need to be shown to improve recognition of early lesions and therefore demonstrate the ability to change patient care. This is impractical to do at this early stage but should be considered when designing future *in vivo* clinical trials in order to demonstrate diagnostic benefits.
- All spectra were acquired from a biased population as all patients had a history of Barrett's oesophagus and/or adenocarcinoma. However, these patients represent the target population for this clinical tool.
- It is possible systematic errors may have affected the dataset, for example systematic drift in the intensity or wave number axes over time. However, careful calibration prior to all sets of measurements was used to obviate these effects. In addition, the data presented showed no evidence of systematic drift and instrument response (green-glass) correction failed to improve the classification performance suggesting that major systemic drift was not a problem.

References

1. National Oesophago-Gastric Cancer Audit. *Third Annual Report*, (2010).
2. A. Watson, R. C. Heading and N. A. Shepherd, *Report of the Working Party of the British Society of Gastroenterology, Principal Authors: Guidelines for the Diagnosis and Management of Barrett's Columnar-lined Oesophagus*. 2005.
3. N. Vakil, S. V. van Zanten, P. Kahrilas, J. Dent and R. Jones, Global Consensus Group, *Am. J. Gastro.*, 2006, **101**, 1900.
4. J. Ronkainen, P. Aro, T. Storskrubb, S. E. Johansson, T. Lind, E. Bolling-Sternevald, M. Vieth, M. Stolte, N. J. Talley and L. Agreus, *Gastroenterology*, 2005, 1291825.
5. J. Jankowski, H. Barr, K. Wang and B. Delaney, *Brit. Med. J.*, 2010, **341**, 4551.
6. L. B. Gerson and S. Banerjee, *Gastro. Endosc.*, 2009, **70**, 867.
7. S. Bhat, H. G. Coleman, F. Yousef, B. T. Johnston, D. T. McManus, A. T. Gavin and L. J. Murray, *JNCI*, 2011, **103**, 1049.
8. F. Hvid-Jensen, L. Pedersen, A. M. Drewes, H. T. Sorensen and P. Funch-Jensen, *NEJM*, 2011, **365**, 1375.
9. M. Sikkema, C. W. Looman, E. W. Steyerberg, M. Kerkhof, F. Kastelein, H. van Dekken, A. J. van Vuuren, W. A. Bode, H. van der Valk, R. J. Ouwendijk, R. Giard, W. Lesterhuis, R. Heinhuis, E. C. Klinkenberg, G. A. Meijer, F. ter Borg, J. W. Arends, J. J. Kolkman, J. van Baarlen, R. A. de Vries, A. H. Mulder, A. J. van Tilburg, G. J. Offerhaus, F. J. ten Kate, J. G. Kusters, E. J. Kuipers and P. D. Siersema, *Am. J. Gastro.*, 2011, **106**, 1231.
10. W. L. Curvers, F. J. ten Kate, K. K. Krishnadath, M. Visser, B. Elzer, L. C. Baak, C. Bohmer, R. C. Mallant-Hent, A. van Oijen, A. H. Naber,

P. Scholten, O. R. Busch, H. G. Blaauwgeers, G. A. Meijer and J. J. Bergman, *Am. J. Gastro.*, 2010, **105**, 1523–1530.

11. B. J. Reid, P. L. Blount, Z. Feng and D. S. Levine, *Am. J. Gastro.*, 2000, **95**, 3089.

12. C. Bennett, S. Green, H. Barr, P. Bhandari, J. DeCaestecker, K. Ragunath, R. Singh, A. Tawil and J. Jankowski, *Cochrane System. Rev.*, 2010, **5**.

13. M. Vieth, C. Ell, L. Gossner, A. May and M. Stolte, *Endosc.*, 2004, **36**, 776.

14. K. B. Dunbar and M. I. Canto, *Gastro. Endosc.*, 2010, **72**, 668.

15. L. Qiu, D. K. Pleskow, R. Chuttani, E. Vitkin, J. Leyden, N. Ozden, S. Itani, L. Guo, A. Sacks, J. D. Goldsmith, M. D. Modell, E. B. Hanlon, I. Itzkan and L. T. Perelman, *Nature Med.*, 2010, **16**, 603.

16. M. A. Kara, M. E. Smits, W. D. Rosmolen, A. C. Bultje, F. J. Ten Kate, P. Fockens, G. N. Tytgat and J. J. Bergman, *Gastro. Endosc.*, 2005, **62**, 671.

17. W. Curvers, L. Baak, R. Kiesslich, A. Van Oijen, T. Rabenstein, K. Ragunath, J. F. Rey, P. Scholten, U. Seitz, F. Ten Kate, P. Fockens and J. J. Bergman, *Gastroenterol.*, 2008, **134**, 670.

18. T. Thomas, R. Singh and K. Ragunath, *Surg. Endosc.*, 2009, **23**, 1609.

19. M. A. Kara, F. P. Peters, W. D. Rosmolen, K. K. Krishnadath, F. J. ten Kate, P. Fockens and J. J. Bergman, *Endosc.*, 2005, **37**, 929.

20. E. L. Bird-Lieberman, A. A. Neves, P. Lao-Sirieix, M. O'Donovan, M. Novelli, L. B. Lovat, W. S. Eng, L. K. Mahal, K. M. Brindle and R. C. Fitzgerald, *Nature Med.*, 2012, **18**, 315.

21. M. A. Kara, F. P. Peters, W. D. Rosmolen, K. K. Krishnadath, F. J. ten Kate, P. Fockens and J. J. Bergman, *Endoscopy*, 2005, **37**, 929.

22. M. A. Kara, F. P. Peters, P. Fockens, F. J. ten Kate and J. J. Bergman, *Gastro. Endosc.*, 2006, **64**, 176.

23. M. Filip, S. Iordache, A. Saftoiu and T. Ciurea, *WJG*, 2011, **17**, 9.

24. W. L. Curvers, R. Singh, L. M. Song, H. C. Wolfsen, K. Ragunath, K. Wang, M. B. Wallace, P. Fockens and J. J. Bergman, *Gut*, 2008, **57**(2), 167–172.

25. J. A. Evans, J. M. Poneros, B. E. Bouma, J. Bressner, E. F. Halpern, M. Shishkov, G. Y. Lauwers, M. Mino-Kenudson, N. S. Nishioka and G. J. Tearney, *Clin. Gastro. Hepatico.*, 2006, **4**, 38–43.

26. M. Goetz and R. Kiesslich, *Anticancer Res.*, 2008, **28**, 353.

27. H. Pohl, T. Rosch, M. Vieth, M. Koch, V. Becker, M. Anders, A. C. Khalifa and A. Meining, *Gut*, 2008, **57**, 1648.

28. L. B. Lovat, K. Johnson, G. D. Mackenzie, B. R. Clark, M. R. Novelli, S. Davies, M. O'Donovan, C. Selvasekar, S. M. Thorpe, D. Pickard, R. Fitzgerald, T. Fearn, I. Bigio and S. G. Bown, *Gut*, 2006, **55**, 1078.

29. C. V. Raman and K. S. Krishnan, *Nature*, 1928, **122**, 501.

30. N. Stone, C. Kendall and H. Barr, *J. Biomed. Opt.*, 2008, **24**, 203.

31. C. Kendall, J. Hutchings, H. Barr, N. Shepherd and N. Stone, *Faraday Discuss*, 2011, **149**, 279.

32. M. C. Grimbergen, C. F. van Swol, C. Kendall, R. M. Verdaasdonk, N. Stone and J. L. Bosch, *Appl. Spectrosc.*, 2010, **64**, 8.

33. J. Horsnell, P. Stonelake, J. Christie-Brown, G. Shetty, J. Hutchings, C. Kendall and N. Stone, *The Analyst*, 2010, **135**, 3042.
34. J. Hutchings, C. Kendall, N. Shepherd, H. Barr and N. Stone, *J. Biomed. Opt.*, 2010, **15**, 660.
35. C. Kendall, J. Day, J. Hutchings, B. Smith, N. Shepherd, H. Barr and N. Stone, *Analyst*, 2010, **135**, 3038.
36. M. Sattlecker, C. Bessant, J. Smith and N. Stone, *Analyst*, 2010, **135**, 895.
37. J. C. Day, R. Bennett, B. Smith, C. Kendall, J. Hutchings, G. M. Meaden, C. Born, S. Yu and N. Stone, *Phys. Med. Biol.*, 2009, **54**, 7077.
38. J. Hutchings, C. Kendall, B. Smith, N. Shepherd, H. Barr and N. Stone, *J. Biopoton.*, 2009, **2**, 91.
39. C. Kendall, M. Isabelle, F. Bazant-Hegemark, J. Hutchings, L. Orr, J. Babrah, R. Baker and N. Stone, *Analyst*, 2009, **134**, 1029.
40. F. Bazant-Hegemark, K. Edey, G. R. Swingler, M. D. Read and N. Stone, *TCRT*, 2008, **7**, 483.
41. N. Stone, R. Baker, K. Rogers, A. W. Parker and P. Matousek, *Analyst*, 2007, **132**, 899.
42. N. Stone, M. C. Hart Prieto, P. Crow, J. Uff and A. W. Ritchie, *Anal. Bioanal. Chem.*, 2007, **387**, 1657.
43. G. Shetty, C. Kendall, N. Shepherd, N. Stone and H. Barr, *Brit. J. Cancer*, 2006, **94**, 1460.
44. P. Crow, B. Barrass, C. Kendall, M. Hart-Prieto, M. Wright, R. Persad and N. Stone, *Brit. J. Cancer*, 2005, **92**, 2166.
45. P. Crow, A. Molckovsky, N. Stone, J. Uff, B. Wilson and L. M. WongKeeSong, *Urology*, 2005, **65**, 1126.
46. M. C. Prieto, P. Matousek, M. Towrie, A. W. Parker, M. Wright, A. W. Ritchie and N. Stone, *J. Biomed. Opt.*, 2005, **10**(4), 440–446.
47. N. Stone, C. Kendall, J. Smith, P. Crow and H. Barr, *Faraday Discuss.*, 2004, **126**, 141.
48. P. Crow, N. Stone, C. A. Kendall, J. S. Uff, J. A. Farmer, H. Barr and M. P. Wright, *Brit. J. Cancer*, 2003, **89**, 106.
49. C. Kendall, N. Stone, N. Shepherd, K. Geboes, B. Warren, R. Bennett and H. Barr, *J. Path.*, 2003, **200**, 602.
50. S. M. Angel and M. L. Myrick, *Appl. Opt.*, 1990, **29**, 1350.
51. M. L. Myrick, S. M. Angel and R. Desiderio, *Appl. Opt.*, 1990, **29**, 1333.
52. A. Mahadevan-Jansen, M. F. Mitchell, N. Ramanujam, U. Utzinger and R. Richards-Kortum, *Photochem. Photobiol.*, 1998, **68**, 427.
53. M. G. Shim, B. C. Wilson, E. Marple and M. Wach, *Appl. Spectrosc.*, 1999, **53**, 619.
54. M. G. Shim, L. M. Song, N. E. Marcon and B. C. Wilson, *Photochem. Photobiol.*, 2000, **72**, 146.
55. J. T. Motz, S. J. Gandhi, O. R. Scepanovic, A. S. Haka, J. R. Kramer, R. R. Dasari and M. S. Feld, *J Biomed. Opt.*, 2005, **10**, 97.
56. L. M. Wong Kee Song, A. Molckovsky, K. K. Wang, L. J. Burgart, B. Dolenko, R. J. Somorjai and B. C. Wilson, *Proc. SPIE*, 2005, **5692**, 140.

57. Z. Huang, S. K. The, W. Zheng, J. Mo, K. Lin, X. Shao, K. Y. Ho, M. The and K. G. Yeoh, *Opt. Lett.*, 2009, **34**, 758.
58. M. S. Bergholt, W. Zheng, K. Lin, K. Y. Ho, M. Teh, K. G. Yeoh, J. B. So and Z. Huang, *Analyst*, 2010, **135**, 3162.
59. M. S. Bergholt, W. Zheng, K. Lin, K. Y. Ho, M. Teh, K. G. Yeoh, J. B. So and Z. Huang, *TCRT*, 2011, **10**, 103.

CHAPTER 9

Volatile Analysis for Clinical Diagnostics

CATHERINE KENDALL,*[a,b] HUGH BARR[a,b,c] AND
NARESH MAGAN[b]

[a] Biophotonics Research Unit, Leadon House, GHNHSFT, Gloucester,
GL1 3NN, United Kingdom; [b] Cranfield Health, Cranfield University,
MK43 0AL, United Kingdom; [c] Gloucestershire Royal Hospital,
Gloucestershire NHS Foundation Trust, United Kingdom
*Email: c.kendall@medical-research-centre.com

9.1 Volatile Diagnostics

The human nose is used as a qualitative analytical tool for discrimination/
detection of odours from food (fish, meat, cheese, wine), perfumes (cosmetics,
soaps) and flavours. Since the time of the ancient Greeks, physicians have been
aware that it is possible to smell disease on the breath of those afflicted. An
obvious example of this is the pear drop like smell of acetone in the breath of a
diabetic or bodily fluids such as urine can also have distinctive smells, which are
specific to disease, such as the honey-like smell of urine in phenylketonuria.

People can be highly skilled at detecting odours, but there may be problems
because of a lack of sensitivity over time and the variation between individuals.
It may also sometimes be unsafe or unhealthy and variable depending on the
individual.[1] To avoid these difficulties, attempts have been made to build an
artificial system that can mimic the olfactory sense of mammals and that can
detect and discriminate the production of volatile compounds from various
sources. Since the 1970s the potential of breath analysis as a way of detecting

RSC Detection Science Series No. 2
Detection Challenges in Clinical Diagnostics
Edited by Pankaj Vadgama and Serban Peteu
© The Royal Society of Chemistry 2013
Published by the Royal Society of Chemistry, www.rsc.org

disease in the human body has been recognised. These early studies were primarily performed using gas chromatography-mass spectrometry (GC-MS). In 1971 Pauling *et al.*[2] demonstrated that several hundred compounds were present in human breath and that some of these could be associated with abnormal physiological states or disease. When considering breath analysis, diseases of the lung and respiratory tract are the obvious choice of pathologies to be investigated as the vast majority of expired breath gas has been produced in the lung. Breath analysis is attractive because it offers the potential for rapid, noninvasive near-patient diagnosis.

All the methods for breath analysis rely on the detection of volatile organic compounds or VOCs. Organic compounds are those that contain carbon and are found in all living things. VOCs are organic compounds that easily become gases or vapours; as such they can be detected in breath and bodily fluids. Many studies, including those using GC-MS have shown that for certain pathological conditions (infection, malignancy, liver and cardiopulmonary disease) specific patterns of VOCs can be detected. These patterns are known as "volatile fingerprints". It is hoped that these fingerprints will provide the basis for accurate breath screening for disease in the future. These same technologies can also be employed to smell the headspace above bodily fluids such as blood and urine, which adds other potential medical applications to the uses of these technologies.

9.1.1 GC-MS and SIFT-MS Techniques

Techniques such as gas chromatography (GC), gas chromatography linked with mass spectrometry (GC-MS) and selected ion flow tube-mass spectrometry (SIFT-MS) have been used to analyse volatile compounds. SIFT-MS was developed in 1976 for real-time quantification of gas analysis of ambient air, exhaled breath and the headspace above liquids.[3,4] Headspace analysis of liquid samples such as urine to detect nephrotoxicity,[5] blood to detect bacterial growth[6] and human sweat to investigate stimulant compounds for mosquitoes causing malaria[7] have also been examined.

A large number of studies have been carried out to identify Microbial VOCs (MVOCs) growing on a diversity of matrices using chromatographic techniques but only a few have investigated these compounds from micro-organisms involved in ventilator associated pneumonia (VAP). Humphreys[8] used GC-MS for patient's breath analysis in order to identify patients with VAP. Scotter *et al.*[9] used SIFT-MS to find the fingerprints of volatiles from medically important fungi (*Aspergillus flavus, A.fumigatus, Candida albicans, Mucor racemosus, Fusarium solani* and *Cryptococcus neoformans*) grown on a variety of laboratory media (SDA, BHIB, Columbia agar, blood agar). Others[10] have identified MVOCs present in diverse building materials and from isolated and subcultured pure mould cultures (*Stachybotrys chartarum, Aspergillus penicillioides* and *Chaetomium globosum*) related to allergic status using GC-MS and solid-phase microextraction (SPME). Finally, there has been interest in monitoring mycotoxigenic fungal strains of *Penicillium roqueforti* using SPME with GC-MS by the detection of some metabolites formed in the biosynthesis of the mycotoxins.[11]

9.1.2 Electronic Noses

In the 1960s the development of instruments to detect odours started using mechanical noses based on redox reactions. It was not until the 1980s when integrated e-nose (electronic-nose) systems began appearing as research tools and commercial devices. The first "electronic nose" was described by Persaud and Dodd in 1981.[12] They attempted to replicate the various stages of the human olfactory process using biochemical sensors. Volatile compounds react on the surface of the sensors, neural networks can be used to analyse the sensor responses and evaluate the main useful components of the data, enabling in volatile odour recognition. The currently accepted definition of an e-nose, regarding the chemical sensor arrays used to detect odorant molecules, was proposed by Gardner and Bartlett in 1994:[13]

"An electronic nose is an instrument which comprises an array of electronic chemical sensors with partial sensitivity and an appropriate pattern recognition system, capable of recognising simple or complex odours".

The odorant molecules that interact with the surface of the sensors in an e-nose (equivalent to the stimuli in the olfactory system) are typically hydrophobic and volatile due to their molecular weight (30–300 Da), strength of the interactions between molecules and molecular shape.[14] Many of the odorant molecules are produced by micro-organisms. The ability of the e-nose to detect volatile fingerprints combined with pattern recognition systems may permit the early detection of specific micro-organisms or groups of micro-organisms.

A wide range of MVOCs are released from both primary and secondary metabolites and obtained from the breakdown of protein and carbohydrates during the metabolic pathways. These compounds can be alcohols, ketones, esters, lactones, aldehydes, sulfur and nitrogen-based products, as well as mycotoxins such as trichothecenes. These MVOCs can be characteristic and different for each species.[15,16] They also can vary depending on temperature, the age of cultures, the nutritional status[17] and the different water availability of the substrates.[16]

Micro-organisms are known to produce a wide range of different organic compounds that are released, with some of them being volatile. These MVOCs may be used as fingerprints of microbial growth and perhaps to discriminate between different species.

Analytical approaches involving more complex devices such as gas chromatography associated with mass spectrometry (GC-MS) and selected ion flow tube-mass spectrometry (SIFT) have also been employed to detect volatile biomarkers in headspace samples for similar applications. However, such approaches are more expensive and require a high level of expertise for analyses and interpretation of data sets.

9.1.2.1   E-Nose Technology

Typically, an e-nose consists of three parts: a sensor array, where the sensors measure electrical changes generated by adsorption of volatiles to the odour-sensitive chemical surface; a transducer that converts signals in readable values and the software for the analysis of the data. This structure has been copied from the olfactory organ of mammals in the sense that the active material of the sensors in an e-nose device (olfactory receptors) interacts with the volatiles creating a sensor response (electric signal or pattern) which is transmitted to the preprocessor (olfactory bulb) where the output is amplified and the noise is reduced. In the final stage the simplified signals (nerve impulses), as patterns of responses, are processed into the data-analysis system (hypothalamus and olfactory cortex in the brain).[18]

9.1.2.2   Types of Odour Sensors

Since the first model of an e-nose reported by Persaud and Dodd,[12] many advances have been made in the development of sensor technology. Some of the most important aspects of sensors are that they must be sensitive enough to the chemicals even at low concentrations and each sensor in the array must have partial sensitivity. Thus, response to a range of different gases rather than to a specific one is an advantage. This variable sensitivity leads to the production of distinct qualitative volatile fingerprints, enabling the identification of different patterns.

   The e-nose sensor technologies can be classified according to the sensor material and its transduction or conversion principle. Table 9.1 provides a summary of the main sensor transducer systems and their characteristics.

Table 9.1   Summary of the sensor technologies for e-noses. (Adapted from refs. 67, 68.)

Category	Measures	Material	Sensitivity	Principle
Electrochemical				
MOS	Resistance	Oxides	5–500 ppm	Changes in current or in
MOSFET	Capacitance	Catalytic metals	ppm	potential
CP	Resistance	Polymers	0.1–100 ppm	
Optical				
Optical fibres	Intensity/	Fluorescent	Low ppb	Changes in light
Fluorescence	Spectrum	dyes		intensity
Chemoluminescence				
Piezoelectric				
SAW	Piezoelectricity	Polymer-coated resonating	1 pg mass change	Changes in mass, density, viscosity
QCM		quartz disc	1 ng mass change	and acoustic phenomena

Electrochemical sensors can be:

a) Conductivity sensors: these exhibit a change in resistance when they are exposed to volatile organic compounds. There are two types of conductivity sensors:

   1. Conducting polymer sensors (CP sensors): This kind of sensor has reversible physiochemical properties in terms of change in conductivity under ambient temperature conditions when the active material (such as polypyrroles, thiophenes) is exposed to the chemicals, which bond with the polymer backbone. They are extremely sensitive and the standardisation of sample presentation is crucial for avoiding problems with humidity and drift over time.

   2. Metal oxide semiconductor sensors (MOS sensors): In this case, the doped materials with which the volatile organic compound (VOC) interacts are oxides that can be made of tin, zinc, and tungsten. Those materials are deposited between two metal contacts over a heating element that operates at high temperatures, 200–650 °C. When the sample interacts with the oxide material, oxidation occurs. The output signal is produced by a change in resistance caused by the reaction between the adsorbed oxygen and the gas molecules. They are quite sensitive (5–500 ppm) and this can be improved by using noble metals or changing the operating temperature. They are also resistant to humidity, but they may have drift problems over time and they may be poisoned by irreversible binding by different compounds such as sulfur or weak acids.

b) Metal oxide silicon field effect sensors (MOSFET sensors): They consist of three layers: a semiconductor, which can be made of silicon (Si) or silicon carbide (SiC); a thick oxide layer (SiO_2) as insulator; and on top a thin catalytic metal layer made of platinum (Pt), iridium (Ir) or palladium (Pd). In these sensors, the contact between the gas molecules and the catalytic gate metal affects the voltage and thus change the capacitance through the transistor. The gas detection is recorded as a voltage change in the sensor signal. The sensitivity and selectivity can be determined by the choice of the operation temperature, gate metal and structure of the gate metal. The sensitivity of the sensors is normally high for low concentrations of the gases, while it becomes saturated for high concentrations of gases. MOSFET and MOS sensors operate at high temperatures and are considered to be less sensitive to humidity with less carryover from one measurement to another.[19]

Piezoelectric sensors can measure temperature, mass changes, pressure, force and acceleration. The e-nose with piezoelectric sensors is designed to detect mass changes. There are two types of piezoelectric sensors:

a) Quartz crystal microbalance (QCM sensors): They consist of a resonating quartz disc that vibrates at a characteristic frequency (10–30 MHz). When the volatiles are absorbed there is an increase in the mass of the disc

reducing the resonance frequency. This reduction is inversely proportional to the mass of the odorant absorbed.

b) Surface acoustic-wave devices (SAW sensors): These sensors measure
mass changes too but differ from those above in that they operate at much
higher frequencies and require waves to travel over the surface of the
device. They can be less sensitive than other piezoelectric sensors.

Optical sensors include, amongst other materials, glass fibres coated with a
thin chemically active material that contains a fluorescent dye. When the
volatiles interact with the dye under a light source of a determinate frequency,
or narrow range of frequencies, the polarity light of the fluorescent dye is
altered and then a shift is produced in the emission spectrum. This change of
colour (or of fluorescence, chemoluminescence, absorbance or reflectance) can
be measured. Optical sensors have wide biological applications though they
present bleaching problems with sunlight.

9.2 Applications of Volatile Fingerprinting

These sensor array systems, so called e-noses, have been examined for potential
applications in many fields such as food quality control and spoilage, cosmetics
and beverages industries, environmental monitoring and more recently in
medical diagnostics.[20]

9.2.1 Food

E-nose technology has been applied in many fields, but the most frequently
considered has been the analysis of raw or manufactured food products;
freshness and maturity monitoring; shelf-life investigations; microbial detection
and control of food packaging material. In this field, many studies have been
performed in order to find the quality of raw materials like grains;[21,22] to
evaluate the olive oil oxidation during storage;[23] for freshness evaluation of
meats or fish; to classify a wide range of beverages such as coffee, beer, wine;[24]
to recognise spoilage bacteria and yeasts in milk;[25] to detect spoilage fungi in
bread[26] and to monitor the ripening of Danish blue cheese.[27]

Today, there is significant interest in detecting the levels of mycotoxins
produced by fungi in many food products. Thus, diverse studies using the
e-nose technology have been carried out to differentiate between mycotoxigenic
and nonmycotoxigenic strains of *Fusarium, Aspergillus, Penicillium* in food raw
materials[28,29] and in ham-based medium.[30] Discrimination between mycotoxin
producers and nonproducers strains is based, as some studies have reported, on
the different volatile fingerprints produced because of the different biosynthetic
pathways for mycotoxin production used.

Food applications have received significant attention because of the ease of
consistent sample presentation, sample size that may be required, it is noninvasive technique and the lack of sample preparation. Thus, potential exists in
using this approach as part of a quality-assurance system.[31]

9.2.2   Medical Applications

If the e-nose is able to recognise volatiles from bacterial and fungal growth on food substrates, it may also be able to detect and diagnose certain illnesses. There is a need for rapid, inexpensive and objective systems for clinical diagnosis. E-nose technology could provide a tool for early detection and reduce the high costs and the time consuming nature of using molecular, microbiological cultures and serological tests in the detection of diseases like *Helicobacter pylori*,[32] and Mycobacterium tuberculosis.[18,33] The use of e-nose systems in medicine for the early diagnosis of noninfectious diseases such as breast and lung cancers[34,35] kidney disorders and infectious diseases has been recognised[33,36–38] and will be discussed further. Clinical studies for rapid real-time diagnosis have proved challenging. We conducted a prospective study to detect the premalignant phenotype Barrett's oesophagus that can progress to oesophageal cancer. Patients attending for upper GI endoscopy who have given written informed consent to participate were examined using "the sniffing endoscope" aspirating the insufflated gas through an e-nose device. The results (Figure 9.1) revealed to complexity of obtaining a real-time *in vivo* diagnosis. Although differences are apparent the results are inconsistent and very individual. We will require technology improvement and very large studies. When moving onto real patients, this very complex project encountered some challenges.

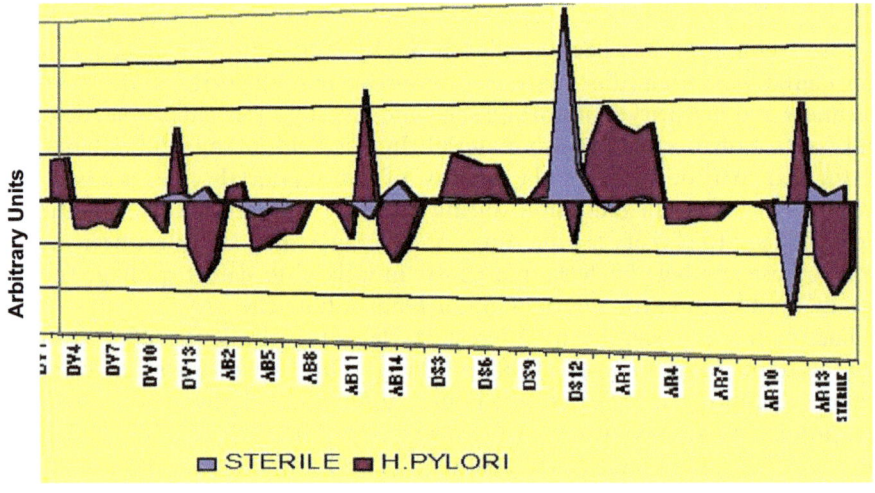

Figure 9.1   This figure shows that using an artificial stomach it is possible to "sniff" the stomach and detect helicobacter infection. There is complete discrimination between *H.pylori* and sterile patterns.[32] The project to develop real-time "sniffing endoscopy" using volatile fingerprints in association with standard visualization is complex.

9.2.2.1 Lung Cancer

In 1999 Phillips *et al.*[39] found a combination of 22 VOCs in the breath of lung-cancer patients that could be used to identify those with the disease. The compounds were identified using GC-MS and were predominantly alkanes, alkane derivatives and benzene derivatives. This was a cross-sectional study that looked at breath samples from 108 patients with abnormal chest X-rays. Patients then underwent bronchoscopy and histological or cytological sampling. 60 patients were confirmed to have lung cancer. Patients underwent breath sampling within 24 h prior to bronchoscopy and the breath samples were analysed using GC-MS techniques. 67 VOCs were common to the breath samples of 62 patients, of these VOCs 22 were selected by discriminate analysis. Using this approach for the detection of stage 1 lung cancer the technique had 100% sensitivity and 81.3% specificity. Crossvalidation of the technique correctly predicted the diagnosis in 71.7% with lung cancer and 66.7% of those without.

Di Natale *et al.*[40] used an electronic nose to identify lung cancer patients. A total of 62 breath samples were taken from 60 individuals by the bag-collection method. 35 patients had lung cancer, 18 were used as reference subjects and 7 patients were postsurgery for lung cancer. 100% of the lung cancer patients were correctly classified, 94% of the controls were correctly classified and 44% of the postsurgical patients were also correctly classified. The misclassified postsurgical patients were classified as healthy controls.

Machado *et al.*[41] studied the exhaled breath from individuals with lung cancer and compared this with subjects with noncancer diseases and a healthy group as noncancer control volunteers. They demonstrated that the e-nose could identify the different characteristics of exhaled breath in patients with lung cancer in a noninvasive way. A total of 330 of 388 breath samples collected from this group were correctly classified; an overall accuracy of 85%. The electronic nose had 71.4% sensitivity and 91.9% specificity. More recently Peng *et al.*[35] were able to differentiate between distinct cancer types by means of patient breath analysis using a nanosensor array.

9.2.2.2 Breast Cancer

In 2003 Phillips *et al.*[42] used 22 C4–C20 alkanes found in breath to identify women suffering with breast cancer from those without breast cancer. Two hundred breath samples were obtained from women with abnormal mammograms who subsequently underwent breast biopsy. Fifty one cases of breast cancer were detected in 198 consecutive biopsies. The breath test distinguished between women with breast cancer and healthy volunteers with a sensitivity of 94.1% (48/51) and a specificity of 73.8% (31/42) (crossvalidated sensitivity 88.2% (45/51), specificity 73.8% (31/42)). Compared to women with abnormal mammograms and no cancer on biopsy, the breath test identified breast cancer with a sensitivity of 62.7% (32/51) and a specificity of 84.0% (42/50) (cross-validated sensitivity of 60.8% (31/51), specificity of 82.0% (41/50)). The negative predictive value (NPV) of a screening breath test for breast cancer was superior to a screening mammogram (99.93% *versus* 99.89%); the positive predictive value (PPV) of a screening mammogram was superior to a screening breath test (4.63% *versus* 1.29%).

Previous *in vitro* studies suggest it could be possible to diagnose microbial pathogens of medical importance such as anaerobic bacteria *Clostridium spp* and *Bacteroides fragilis*, and even to discriminate between different *Clostridia* based on volatiles produced.[43]

9.2.2.3   Microbial Infections

Furthermore, there has been an increasing need for early detection of microbial infections in the hospital environment. Dutta *et al.*[44] distinguished between different types of *Staphylococcus aureus* species: MRSA, MSSA and coagulase-negative *staphylococci* (C-NS) from swab samples of infected patients using a sensor array. Other acquired diseases of great importance in hospitals are those related to prolonged mechanical ventilation like sinusitis, bronchitis and pneumonia.[38,45]

In 2004 Hockstein *et al.*[46] used e-nose analysis of the breath of ventilated patients in an intensive-care unit to diagnose ventilator associated pneumonia (VAP). He compared breath analysis to computed tomography (CT) of the chest. They report an accuracy of at least 80% when comparing e nose analysis to chest CT.

Adrie *et al.*[47] have reported increased levels of nitric oxide in the breath of patients with pneumonia compared to those without pneumonia.

Humphreys *et al.*[38] investigated the diagnosis of VAP by detecting micro-organisms in bronchoalveolar lavage (BAL) fluid in a prospective comparative study of e-nose and microbiology using LDA for the data analysis. 77% of the samples were correctly classified and 68% when patients on antibiotics were included.

Thaler *et al.* (2006)[48] was able to identify correctly biofilms *versus* nonbiofilm producing *Pseudomonas* and *Staphylococcus* species with accuracy ranging from 72.2% to 100% in patients suffering with rhinosinusitis.

9.2.2.4   Bacterial Infections

Pavlou *et al.*[36] used an e-nose to identify samples of urine that were infected with bacteria. They were further able to identify the species of bacteria causing the infection. Shyknon *et al.* used e-nose technology to differentiate between different bacterial causes of otits media.[49]

Some recent studies have demonstrated the capacity of a hybrid metal-oxide/metal-ion-based sensor systems to discriminate between different species of Trichophyton that cause dermatophytosis.[50] At present, the diagnosis of these skin diseases takes up to 1–2 weeks. They were able to differentiate between species and distinct concentrations within 4 days of growth. This would permit a rapid diagnosis, the appropriate use of antifungal drugs and also monitoring their activity during the treatment. Naraghi *et al.*[51] studied the potential of using an e-nose for discriminating between five antifungal treatments against two dermatophytes species within 96–120 h. This approach is important due to the lack of accurate susceptibility testing, especially for dermatophytic fungi.

9.2.3 Environmental

Pollution has become one of the main concerns among public opinion and for local governments. The water supply can be contaminated with faecal micro-organisms and with chemical products discharged from industrial, agricultural and animal farm activities. The traditional methods for detecting indicator micro-organisms of pollution, such as microbiological cultures, can underestimate their presence in water. Sensor arrays have been used to monitor wastewater in treatment plants, sewage treatment works and to discriminate between different products from sources as diverse as leather and automotive industries.

A wide spectrum of organisms, both bacteria and fungi, and chemical components such as pesticides and insecticides can contaminate water supplies.[52] E-noses have also been used to monitor the quality of potable water for testing cyanobacterial species[53] or detecting and differentiating between *Streptomyces* spp spores in reverse osmosis and tap water[54] (Bastos and Magan,[54]). Recently, they showed that soil quality and microbial status could be differentiated using volatile production patterns under different environmental regimes.[55] Other interesting applications such as control of health and safety air quality in the aircraft-cabin environment have gained special interest.[56]

9.3 Ventilator-Associated Pneumonia (VAP)

Ventilator-associated pneumonia (VAP) is a significant infection and has become of great medical importance because it is associated with antibiotic-resistant pathogens and might result in high rates of morbidity, mortality and increased medical care costs. The diagnosis of VAP is still difficult since there is no gold standard diagnostic test and the specificity and sensitivity of all existing diagnostics are not very high. Clinical signs of VAP are not specific and the scoring systems do not give high accuracy for VAP diagnosis.

VAP is the acquired pneumonia that occurs in patients who require mechanical ventilator support by endotracheal tube or tracheostomy for ≥ 48 h.[57] Therefore, VAP is considered a nosocomial disease because it is acquired in the health care environment with no evidence that it is present or incubated at the time of hospital admission.

VAP can be caused by a range of individual or mixtures of micro-organisms. Gram-negative enteric rods, *Staphylococcus aureus* and *Pseudomonas aeruginosa* are the predominant organisms responsible for this infection. However, many other micro-organisms can cause VAP including *Streptococcus pneumoniae*, *Haemophilus influenzae* and *Acinetobacter baumanii* depending on the onset of the illness.[58] There are other opportunistic micro-organisms such as *Candida albicans* and *Aspergillus fumigatus* that can play a role in immune-compromised patients. Polymicrobial infections appear to be especially high in patients with acute respiratory distress syndrome (ARDS).

Identification of the micro-organisms that may have the most clinically impact on a patient, depends on the individual patient, the type of intensive-care
unit (ICU), duration of stay and on any previous antibiotic treatment.[58] Several
studies suggest that the time of onset determines that pathogens will infect and
the prognosis of the disease.[59,60] For the latter authors *H.influenzae, S.pneu-
moniae, methicillin sensitive S.aureus (MSSA)* and some *Enterobacteriaceae*
were common in early-onset VAP, while *P.aeruginosa, Acinetobacter spp and
methicillin resistant S.aureus (MRSA)* were typically more common in late-
onset infections. The latter species belong to the group of micro-organisms that
potentially can be multidrug resistant (MDR). These pathogens have a high
and very variable range of mortality, between 0–50%, due to their specific risk
factors, patterns of clinical resolution and resistance.[61]

The incidence of VAP is estimated at between 8–28%[59,60] and in all cases
there has been a close relationship between VAP and the length of mechanical
ventilation. Mortality rates due to VAP may reach values of up to 50%, especially in elderly populations, in severely ill patients, and those who may have
received inappropriate previous drug therapy. However, some studies have not
found any relationship between the high rates of mortality and VAP, but are
mostly due to the underlying diseases.

The diagnosis of VAP is difficult because there is no specific and single
clinical criterion. A large number of diseases, even noninfectious, may present
similar clinical signs such as ARDS, nosocomial tracheobronchitis, congestive
heart failure, atelactasis, pulmonary thromboembolism and pulmonary
haemorrhage. Likewise, the lack of a gold standard in diagnosis may result in a
nonspecific and sometimes unsuccessful treatment, which exposes the patients
to high levels of toxicity, longer stays in hospital, and elevates the costs and
risks of having resistant micro-organisms.[62]

There have been various clinical strategies for VAP diagnosis such as CPIS
(clinical pulmonary infection score) developed and modified by Pugin *et al.*[63]
although some consider CPIS of little usefulness for daily evaluation due to the
need of a waiting period for culture results and for its low sensitivity and
specificity.[64] Also, CPIS has not been validated in immunosuppressed patients.

Regarding VAP treatment, the decisive points to be considered once the
therapy for VAP is chosen are the antibiotic selection and its duration. The
empiric antibiotic therapy has to be based on knowledge of the most common
micro-organisms and its local resistance pattern within the ICU,[65] on the
duration of mechanical ventilation and on the previous treatment, if there was
one. The duration of the therapy has been lately reduced due to the obser-
vations that prolonged therapies lead to colonization with antibiotic resistant
bacteria such as *P.aeruginosa, S.aureus* or *Enterobacteriaceae*. This can result in
a relapse of VAP, mainly in the second week of treatment.[66] An early effective
diagnosis and therapy are crucial for good outcomes with less antibiotic use
and therefore for reducing mortality.

Work has been undertaken to examine the potential for using qualitative
volatile fingerprints of specific micro-organisms involved in VAP infections
using the hybrid sensor e-nose to discriminate between different species; to

build an *in vitro* model based on these data sets to interpret samples from VAP patients to enable better and more accurate treatments to be applied. Humphreys *et al.*[38] investigated the diagnosis of VAP by detecting microorganisms in bronchoalveolar lavage (BAL) fluid in a prospective comparative study of e-nose and microbiology using LDA for the data analysis. 77% of the samples were correctly classified and 68% when patients on antibiotics were included.

Acknowledgements

The following people are kindly acknowledged for their help: Neus Planas Pont, Napoleon Charaklias, Martyn Lee Humphries, Natasha Sahgal, Kamran Naraghi.

References

1. H. T. Nagle, R. Gutiérrez-Osuna and S. S. Schiffman, The how and why of electronic noses, *Spectrum, IEEE*, 1998, **35**, 22–31.
2. L. Pauling, *et al.*, Quantitative analysis of urine vapor and breath by gas-liquid partition chromatography, *Proc. Natl. Acad. Sci. USA*, 1971, **68**(10), 2374–6.
3. P. Španěl and D. Smith, Selected ion flow tube mass spectrometry for on-line trace gas analysis in biology and medicine, *Eur. J. Mass Spectrom.*, 2007, **13**, 77–82.
4. P. Španěl and D. Smith, The challenge of breath analysis for clinical diagnosis and therapeutic monitoring, *Analyst*, 2007, **132**, 390–396.
5. K. J. Boudonck, M. W. Mitchell, L. Német, L. Keresztes, A. Nyska, D. Shinar and M. Rosenstock, Discovery of metabolomics biomarkers for early detection of nephrotoxicity, *Toxicol Pathol.*, 2009, **37**, 280–292.
6. J. M. Scotter, R. A. Allardyce, V. S. Langford, A. Hill and D. R. Murdoch, The rapid evaluation of bacterial growth in blood cultures by selected ion flow tube-mass spectrometry (SIFT-MS) and comparison with the BacT/ALERT automated blood culture system, *J. Microbiol. Methods*, 2006, **65**, 628–631.
7. J. Meijerink, M. A. H. Braks, A. A. Brack, W. Adam, T. Dekker, M. A. Posthumus, T. A. Van Beek and J. J. A. Van Loon, Identification of olfactory stimulants for *Anopheles gambiae* from human sweat samples, *J. Chem. Ecol.*, 2000, **26**, 1367–1382.
8. M. L. Humphreys, Volatile diagnostic techniques for ventilator associated pneumonia. *DM Thesis*, 2010, Cranfield University.
9. J. M. Scotter, V. S. Langford, P. F. Wilson, M. J. McEwan and S. T. Chambers, Real-time detection of common microbial volatile organic compounds from medically important fungi by Selected Ion Flow Tube-Mass Spectrometry (SIFT-MS), *J. Microbiol. Methods*, 2005, **63**, 127–134.

10. L. Wady, A. Bunte, C. Pehrson and L. Larsson, Use of gas chromatography-mass spectrometry/solid phase microextraction for the identification of MVOCs from moldy building materials, *J. Microbiol. Methods*, 2003, **52**, 325–332.

11. J. C. R. Demyttenaere, R. M. Moriña and P. Sandra, Monitoring and fast detection of mycotoxin-producing fungi based on headspace solid-phase microextraction and headspace sorptive extraction of the volatile metabolites, *J. Chromat. A*, 2003, **985**, 127–135.

12. K. Persaud and G. Dodd, Analysis of discrimination mechanisms in the mammalian olfactory system using a model nose, *Nature*, 1982, **299**(5881), 352–5.

13. J. W. Gardner and P. N. Bartlett, A brief history of electronic nose, *Sens. Actuators B*, 1994, **18**, 210–211.

14. J. W. Gardner and P. N. Bartlett, *Electronic Noses: Principles and Applications*. Oxford University Press, Oxford, 1999.

15. T. O. Larsen, Identification of cheese-associated fungi using selected ion monitoring of volatile terpens, *Lett. Appl. Microbiol.*, 1997, **24**, 463–466.

16. N. Magan and P. Evans, Volatiles as an indicator of fungal activity and differentiation between species, and the potential use of electronic nose technology for early detection of grain spoilage, *J. Stored Products Res.*, 2000, **36**, 319–340.

17. K. Fiedler, E. Schütz and S. Geh, Detection of microbial volatile organic compounds (MVOCs) produced by moulds on various materials, *Int. J. Hyg. Environ. Health*, 2001, **204**, 111–121.

18. A. P. F. Turner and N. Magan, Electronic noses and disease diagnostics, *Nature Rev.: Microbiol.*, 2004, **2**, 1–6.

19. E. Schaller, J. O. Bosset and F. Escher, Electronic nose and their application to food, *Lebensmittel-Wissenschaft und-Technologie*, 1998, **31**, 305–316.

20. J. W. Gardner, H. W. Shin and E. L. Hines, An electronic nose system to diagnose illness, *Sens. Actuators B*, 2000, **70**, 19–24.

21. A. Jonsson, F. Winquist, J. Schnürer, H. Sundgren and I. Lundström, Electronic nose for microbial quality classification of grains, *Int. J. Food Microbiol.*, 1997, **35**, 187–193.

22. P. Evans, K. C. Persaud, A. S. McNeish, R. W. Sneath, N. Hobson and N. Magan, Evaluation of a radial basis function neural network for the determination of wheat quality from electronic nose data, *Sens. Actuators B*, 2000, **69**, 348–358.

23. S. Buratti, S. Benedetti and M. S. Cosio, An electronic nose to evaluate olive oil oxidation during storage, *Ital. J. Food Sci.*, 2005, **17**, 203–210.

24. C. Di Natale, A. Macagnano, F. Davide, A. D'Amico, R. Paolesse, T. Boschi, M. Faccio and G. Ferri, An electronic nose for food analysis, *Sens. Actuators B*, 1997, **44**, 521–526.

25. N. Magan, A. Pavlou and I. Chrysanthakis, Milk-sense: a volatile sensing system recognises spoilage bacteria and yeasts in milk, *Sens. Actuators B*, 2001, **72**, 28–34.

26. G. Keshri, P. Voysey and N. Magan, Early detection of spoilage moulds in bread using volatile production patterns and quantitative enzyme assays, *J. Appl. Mycol.*, 2002, **92**, 165–172.

27. J. Trihaas, L. Vognsen and P. V. Nielsen, Electronic nose: New tool in modelling the ripening of Danish blue cheese, *Int. Dairy J.*, 2005, **15**, 679–691.

28. G. Keshri and N. Magan, Detection and differentiation between mycotoxigenic and non-mycotoxigenic strains of two *Fusaria* spp. using volatile production profiles and hydrolytic enzymes, *J. Appl. Mycol.*, 2000, **89**, 825–833.

29. N. Sahgal, R. Needham, F. J. Cabañes and N. Magan, Potential for detection and discrimination between mycotoxigenic and non-toxigenic moulds using volatile production patterns: A review, *Food Additives Contamin.*, 2007, **24**, 1161–1168.

30. M. Camardo Leggieri, N. Planas Pont, P. Battilani and N. Magan, Detection and discrimination between ochratoxin producer and non-producer strains of *Penicillium nordicum* on a ham-based medium using an electronic nose, *Mycotox Res.*, 2010, DOI: 10.1007/s12550-010-0072-5.

31. N. Magan and N. Sahgal, Electronic noses for quality and safety control, in *Advances in Food Diagnostics*. Eds. L. M. L. Nollet, F. Toldra and Y. H. Hui, Blackwell, Iowa, USA, 2007.

32. A. K. Pavlou, N. Magan, D. Sharp, J. Brown, H. Barr and A. P. F. Turner, An intelligent rapid odour recognition model in discrimination of *Helicobacter pylori* and other gastroesophageal isolates in vitro, *Biosens. Biolectron.*, 2000, **15**, 333–342.

33. A. K. Pavlou, N. Magan, J. Jones, J. Brown, P. Klatser and A. P. F. Turner, Detection of *Mycobacterium tuberculosis* (TB) in vitro and in situ using an electronic nose in combination with a neural network system, *Biosens. Bioelectron.*, 2004, **20**, 538–544.

34. A. D'Amico, G. Pennazza, M. Santonico, E. Martinelli, C. Rocioni, G. Galluccio, R. Paolesse and C. Di Natale, An investigation on electronic nose diagnosis of lung cancer, *Lung Cancer*, 2010, **68**, 170–176.

35. G. Peng, M. Hakim, Y. Y. Broza, S. Billan, R. Abdah-Bortnyak, A. Kuten, U. Tisch and H. Haick, Detection of lung, breast, colorectal, and prostate cancers from exhaled breath using a single array of nanosensors, *Brit. J. Cancer*, 2010, **103**, 542–551.

36. A. K. Pavlou, Magan, C. McNulty, J. M. Jones, D. Sharp, J. Brown and A. P. F. Turner, Use of an electronic nose system for diagnoses of urinary tract infections, *Biosens. Biolectron.*, 2002, **17**, 893–899.

37. P. Boillot, E. L. Hines, J. W. Gardner, R. Pitt, S. John, J. Mitchell and D. W. Morgan, Classification of bacteria responsible for ENT and eye infection using the Cyranose System, *IEEE Sens. J.*, 2002, **3**, 247–253.

38. L. Humphreys, R. M. L. E. Orme, P. Oore, N. Charaklias, N. Sahgal, N. Planas Pont, N. Magan, N. Stone and C. A. Kendall, Electronic nose analysis of bronchoalveolar lavage fluid, *Eur. J. Clin. Invest.*, 2011, **41**, 52–58.

39. M. Phillips, *et al.*, Volatile organic compounds in breath as markers of lung cancer: a cross-sectional study, *Lancet*, 1999, **353**(9168), 1930–3.

40. C. Di Natale, *et al.*, Lung cancer identification by the analysis of breath by means of an array of non-selective gas sensors, *Biosens Bioelectron.*, 2003, **18**(10), 1209–18.

41. R. F. Machado, O. Deffenderfer, T. Burch, S. Zheng, P. J. Mazzone, T. Mekhail, C. Jennings, J. K. Stoller, J. Pyle, J. Duncan, R. A. Dweik and S. C. Erzurum, Detection of lung cancer by sensor array analyses of exhaled breath., *Am. J. Resp Crit. Care Med.*, 2005, **171**, 1286–1291.

42. M. Phillips, R. N. Cataneo, C. Saunders, P. Hope, P. Schmitt and J. Wai, Volatile biomarkers in the breath of women with breast cancer, *J. Breath Res.*, 2010, **4**, DOI: 10.1088/1752-7155/4/2/026003.

43. A. K. Pavlou, A. P. F. Turner and N. Magan, Recognition of anaerobic bacterial isolates *in vitro* using electronic nose technology, *Lett. Appl. Microbiol.*, 2002, **35**, 366–369.

44. R. Dutta, D. Morgan, N. Baker, J. W. Gardner and E. L. Hines, *Sens. Actuators B*, 2005, **109**, 355–362.

45. M. J. Rumbak, Pneumonia in patients who require prolonged mechanical ventilation, *Microbes Infect.*, 2005, **7**, 275–278.

46. N. G. Hockstein, *et al.*, Diagnosis of pneumonia with an electronic nose: correlation of vapor signature with chest computed tomography scan findings, *Laryngoscope*, 2004, **114**(10), 1701–5.

47. C. Adrie, *et al.*, Exhaled and nasal nitric oxide as a marker of pneumonia in ventilated patients, *Am. J. Respir. Crit. Care Med.*, 2001, **163**(5), 1143–9.

48. E. R. Thalen and C. W. Hansen, Use of an electronic nose to diagnose bacterial sinusitis, *Am. J. Rhinol.*, 2006, **20**, 170–172.

49. M. E. Shyknon, D. W. Morgan, R. Dutta, E. L. Hines and J. W. Gardner, Clinical evaluation of the electronic nose in the diagnosis of ear, nose and throat infection: a preliminary study, *J. Laryngol. Otol.*, 2004, **118**, 706–709.

50. N. Sahgal, B. Monk, M. Wasil and N. Magan, Tricophyton species: use of volatile fingerprints for rapid identification and discrimination, *Brit. J. Dermatol.*, 2006, **155**, 1209–1216.

51. K. Naraghi, N. Sahgal, B. Adriaans, H. Barr and N. Magan, Use of volatile fingerprints for rapid screening of anti-fungal agents for efficacy against dermatophyte *Trichophyton* species, *Sens. Actuators B*, 2010, **146**, 521–526.

52. O. Canhoto and N. Magan, Electronic nose technology for the detection of microbial and chemical contamination of potable water, *Sens. Actuators B*, 2005, **106**, 3–6.

53. J. W. Gardner, H. W. Shin, E. L. Hines and C. S. Dow, An electronic nose system for monitoring the quality of potable water, *Sens. Actuators B*, 2000, **69**, 336–341.

54. A. C. Bastos and N. Magan, Potential of an electronic nose for the early detection and differentiation between *Streptomyces* in potable water, *Sens. Actuators B*, 2006, **116**, 151–155.

55. A. C. Bastos and N. Magan, Soil volatile fingerprints: Use for discrimination between soil types under different environmental conditions, *Sens. Actuators B*, 2007, **125**, 556–562.

56. J. D. Spengler and D. G. Wilson, Air quality in aircraft, *J. Process. Mech. Eng.*, 2003, **217**, 323–335.

57. J. D. Hunter, Ventilator associated pneumonia, *Postgrad. Med. J.*, 2006, **82**, 172–178.

58. E. Bouza, C. Brun-Buisson, J. Chastre, S. Ewig, J. Y. Fagon, C. H. Marquette, P. Muñoz, M. S. Niederman, L. Papazian, J. Rello, J. J. Rouby, H. Van Saene and T. Welte, Ventilator-associated pneumonia, *Eur. Respir. J.*, 2001, **17**, 1034–1045.

59. J. Chastre and J. Y. Fagon, Ventilator-associated pneumonia, *Am. J. Respir. Crit. Care Med.*, 2002, **165**, 867–903.

60. M. H. Kollef, What is ventilator-associated pneumonia and why is it important?, *Respir. Care*, 2005, **50**, 714–724.

61. J. Rello, E. Díaz and A. Rodríguez, Etiology of Ventilator-Associated Pneumonia, *Clin. Chest Med.*, 2005, **26**, 87–95.

62. J. S. Solomkin, Ventilator-associated pulmonary infection: the germ theory of disease remains viable, *Microbes Infect.*, 2005, **7**, 279–291.

63. J. Pugin, R. Auckenthaler, N. Mili, J. P. Janssens, P. D. Lew and P. M. Suter, Diagnosis of ventilator-associated pneumonia by bacteriologic analysis of bronchoscopic and nonbronchoscopic "blind" bronchoalveolar lavage fluid, *Am. Rev. Respir. Dis.*, 1991, **143**(1), 1121–1129.

64. R. P. Baughman, Microbiologic diagnosis of ventilator-associated pneumonia, *Clin. Chest Med.*, 2005, **26**, 81–86.

65. S. M. Koenig and J. D. Truwit, Ventilator-associated pneumonia: Diagnosis, treatment and prevention, *Clin. Microbiol. Rev.*, 2006, **19**, 637–657.

66. P. J. W. Dennesen, J. A. M. Van der Ven, A. G. H. Kessels, G. Ramsay and M. J. M. Bonten, Resolution of infectious parameters after antimicrobial therapy in patients with ventilator-associated pneumonia, *Am. J. Respir. Crit. Care Med.*, 2001, **163**, 1371–1375.

67. T. C. Pearce, S. S. Schiffman H. T. Nagle and J. W. Gardner, *Handbook of Machine Olfaction*. Wiley-VCH, Weinheim, 2003.

68. E. Mahmoudi, Electronic nose technology and its applications, *Sens. Transduc. J.*, 2009, **107**, 17–25.

Subject Index